REALITY

40 REVELATIONS

BOOK 1

Real Answers for Everything

ASHLAY VAZ

This book contains 11 out of 40 revelations elucidating existences or physical reality

CAUTION **Read at your own risk. Information contained herein may alter your minds, which may change your life, lifestyles and/or health.**

ISBN 0-7414-3320-6

Cover illustration: The formation of a star, from where every existence engenders.

Published by:

INFINITY
PUBLISHING.COM

1094 New DeHaven Street, Suite 100
West Conshohocken, PA 19428-2713
Info@buybooksontheweb.com
www.buybooksontheweb.com
Toll-free (877) BUY BOOK
Local Phone (610) 941-9999
Fax (610) 941-9959

Printed in the United States of America

Printed on Recycled Paper

Published January 2008

ACKNOWLEDGMENT

During the past hundreds of years, philosophers and scientists in various disciplines have toiled, seeking to find the fundamental blocks of existences, or reality. Some have spent their entire lives, and some continue to spend, painstakingly researching-thinking, analysing, probing, experimenting, and testing-hypothesis and theories. Some are well known, but there are many others, who have contributed vast, and significant knowledge, or information on multifarious phenomena. This has facilitated these revelations to fruition.

I cannot acknowledge each one of them, and they have my deepest sincere acknowledgement, appreciation and gratitude.

Ashlay Vaz

CONTENTS

INTRODUCTION

We consider ourselves humans (our concept), as we arrogate our thoughts, feelings/emotions and actions/behaviour to be different from those of other animals, and/or our closest mammal ancestors-the primates. Humans 'evolved' from primates through many events (revelation 33). One remarkable event was the increasing utterances of self-generated sounds, or words, which eventually unfolded into a language. Our language resulted from our own engendered sounds or words, relating, interpreting, describing or explaining objects or things, events or phenomena. With and within this language humans communicate, and understand each other.

As we think/ideate within this self generated language, our lives (thoughts, emotions, action or behaviour, which are not real), or our world is entirely an illusion. As our language is our own creation, we cannot understand reality, or the real world. As humans live (think, feel and action) within their new language world, we are only partly psychologically different, but physiologically similar to/as primates. Revelations 33-40 details how humans live within their language world.

Querying within an illusory language we are forced to theorize, and/or hypothesize phenomena or nature, and therefore, answers/explanations are inaccurate and/or incomplete, exacting yet further questions. As there are unanswered questions, more illusory knowledge keeps accumulating. Accumulating illusory knowledge continuously exacts yet further meanings and explanations to sounds/words and concepts. This explains, why even today, we still cannot accurately and completely answer our why questions, and/or solve most of our problems.

As we live (think, feel and action/behave), and continue to be educated, and learn within our own generated knowledge, everything-our entire lives-continues to be illusory. As humans continue to live within their illusory world, there were, and are, and there would always have been-if not for these revelations-further continuing queries and problems.

Imagining and subsequent querying within sounds or words, or within our illusory language (necessitating explanations), originated religions, philosophy and/or science. Religious authorities with their creed, philosophers with their doctrines, and scientists through their scientific methods are continuously trying to answer questions of the real world. Within our illusory language they face great difficulties, always failing to explain existential objects/beings, stimuli, and phenomena.

As we ideate, query and create within an illusory language, or our world, we fail to describe the real world. We conceptualize and create an existence we call 'universe', and to explain our concepts and creation we build theories. For example, the concept 'universe' ushered in cosmology, and even today, cosmologists cannot explain their theories, continuing to speculate, postulate and hypothesize. Failing to present appropriate explanations, they eventually present more theories. The Big bang theory, the Steady state theory, dark matter, dark energy, membranes, parallel universes, the Theory of Everything in cosmology and theoretical physics are some of the circulating concepts, generating more theories, more illusory knowledge, more questions and more problems.

We create energy particles, quarks, atoms, genes and cells, but cannot finite them, as we fail to find the beginnings or the origins of anything. We also cannot explain why existences spontaneously and randomly appear, alter, or grow or decay, or exist or endure, and/or eventually terminate.

Philosophers and/or scientists in all disciplines work with the concept of causation. Postulating everything existing is the consequence of causation, we always face the chicken and egg, or the DNA/RNA problem-what came first? We are bewildered whether RNA or DNA originated first, and how DNA can create proteins through amino acids, and vice versa. Did life create DNA, or did DNA create life? As there is no such thing as causation, all our scientific theories are incorrect. Shockingly, all our beliefs in divinities or creators have no justification.

Science investigates to try and explain its own questions, and solve problems within our illusory world. Questions about the real world are generated from within our illusory language, and as the questions per se are illusory, we further generate more illusory or incorrect answers, and unsolved problems. As we ideate within an illusory language, all our theories result in problems, incapable of explaining observed phenomena or experimental results. If we do not know why or how things originated, or began existing, trying to find out how things work, in order to be able to find out how they began existing are exercises in futility.

Through the years, we are all aware of scientific research, periodically presenting us with new inventions, and discoveries. These publications ephemerally excite, and elate our minds. Some promote false hopes, expectations, and some delve us in awe. With every scientific publication of discoveries there are confrontations, arguments, refutations, disagreements, more theories and speculations.

These fascinating revelations corroborate our very own observations and experiences, rightly, soundly, and courageously challenging our existing scientific knowledge, and conventional scientific thoughts. Each revelation is intriguing, having far-reaching implications, doing away with highly troubled theories from the past and present, and also demystifying mysteries and paradoxes.

Upsetting and as it is amazing, there is an abundance of provocative, shocking, prodigious, and breakthrough information on varied subjects, challenging some stringent and fundamentalist societies, religions, and world-views. These revelations will fly in the face of the scientific establishments, and religious institutions. Information presented will have serious repercussions on past, present and future influences, beliefs, and controls that some authorities, governments, and religious bodies exercise on humanity. These compelling and controversial read revelations will set a new era in the scientific world, revolutionizing doctrines of several disciplines of science-cosmology, theoretical physics, philosophy, physiology, psychology, evolution and behavioral sciences. Every physical law and scientific belief are reconstructed and replaced, as every revelation presents new explanations for the mysteries of the phenomena seen around us.

Philosophers/scientists will now be able to easily completely analyze, and understand physical phenomena, virtual or latent processes. We will look at existences with a fresh, genuine, and different perspective, and in a different light. If these revelations were not presented, humanity would continue with unanswered questions, and unsolved problems.

Each revelation will shock us, as we will learn about our physiological and psychological existence. We will accept the reality that we are alone, and there are no other types of lives, or equals, or masters, or creators. We will never find aliens, or extra terrestrials. These revelations will have mind-boggling theological and philosophical consequences, confirming that life as we know it is unique only to/on earth. Information within these revelations will have a profound impact on religions, completely shattering the traditional perspective of god's relationship with humans. Humans are not created by divinities, and obviously cannot be uniquely favored. As we conceptualized the 'universe', we also likewise created 'god' (revelation 34). As philosophers and scientists have failed to explain 'beginnings', we are made believers.

People all over the world buy non-fiction books hoping to seek solutions to their problems, or answers to their questions. Reading through, they finally end up with further questions, and more uncertainties, confusions and complexities. Information herein will end human perplexed minds. Readers will immediately introspect with the new information, corroborating their experiences and observations, unraveling and self-solving many of their puzzles, mysteries and myths. We will now understand the purpose and meaning of everything, including the mechanisms of nature. Furthermore, there are proper solutions to persisting problems, anxieties, and confusions. These revelations give accurate and complete answers, especially to the 'why' questions.

There has never been, and there will never be revelations like these that will revolutionize thinking, and besides flummoxing minds will also volte-face lives, flipping conventional wisdom on its head. Information contained within will dazzle your minds, and have profound implications on how you comprehend your existence, altering your lifestyle. These revelations will elucidate, and enlighten you on your choices and beliefs. You will reconsider your hopes and fears, alter or lessen your uncertainties, anxieties and sufferings. The information will assist you to review your underlying requirements of gratifying your minds, and/or body, your relationship with others, and your possessions.

Over the years, I sought answers to some of the profound existential questions; on the gods, energy, the universe, space, galaxies, stars, the solar system, force, light, matter, life, consciousness, brains, minds, volition, body, soul and death. Continuous probing and introspection of my two language minds (revelation 29), finally brought about few profound thoughts/ ideas, manifesting reality. Now we will finally understand the simple common sense world that we inhabit.

These 40 revelations-each a bombshell-explains the great problem of being; its source, nature and destiny, or in physical terms, the origin, evolution and fate of existences, and or beings, invisible or visible. Forty separate books could have been written detailing each revelation, but each of them would have been abstruse. Summarized and conjoined in a nutshell, each presented revelation, is/as a piece of a jigsaw puzzle, substantiates and corroborates each other. Collectively, they skillfully elaborate, and elucidate the existence of energy, as to why and how its transmutation per se engenders, and endures duration existences, and/or the reality picture. These revelations present a thought provoking account of understanding existences, and/or reality. There are hundreds of questions (most of them that have never been imagined or queried) answered within these revelations divulging incredible, sensitive, and profound information, which would probably never have come to fruition. The answers to questions are indeed solid, complete and accurate, yet simple and rational.

The authentic information contained in these revelations need no corroboration. Philosophical doctrines or scientific research cannot present any contrary arguments, or refutations. These are not theories but facts, and this is why they are called revelations.

We kept producing and accumulating illusory knowledge by, through, and within ourselves, but now these revelations will finally terminate further non-constructive information. Suddenly, humanity would be unprepared to cope with the social, economic, political and ethical revolution of such information. However, these revelations will bring people closer together, as humankind will wake up to reality and realization, amalgamating their world with the real world.

These revelations are written in simple non-technical format for readers of all ages, non-scientists and scientists, as the contents are not tough to grasp, with/and no complex mathematical hurdles to jump. Not only are they so simple that the average person can understand, but they are so complete and workable, and can withstand any scientific scrutiny and refutation. These revelations are strategically listed in a certain order, enabling readers to progressively better their understanding of reality. Understanding reality is real freedom. Reality or the truth will now set you 'free'. Henceforth, there will be a new way of thinking, and a new way of living. Choices or decisions will either be very easy, or extremely difficult.

As energy or energy structural processes let us ‘know’ ourselves, and know everything. Knowing reality is to comprehend everything. With knowing and/or understanding everything/ reality, there will be no more questions or problems.

Let us turn the pages, face reality and get real!

PART ONE

REALITY AS/IS EXISTENTIAL ENERGY OR EXISTENCES

A - Nothingness

REVELATION ONE

ENERGY IS/AS EVERYTHING EVERYWHERE

Nothingness or illusions-No such thing as causation-Terminating the concept creation-Everything is the same thing

Profound information facilitating the understanding of these revelations on physical reality

THE 'ORIGIN' OF THOUGHTS

Revelation 33 details how primates transformed into humans, or how humans created themselves. However, the following information is sufficient to understand the eleven revelations on physical reality.

Thousands of years ago, for whatever reasons, groups of primates migrated out of habitats, having lived there for millions of years. Migration necessitated the utterance of sounds/words to describe observed, and experienced things/objects, stimuli and phenomena. This facilitated communications between individuals in a group. Words or sounds accumulating into a vocabulary or language, concurrently commenced thinking/ideation, and imagination, and these processes engendered humanity.

Our human language is our own unintentional creation. Every sound or word is an idea. These sound or words describing objects/things and phenomena, or existences have meaning only within the domain of our language world. Our thoughts are ideas, and ideas are our thoughts, as we think concurrently as we ideate, and vice versa. We conceptualize everything. As our language is our own creation, it is illusory.

As we live within our language world, our thoughts, and our ideation of our feelings/emotions, and of our actions/behaviour, or in other words, our entire lives is an illusion. As we live (think feel and action) within this illusory language world, we never understood/understand the real world. Within this illusory language, everything is incorrectly, or incompletely conceptualized, and explained. Our own queries result in incorrect and incomplete answers, and more questions. We cannot explain reality or the real world, and this explains why we have to constantly introduce new concepts to describe or explain previous speculations, theories, and ideas. Querying further within this accumulating knowledge, our illusory language accumulating information world gets larger, with unsolved problems, and increasing complexities and confusions. From the dawn of humanity, we are still trying to understand explanations to our own questions.

UNSUCCESFUL QUEST OF SCIENCE

As we think within our illusory language world, we cannot describe the other real world, or reality. We need to hypothesize or theorize, and then try explaining existences. Without correct answers we console ourselves with yet other new models, and theories, which explain some phenomena, but fail to account for others, which require further theories generating yet more illusory information, contradictions, paradoxes, confusions, problems, law violations, anomalies, paradoxes and sleepless nights.

Today's scientific knowledge are concepts from simpler times, when there was far less understanding. These ideas survived as they gave us some understanding of our world, and some functional models to work with. Science today is a collection of abstract theories, offering no deep, or unifying physical explanations, and giving no reason, or justification from observations and experiences, explaining why questions and theories abound. Our world today requires real answers, and clear understanding.

Our 21st century physics presents bizarre ideas, as multiple dimensional strings, mysterious quantum effects, special relativities, super strings in 10 dimensions, warped space time, quantum gravity, gluons, ghostly neutrinos, and a host of other concepts, none sensible, or workable, which only some scientists understand.

Science has been speculative describing existences through its own laws, and therefore, we have an unclear picture of reality within our illusory imagination. Science continues in its quest with scientific faith. There are statements made with discoveries, but they end with 'whose details are uncertain', or ‘there is still more research to be done’. However, science continues with its research, as we trust in what is missing, or not completely understood, as this is scientific faith. Today’s science is given the benefit of faith, as we all tend to believe that the explanations are true. However, when the true picture is inaccurate or incomplete, we believe in strange things, for example, presuming the hypothetical evasive, non-localized, invisible electron orbiting the postulated atom’s nucleus.

GENERATING CONFUSION

Science describes what we observe, but never explains why and how existences or things engender, why they exist, endure, and eventually terminate. These revelations answer the 'why' questions.

We cannot describe the real world from within our language world. How do we define/describe the planet earth? The atmosphere, oceans, mantle, the inner and the outer core are all 'different' 'existences'. Yet within our thinking they comprise planet earth. How can we describe its boundaries? Is it the beginning, or the end of the Van Allen Zones, or the end of each layered atmosphere, or the surface of the earth, with or without its oceans? Why?

We are made to understand our world with the theories of relativity and quantum physics, but these incompatible probable stories are per se troubled by many problems. The postulated ‘universe’ was brought into existence by the Big bang theory, but we shall learn there is no such thing as a 'universe'. We conceive mathematical space in our illusory minds, but these ideas per se interact and build the mind itself, so obviously we run into numerous difficulties and problems.

NO SUCH THINGS AS 'BEGINNINGS' OR 'ENDS'

Revelation 33 shockingly elucidates how humanity dawned. Humans inadvertently created themselves, concurrently with their gods. From the 'beginning' of humanity, we postulate everything must have a cause, or be created, and concluded that there must be a creator. This 'began' thinking or ideation, imagination and querying, and humans were now different from other mammals. Thinking/conceptualizing within our illusory language, we ask who, how, and what created god. What is energy? Who created energy?

As we observe and experience objects, or things, or phenomena, conceptualizing them within our illusory minds, we describe them with sounds, or what we call words. Words and/or concepts, and theories from within an illusory language, trying to describe physical phenomena, or existences (our concepts) of the real world, generate all the difficulties, confusion and problems. There are incomplete and inaccurate answers to our questions of the real world, or reality, as we conceptualize, and create existences from within our own engendered language, which is illusory. Within an illusory language we cannot ask questions, or even try to describe phenomena or existences. This explains why we are unable to answer the WHY questions. Why things exist? Why are we here?

As we observe things spontaneously engendering, and existing, enduring, and eventually terminating, we postulate everything is 'caused', or must have a source, or an origin, or a beginning, or a creator. Within human imagination there must be a 'beginning', source, or origin for anything including energy to exist. From the 'beginnings' of humanity, envisaging divinities to be the cause of everything, we have always tried to answer questions, and solve problems with causation. This explains why science can only answer some 'how' questions, and not a single 'why' one, and why problems generate more confusions, and anxieties, some without any solutions.

'God', 'universe', 'infinity', 'continuity', ', 'eternally', 'perpetually', 'creation', 'source', 'origins', 'engender', 'beginnings', 'ends', 'time', 'space', 'duration', 'permanence', 'boundaries', 'edges' 'existences', 'entities', and even 'energy' and 'reality' are human concepts, and creations. Shockingly, if there are no beginnings, ends and no 'entities', there is no causation, and therefore, no divinities.

Hypothetical atoms, molecules and elements are structures of transmuting energy. Fusion, integration, disintegration, collapsing, terminating, all forms of radiation, temperatures, or light, concentrations, condensation, momentum, spin, pressures/force, expulsion, extraction are forms, states, processes of transmuting energy.

Existential energy was not 'created', nor does it have a 'beginning', and therefore, reality, or existential energy cannot be conceptualized, queried or explained.

As we read through these revelations, we will understand these statements.

WHAT IS ENERGY OR EXISTENTIAL ENERGY?

Energy is a human word/construct/concept, made up of/from three phonemes, or units of sounds. As we are constituted of 'something', we are also empirically aware of its existence, what humans named 'energy'.

We ideate, imagine and create 'energy', and existences. We query how and why energy exists, and cannot explain it. How, when, why and where are also our concepts. We keep on querying our creations and concepts and therefore, all our accumulating information generated by our questions is illusory. As our language created by us is illusory, we cannot query. We are energy processes and structures, querying about energy that engendered ourselves, and everything around us.

Existential energy is that something existing, as/in everything everywhere. As energy continuously transmutes, there are existences as structures, processes, state or forms, continuously transmuting into yet other structures, processes, states or forms. Therefore, as energy continuously transmutes, all existences living and non-living exist for duration, eventually disintegrating or terminating into, or within/as the fabric of space/energy.

The cloud is a structure of energy comprised of atmospheric energy particles, existing as/within the fabric of space/atmosphere (energy). It spontaneously 'begins' existing, exists for a certain period of time, eventually disintegrating or terminating. All existences and us are like that cloud, and constituted of energy particles, engender, or exist/endure, but eventually disintegrate, or terminate, continuously transmuting into yet other forms, states, processes, or structure of the atmosphere, or space/energy.

Why does, and how can, existential energy exist perpetually, or without boundaries?

As we ideate or think within our illusory language, we create 'beginnings', 'ends', 'infinity' and 'boundaries', as with all other phenomena.

Existences cannot have a beginning, as everything everywhere is energy. If we think of existences we create them, as everything exists within our thoughts. If we do not think, then there are no such things as energy, or existences, and also no such thing as 'beginnings' and 'boundaries'. If we do not think, there are no questions and therefore, no need for answers-nothingness.

Conceptualizing, imagining and querying within our own generated language, which is illusory, all questions are incorrect, incomplete and meaningless, and can therefore, never be answered or explained correctly, and completely.

Why is energy nothingness? When is energy something? When does energy exist? Why does energy exist? Why does anything exist?

We created the concepts god, energy, evolution, gravity, matter, and everything observed and experienced. Creating things they exist, and then we have to explain their existence. When we are asleep, drugged, or intoxicated, we are not thinking or ideating, and existences do not exist-nothingness.

Energy exists, as we think or ideate, observe, and experience existences, or phenomena. As we can feel, and observe force/momentum, we explain energy to be 'energetic'.

Energy is something, as we can think/ideate, imagine, query, create, and explain our creation, and concepts to ourselves. We assume we are 'knowledgeable', but information gathered from illusory ideas, incorrect and incomplete answers, and explanations is all an illusion. Other creatures can somewhat ideate, but without a vocabulary of sounds, or words, or language, cannot imagine, or query, and their world, or existential energy, is nothingness. Light, heat, temperatures pressures/force, and sounds, manifest stimuli, phenomenon, and environment changes to/in organisms (revelation 14), explaining why, and how all living things can 'know', or 'sense'.

If a surgeon removes a few thought neurons, or memory tissues from under our cranium, we will immediately turn into a chimpanzee or an ape. However, we will have to first find out where thoughts or memory lie, as they exist similar to the postulated unpredictable, elusive electron-here, there and everywhere.

RETHINKING EVERYTHING

NINE LAWS OF REALITY OR EXISTENTIAL ENERGY OR EXISTENCES

1. Energy is everything and everywhere.

2. Existential energy in continuous momentum continuously transmutes. Energy is static (a structure/form of energy-as humans understand it) if it is contained by existences (transmuting energy in states processes structures or forms).

3. There are existences (invisible or visible-as humans create/understand them), as processes, structures, forms or states of energy, as it transmutes

4. As energy continuously transmutes, there is flux, and as everything is energy, existences/beings are continuously altering, or growing, or decaying.

5. As energy continuously transmutes, there is flux, and existences/beings exist/endure, but eventually terminate.

6. As energy is everything, everywhere, all existences or beings, as forms, states, processes or structures of transmuting energy exist, as/within the fabric of existential energy's

expanse. There are no entities, as space is not a void, but a form of energy.

7. An existence can only engender/exist/endure, concurrently with another existence (each with different characteristics), as parts of the same cycle sustaining the existence/endurance of each other. Every cycle is part of other cycles.

8. Existences engender/endure (continuously altering and eventually terminating), as a result of the energy radiation/transmuting cycle within a star field.

9. As there are no entities, and as everything exists as part of an energy transmuting cycle, there is no causation, or creation.

ELUCIDATING REALITY OR EXISTENTIAL ENERGY

1. Energy is everything everywhere.

'Within' existential energy's 'expanse' there are no dimensions, and there are no such things as boundaries, edges or limits. Human words/concepts such as great or small, superior or inferior become meaningless, as nothing can be compared, as there are no standards. There is no order or disorder, precedence or sequence, precursor or sequel, beginnings or ends, newness or oldness. There are no directions, progressions or regressions. 'Duration' and time have no meaning, as energy continuously transmutes, and therefore, there is no permanence, stability and destiny. Energy in its entirety or its totality (our words/concepts) cannot expand or contract, increase or decrease, or be created or destroyed. There is no such thing as a part or a whole, as nothing changes in quantity or quality, or nothing increases or decreases. Distances, nearness, lengths, breadths, density, thickness, thinness, height, depth, front, back, up, down, east, west, south, north, right, and left are human notions, assisting, and or enabling us to communicate, plan and to live within our language world. These concepts have no meaning in the real world.

The first law of thermodynamics is correct, but not the second. As everything everywhere is energy, it cannot be created or destroyed. It does not 'go' from 'more useful forms' to 'less useful forms'. Some states, processes, structures or forms of energy are more or less 'useful', depending on how we want to comprehend it. Burning matter is transmuting energy, and this change of state, or structure, or process of matter/energy gives away (transmutes) heat (another form of energy). This heat (a form of energy) disperses, as it amalgamates (transmutes) into space-another form of energy. As we do not have the technology to utilize energy from atmospheric space, we postulate that the heat dissipating into space (a form of energy) is a 'less useful form' of energy (as we cannot recover it to perform 'useful' work). Energy is matter in one form, and the transmutation of matter (a form of energy) to energy performing work as heat is also a form of energy.

Condensation is the transmutation of vapor into liquid-both forms of energy. Condensation per se is an energy transmuting process. These words describing states, processes, structures or forms of energy, are concepts enabling us to 'understand' phenomena (within our illusory language or world) and live our lives.

We were recently informed of a table experiment supposed to have created energy. As everything is energy, anything 'giving out' energy or 'creating' energy was/is, and will always be a form, state, process or structure of energy transmuting. You cannot 'create' or 'give out' energy, as everything everywhere is energy. There are no 'entities'. We assume we 'obtain' and 'use' energy, but in reality, we are only transmuting energy.

Fuels like coal and gas are forms of energy. We ‘obtain’ ‘energy’ heating them. Heat is another form of energy. We obtain energy from fission and fusion of nuclei. Nuclei or the myriad types of matter are all forms of transmuting energy.

Hydrogen cell cars now being produced use electro-chemical reactions to operate motors, and the end products are heat and water. We postulate, that we 'obtain' energy in the form of ethanol from corn, propane, liquid petroleum, gas and electricity. Hydrogen and all the elements, air, chemicals, heat, water, corn, propane, liquid petroleum, electricity, the copper wire, methane, coal or oil and everything everywhere are states, processes, structures and forms of transmuting energy.

We have been informed, that hydrogen is an energy carrier, and not an energy source, meaning that it stores, and delivers energy in a usable form. This is preposterous. Everything everywhere is energy. Hydrogen is always susceptible, as it is continuously extracted within/by the energy transmuting cycle (revelation 6).Therefore it is the lightest element, and can be utilized easily to heat, or pressure/force, and other uses.

We have to transmute some form, state, process, or structure of energy to momentum, or to force, or to action. Momentum, or force, or actions are per se forms of energy. These physical phenomena manifest as energy transmutes.

Existences are states, processes, structures or forms of transmuting energy, existing within/as the fabric of existential energy. Reality is simply transmuting energy. Nothing other than energy exists, as/in anything, anywhere.

2. Existential energy in continuous momentum continuously transmutes. Energy is static if it is contained by existences (transmuting energy in/as states/processes/structures or forms).

Existential energy exists in continuous momentum (energetic state-human concepts). This is, has always been, and will always be the state of energy. There is no creation, determinism, or intelligent design.

Revelation four describes 'free' energy, and the engendering of star fields. Briefly, energy transmuting as/into temperatures, light and all forms of radiation, pressure/force, and condensing into particles/plasma within/as star fields is not 'free'. Energy 'becomes' 'free', as star fields terminate, and/as energy transmuting cycles end, drowning/amalgamating as black holes. A black hole continues the whirl of a collapsing star field, and concurrently energy ‘becomes’ free increasing the rotation velocity. From swirling black holes 'free' energy transmutes/condenses into a dense region, and concurrently generating star field space commences energy transmuting cycles between bodies within-a functioning star field.

Conjoining star fields form a galaxy. However, star fields cannot continue to conjoin, and/or galaxies can not grow any larger, as they are confined by dark (static) energy, explaining why galaxies exist as islands within/as existential energy's expanse (revelation four).

As star fields form (to only a certain size, as temperatures and pressures reach their extreme limits) conjoining to form galaxies, other regions of existential energy are static, and this energy is named 'dark' energy. Therefore, existential energy exists, parts of it transmuting as star fields/galaxies, and parts of it as dark, or static energy.

Every existence engenders, and endures concurrently with another existence, with different characteristics sustaining each other. Even existential energy exists, both static and transmuting. Transmuting energy generating existences, concurrently influences static or dark energy, and vice versa.

As humans ideate, query and create existences, energy possesses its transmuting characteristic. Energy is explained as something that is 'energetic', that 'actions', moves things, or is in momentum. However, these intrinsic 'energetic' characteristics (states/forms) are manifested only as energy transmutes.

We incorrectly set standards for comparison. Energy can exist deadly dull, or as it transmutes, enormously 'energetic'. We assume dark energy to be deadly dull, or 'ineffective' as its static state. However, dark energy confining galactic space, and galaxies, is as 'energetic', as transmuting energy. This is so, as if it can (as energy is in continuous momentum) it will transmute. It is static, as it cannot transmute. Both 'energetic' and 'static' are human concepts.

Without this peculiar intrinsic transmuting characteristic, or if existential energy existed only in a static state, there would have been no transmutations, and there would be nothingness-no existences, and/or humans to think, and ask questions.

How did existential energy 'begin' the first star fields?

Nothing 'begins', and therefore, star fields have 'always' existed. Energy cannot motion in any one direction three dimensionally. There are no 'boundaries' or 'edges' to existential energy, as these are human concepts. Existential energy is continuously in momentum, as energy possesses this intrinsic condensing, or transmuting characteristic. In other words, this is why energy is in momentum, or 'energetic'. If existential energy remained static, or as dark energy, it would not possess its transmuting (energetic) characteristics, and there would then be no existences (human understanding). Therefore, existential energy is in continuous momentum, as parts, or areas, or volume of it revolves. As it revolves there is momentum, and rotational velocity increases. In other words, energy is in continuous momentum, as it revolves, and as it can revolve, it transmutes. This explains the continuous condensing characteristic of energy (revelation four). With this intrinsic continuous condensing characteristics of energy (in momentum), dense regions, or star fields, or existences engender.

Existential energy exists, has always existed, and will always exist full of galaxies. We cannot query 'beginnings', or the time/era when star fields first formed, as 'beginning' and 'time' are human concepts.

What and who made existential energy to possess its momentum or transmuting characteristic?

Creating existences, and experiencing, and observing the characteristics of energy, it is described with momentum, or as transmuting. Without human minds, energy is nothingness.

Why should it matter whether energy could, or should have remained static, or in continuous momentum? It matters only when humans think/ideate, imagine and query. This question is asked only during a human's lifetime. Before we are born, and after we die, this question is meaningless.

3. As energy transmutes, there are invisible or visible existences, as processes, structures, forms or states.

What are existences? How are existences 'created' with transmuting energy?

Without a language we would not be able to think/ideate, and query, as we would exist as other creatures-not 'knowing' in a world of nothingness, or in eternal bliss. Energy or existences would then be nothingness.

Querying about existences (human concepts) we create them. We therefore, create existences within/with our thoughts. We query how things can exist if there is no 'beginning', source, or divine creation (see law 9). After we create existences within our language, we try to explain, and or understand our own concepts and creations. To explain our created existences, we then have to speculate, and/or theorize what we create. To explain what we create, we have to query, imagine, ideate/think-back at the start of another circle.

As we created the concept god, we have to describe the existence of god. To explain the existence of god, we have to query who created that god we created. We could go on with the same question 'eternally'.

We cannot explain the phenomenon 'light', as humans created it. Light is a form of energy transmutation. We cannot state that light propagates, as everything everywhere is energy. There is no causation, and therefore, no entities. We cannot also state that light, or energy propagates from the star/sun. As there are no entities, in reality, light does not propagate, but space as a form of energy transmutes to light, and light is apparent. As light constitutes heat and pressure or force, energy in momentum at/as the star/sun continuously transmutes into another form, state, process or structure of energy and light is apparent.

In reality, light, the sun and everything does not exist, but only energy is transmuting. As we created light, the sun and existences, we have to explain their existence and 'characteristics'. Energy like every other phenomenon visible or virtual, living or non-living thing is an existence (within human understanding). Our thoughts from within an illusory language are an illusion, and so existences are in reality an illusion.

As free energy transmutes into a dense region, concurrently engendering star field space-a star field-there are existences. Existences randomly engender without 'creation', design or purpose, as all existences are the same thing-transmuting energy. An existence concurrently engenders/endures with another existence, as parts of the same cycle sustaining each other. If an existence alters or terminates, the other existence concurrently will also alter or terminate.

With processes at/as a star, space, force or pressures, light, temperatures, heat, all forms of radiation are continuously generated. Most of these existences are invisible. We cannot see light, but we see with/by, and through light. So observable transmuting energy, as well as invisible transmuting energy in/as states, processes, structures or forms are existences.

Power, force, spin, momentum or motion, temperatures, heat, light, color, sound, electricity and magnetism are all existences, as forms, structures, states, or processes of continuously transmuting energy. Even 'space' containing a particle, nuclei or a dense region, and objects/bodies within star fields, between star fields, and confining galaxies is a form of energy, and an 'existence'. Dark energy which cannot transmute, and is 'static' is also a form of energy, and an existence. Invisible particles and their space (even though they cannot be observed or sensed/experienced) are also existences.

Existences can be invisible particles, hypothetical nuclei/elements, molecules, compounds, chromosomes, cytoplasm, organelles, cells, organisms including humans. Existences can be light, sound, color, temperatures, forces, electricity or magnetism. Existences can be matter as solids, liquids, or gases whether they be in air, or in ether, or in neutrinos, or the in dust, or clouds, or nebula, or in stars, planets, meteors, meteorites, asteroids, galaxies or whatever.

Each existence is conceptualized, and described or explained differently, facilitating humans understand physical phenomena. In reality, there are no such things as 'existences', as only energy exists, and is everything everywhere. Energy is simply nothingness.

4. As energy continuously transmutes, there is flux, and as everything is energy, existences/beings are continuously altering, growing or decaying

Why is everything in flux? Why does everything continuously change?

In reality, as everything everywhere is energy, there is no flux. As we describe every phenomenon, object or thing, observed and experienced, we create existences, or entities. As existences engender, exist, endure, alter and eventually terminate, we conclude everything is flux.

As energy is everything everywhere, and as energy transmutes, there is flux. Structures, processes, states or forms, continuously transmute into yet other structures, or processes, or states or forms, as other existences. Existences engendering, existing and/or enduring are continuously altering, or changing, decaying, growing and deteriorating. Therefore, all existences are of certain duration, continuously altering at every moment of time.

Anything, anywhere, anytime can spontaneously appear, erupt, 'begins', 'originate' or engender. Everything everywhere can either be growing, or deteriorating, or corrupting, as energy transmutes. As energy transmutes, there are existences as parts of a cycle with other existences, and each cycle parts of other cycles. As something, somewhere corrupts or decays, or transmutes in a cycle or system, another thing somewhere, concurrently will engender, appear, grow or change.

The word/concept 'decay' is a phenomenon denoting a thing declining in state, quantity, quality or value. As everything is energy, and in continuous flux, anything (phenomenon/existence) decaying in something somewhere, is concurrently transmuting or transforming, or there is a change, or a growth in something else, somewhere.

The word/concept 'annihilation' is also incorrect, as energy is everything everywhere, and nothing (there is no entity) can be destroyed or created. Existences only continuously transmute into another form of energy.

All existences visible, or invisible, are states, structures, processes and forms of transmuting energy, existing as/within the fabric of energy. Every existence engenders, exists, endures, alters, or terminates, as a result of continuing changing temperatures and pressures/force, as everything is flux, as energy continuously transmutes.

5. As energy continuously transmutes, there is flux, and existences/beings exist/endure, but eventually terminate.

With flux, every existence is continuously altering, eventually terminating, or disintegrating into yet other states, forms, structures or processes as/of energy. Everything is of certain duration, existing for an instant, a minute, hours, days, months, years, or for millions of years. No existence/being possesses the same characteristics at any given moment.

'Stars' like everything else are/as energy transmuting structures are born to die. Non-living things on the surface of earth, shielded from energy radiation by our many layered atmosphere endure for long periods of time. They gradually decay (altering), and eventually disintegrate. All living things continuously alter or grow, exist for duration, deteriorate, and eventually terminate.

All existences, as star fields, stars, astronomical bodies, objects, organisms including humans, nuclei and particles, are duration structures, or processes, analogous to a cloud forming, and eventually disintegrating. Clouds constituted of atmospheric nuclei, endure for a period of time, and depending on changing environments, or temperatures and pressures, eventually terminate. All existences are also analogous to bubbles in a boundless vast ocean, interdependent and interlinked, and existing as/within the fabric of water, enduring/existing for certain duration, and when disintegrated, amalgamate into/as the fabric of water, or energy. Comprised of integrated particles we can disintegrate anytime, back into particles, eventually as part of another cycle, or float aimlessly within/as the fabric of space, or existential energy. We are insignificant and mere specks, interlinked and interdependent in the vast boundless 'eternal' expanse of existential energy.

There is only change, and everything is chance, or contingent. Nothing is foreseen. Every existence is extemporaneous.

6. As energy is everything everywhere, all existences or beings as forms, states, processes or structures of transmuting energy exist, as/within the fabric of existential energy's expanse. Therefore there are no entities, as space is not a void, but a form of energy.

We will learn how free energy condenses into a dense region, concurrently engendering space-a star field. The dense region and its containing space (a star field), occupies/supplants the same area/volume of energy from which it engendered. As everything is energy, and as energy transmutes, the space engendered is not a void, but a form of energy.

Space or a void, or a vacuum are human concepts/constructs. Within existential energy there is no such thing as space (as a void), or there is no place as a void, and a vacuum cannot be created. Any space anywhere, postulated to be a void is a form of energy. However, we have to continue to use the existence 'space'-as a form of energy, but not the concept 'space' as a void.

As 'virtual' particles/plasma integrate into neutrons/protons, nuclei, elements, molecules, objects, planets, bodies, stars, and star fields into galaxies, their space confining them concurrently amalgamates, getting proportionately larger. This explains why we see 'space' around bodies. These existences/bodies are interlinked and interdependent, and as structures, states, processes or forms of transmuting energy with their surrounding 'space' exist (like bubbles within/as water), as/within the fabric of existential energy. Therefore, there are in reality, no entities, and no space-as a void- that we assume. This space we assume to exist is a form of energy.

As energy is everything everywhere, every single particle constituting our body, and everywhere is intimately interdependent, and interlinked, and its identity merged and blended, and absorbed into the totality of existence. As we are part of the boundless, three dimensional existential energy, we see and feel the wholeness with every body, planet, star field and galaxy. The Brahmins of India were correct in expressing and corroborating this reality in their sacred mystic Sanskrit formula, which is yet really so simple, and so clear- "TAT TVAM ASI". This (everything) is you, or 'That thou art', or 'This is who you are', or 'You are the unchanging reality', or 'You are in the east, and in the west, north, south, above and below'. You are this existential energy, in/as everything, everywhere.

7. An existence can only engender/exist/endure, concurrently with another existence (each with different characteristics), as parts of the same cycle sustaining the existence/endurance of each other. Every cycle is part of other cycles.

As space concurrently engenders with the dense region (a star field), possessing different characteristics, facilitating sustaining the existence, and endurance of each other, every particle, proton, nuclei, element, molecule, object, star field and galaxy exist with confining space, and vice versa. As a result, every existence engenders, exists and/or endures, concurrently as another existence engenders, exists and/or endures. As parts of the same cycle they possess different characteristics, sustaining each other.

All cycles are part of yet other cycles, and as each cycle is further interdependent and interlinked with yet other cycles, their existence is subjected to the continuation, or endurance of other cycles.

Two or more existences cannot exist-as parts of the same cycle-possessing the same characteristics. They are of intrinsically different characteristics, to give or take, expend or absorb, attract or repel, act or react, to and from the other, thus sustaining each other's existence/endurance. Receiving or giving, expending, or absorbing, and so on, are processes, states, structures or forms of continuously transmuting energy, and are human concepts.

Every event, occurrence, phenomenon, being or existence is part of a cycle, and exists/endures, concurrently with another existence, as different states, processes, structures or forms of continuously transmuting energy. A star will not engender, if there is no other existence within its star field to support/maintain their processes. Therefore, there is no such thing as a star, but a star field. There is no such thing as a planet, but a solar system. There are no binary stars, but star fields orbiting each other, each with its own energy transmuting cycle within its star field.

A gene in one chromosome exists, supporting and interdependent with another gene, or group of genes in the same, or another chromosome, and vice versa. A chromosome exists with another chromosome. A nucleus exists with the cytoplasm of the cell, and vice versa. A cell will exist interdependent, and supporting another cell, or a group of cells, or tissues, or organs, and vice versa. A single cell, or a unicellular organism, or a multi cellular organism (as the environment-see revelation 20) is in continuous existence with the environment, and vice versa. There is no such thing as an organism, but the ecological cycle, as an organism exists, interdependently supporting part of the environment, and vice versa. Every living and non-living thing within the biosphere, or anywhere, is part of a system, or a cycle. The carbon dioxide-oxygen cycle, the hydrological cycle, and the nitrogen cycle are few of the many cycles operating within our atmosphere. Our atmosphere is a part of the cycle with our oceans, sustaining each other. Even earth has a southern and northern hemisphere, each with different phenomena sustaining each other, as parts of the same cycle. The atmosphere, the oceans, or earth's environment or earth exists, sustaining other cycles existing within the solar system. The cycles between the planets, and the sun, sustain or replenish the 'fuel' expended from the sun enabling it to continue shining/burning. Star fields exist with another companion star field, or a group of star fields. Star fields sustain the existence of others, enduring the galaxy. Galaxies exist, as part with their confining space as a cycle. Galaxies and their space exist with dark energy confining them, and vice versa, as parts of the same cycle.

If any one of these existences disintegrates, the other supporting, and or interdependent part/existence will also concurrently disintegrate.

8. Existences engender/endure (continuously altering and eventually terminating) as a result of the energy radiation/transmuting cycle within a star field.

Revelation four describes how free energy in momentum condenses into a dense region, concurrently engendering star field space-a star field. Existences within star fields engender, exist and endure, concurrently as energy transmuting cycles engender. At/as a star, energy is continuously generating temperatures/all forms of radiation and pressure/force.

Astronomical bodies, planets, satellites, atmospheres, biospheres and organisms-plants and animals, and myriad varying forms of matter within star fields engender, exist and endure, as long as energy transmutes within star fields.

As energy transmutes, there are interactions, integrations and disintegrations of all existences within star fields. Every existence engenders, or endures, or alters, and terminates, as a result of being subjected to altering temperatures pressures/force, and or densities, as energy continuously transmutes within/as a star, and within star fields.

As a result of the intrinsic transmuting property of energy, there are star fields. Without star fields there would be no existences-only static existential energy, as/or nothingness.

9. Everything exists, as part of an energy transmuting cycle. There are no entities, no causation, or divine creations.

There is only randomly, and spontaneously forming, enduring, and eventually terminating existences, of/by/from to the same, and only thing existing-energy.

In reality, 'nothing' is a result of a 'cause', as there are no 'entities'. Without causation, and no ‘entities’, there is no determinism, design, order or purpose. There is no such thing as an 'existence', or 'something', or even 'nothing'. Entities and existences are human concepts to understand phenomenal reality.

The word/concept 'create' is explained, as to bring something into existence, or cause to exist, or make, or produce. Everything is energy everywhere, and in order to create, or produce an existence, and even for it to exist, and or endure, energy continuously transmutes from one state, process, structure or form to yet another state, process, structure or form. Nothing can exist, or be created, or produced from something other than energy. Therefore, there is no creation, and no causation.

Without ‘entities’ or existences, or causation, there is no such thing as creation, and therefore, there are no creators. There is no room for any creative act to occur. Moreover, a creator has to be created, and we could go on looking to 'infinity' (another concept) for the first creator, who created our postulated creator, and on, and on. Why should there be creators or causation? For a creator to exist (if there is such an existence), or for any human to create something, a form state, process, or structure of energy transmutes into another state, process, structure or form of energy. Energy cannot be created or destroyed.

As there is no causation, most of our scientific theories are incorrect. Furthermore, all our beliefs in divinities and/or creators have no justification. Postulating 'entities', 'causation', and 'creators' explains why many profound questions, or problems, cannot be answered or solved. Any philosophical, scientific or religious explanation to any question or problem using causation, and/or creation, is not only incorrect, and/or incomplete, but preposterous.

B - Misconceptions or Illusory Concepts

REVELATION TWO

NO SUCH THING AS SPACE

No such things as entities-Everything is interlinked and interdependent-Revolutionizing philosophical doctrines and scientific thinking/methods

SPACE AS A FORM OF ENERGY

Space engenders concurrently as energy condenses. As both the dense region and space occupy, or supplant the area/volume from which they engendered, space is a form of energy. As energy as/at a dense region, or a star continuously condenses energy, concurrently producing, and propagating heat and pressure or force, space confines the dense region, and vice versa. Therefore, as heat and pressure/force propagate within/as space, other forms of energy, and force, concurrently and continuously propagate within/as space toward the dense region, or the star, within the same pathways, confining the dense region. This explains why all bodies within star field or interplanetary space are forced onto their star, but they cannot 'fall' on to their star, as they exist within/as space orbiting the star. This also explains the energy transmuting cycle between the sun and the planets, expelling and extracting energy detailed within revelation six. Furthermore, this also explains why there is evaporation, and extraction of protons from earth's oceans, surface and atmosphere detailed within revelation eight.

Our interplanetary space is an integral part of the interiors, and the atmosphere of the sun, or the sun collectively as a whole. Therefore, interplanetary space, and the bodies as/within it, as parts of a cycle, sustains processes at/of the sun, and vice versa.

Space engenders/exists/endures, as energy continuously transmutes, and vice versa

As/at a star, as energy continuously transmutes, space concurrently engenders/exists and endures. Within a star field we have a dense region as a star, and other nebulae, or bodies within the star field space. Within our solar system the sun is the dense region, and all the planets exist, within/as interplanetary space. This explains why all astronomical bodies are observed within space, or space surrounding them. There is space around a star, and star field bodies, or space within a star field or interplanetary space, space in between star fields, or interstellar space, and space containing galaxies, or intergalactic space. As the dense region forms, concurrently engendering star field space, the space within star fields, space around bodies within a star field, interstellar space, and (as star fields conglomerate to form a galaxy) inter-galactic space are all transmuting forms of energy.

As star fields conglomerate together to form a galaxy, the galaxy is confined by amalgamating space. Galactic space confines the galaxy, and vice versa. Dark energy confines galactic space, and vice versa. Therefore, there is no space-as a void-anywhere within existential energy's expanse. Space is an integral part of any existence. No existence engenders, exists, or endures without its confining space.

Space confining matter

A neutron star conjoins protons and neutron plasma (revelation 4). As fragments of an exploding neutron star blast through a nascent star field space, nuclei engender. In other words, disintegrating protons and neutron plasma are confined by space, engendering nuclei. This explains why every proton, nuclei/element, molecule and matter is confined by space. No protons, nuclei, element or molecule will engender, exist or endure without its confining space.

Forms, states, processes or structures of energy transmutes, or propagate within/as space-a form of energy. Space transmutes (as it is a form of energy) into yet another form, state, process or structure of energy, manifesting an existence. Without space energy cannot transmute. In other words, without space concurrently engendering/transmuting, energy cannot transmute.

If energy was not 'energetic', or not continuously in motion, there would have been no transmutations, and there would be no existences/beings, as structures, processes, states or forms of energy. As energy continuously transmutes, it is continuously in momentum, or ‘energetic’.

If there was no space generated, as energy transmutes, and vice versa, existential energy would remain static, or motionless. There would be no flux, no momentum, no pressures/force, no forms of radiation/ temperatures, no phenomena, no existences or beings, and no humans to think, and query. Existential energy would be nothingness.

There is no space-as a void-anywhere

As we observe existences within space, we posit existences/beings as 'entities' within space-a void. Space is postulated as a three dimensional region being empty of anything, vacant, a void or a vacuum, but as a form of energy, it is an existence, or 'something', and not a 'nothing'. Space-as a void-is a human concept/creation, explaining physical reality for human planning and living.

There is no such thing as space-a void-as 'space' is a form of energy. We cannot produce a vacuum or space-as a void-as energy cannot be created or destroyed, as only energy exists, as/in everything, everywhere. Space concurrently engenders with existences, or what we call 'entities', and therefore, all existences and space exist, within/as the fabric of existential energy. Space and existences (star fields, galaxies and dark energy) take up all of existential energy.

As the word/concept space is deep-rooted in our lives, we will continue to use the phenomenon space, but as an existence, and importantly, as a form of energy-not a void. Space is also an existence, as it is a form of energy.

Existences, as beings, objects, and astronomical bodies are observed existing 'individually' within 'space'-a postulated void. These existences can be observed, as they comprise of condensed energy (energy in/as structures), or what we call matter. There are other existences, for example, light, temperatures, all forms of radiation, pressure/force and sound, which are not observable, but are apparent transmuting within/as space-a form of energy.

It is incorrect to posit there was a void or vacuum, before the conceptualized creation, or before the theorized big bang, that generated the postulated 'universe' (please see revelation 3).

Einstein and Newton's theories are incorrect without 'entities'.

Einstein surmised that even the emptiest space, devoid of matter and radiation might still have dark energy-this is called a cosmological constant. He failed to note that space, radiation and matter are forms of energy, or every existence virtual/invisible, and observable is a form of transmuting energy.

As there is no space-as a void-anywhere, existences are therefore, interdependent, and or interlinked. As there is no such thing as 'space' (a postulated void), there are in reality, no entities. As there are no entities, Newton's dynamics, and Einstein's theories on relativities and space-time are incorrect.

Newton's gravity mechanisms, Faraday's force field concepts, Einstein's relativity theories, and all modern physics are all space-as-a-void theories. If we do not consider space a form of energy, all our past, and present physical theories are incorrect, and this means that physics has serious scientific, and inconsistency problems. As we accepted-space as a void-theories for centuries, science and its societies have taken the wrong directions and therefore, most of our theories are wrong. This explains why there are uncertainties, paradoxes, problems and unanswered questions.

THE DIFFERENT REGIONS OF SPACE-A FORM OF ENERGY

Within existential energy's expanse there is no space-as a void-anywhere.

Intergalactic space

Star fields conjoin forming a galaxy. The amalgamating space of star fields, or galactic space confines the galaxy, and vice versa. This space is further confined by dark energy, which cannot transmute. As dark energy is invisible we assume 'space' (as a void) between galaxies. Intergalactic space is not a void, or a vacuum, as dark, static, latent, virtual, 'energetic' energy exists, confining galactic space, and vice versa.

Inter-stellar space

The boundaries or the outskirts of star fields (or between star fields) is filled with dark matter, as plasma/protons expended from stars, and/or nuclei from disintegrating bodies settle at this neutral zone. The space between star fields or the interstellar medium is not a void. Vast stretches of gas nuclei, and tiny solid particles also called cosmic dust forming clouds, exist within this space. Inter-stellar space is dark and cold, as light, or radiated heat from stars dwindle with distance.

Inter-planetary space or star field space

Star field space is occupied by a star, and bodies, or nebulae, or planets in/as a solar system. This space propagates/transmutes expended plasma, all forms of radiation/temperatures, and force/pressures from a star/sun. The energy cycle (revelation 4/6) shuttles to-and-fro, transmuting energy 'extracted' toward, and 'expelled' from the star/sun. 'Extract', 'expel' are human concepts.

Earth's atmospheric space

Earth's atmosphere and oceans are continuously being interacted, disintegrated and integrated, over and over again, within the energy transmuting cycle, detailed within revelation four and eight. Our atmospheric space is filled with gaseous elements, molecules, water vapor and other substances. Space surrounds each nucleus/element or molecule.

Space containing protons, and nuclei, or matter

A neutron star conjoins protons and neutron plasma (revelation 4). As fragments of an exploding neutron star blast through a nascent star field space, nuclei engender. In other words, disintegrating protons and neutron plasma are confined by space, engendering nuclei. This explains why every proton, nuclei/element and molecule is confined by space. No protons, nucleus, element or molecule will engender, exist or endure without its confining space.

EVIDENCE THAT SPACE-AS A VOID-DOES NOT EXIST

Free energy transmutes engendering space. Energy transmutes as space exist

Energy transmutes shuttling to-and-fro within/as space, between the star and bodies-a functioning star field. Without space, energy cannot transmute into/as varying forms, states, processes or structures, or existences.

Creating space-a void

There is no such thing as empty space-a void. All experiments trying to create a vacuum are incorrect, as the space postulated to being contained, or trying to be removed to create a 'vacuum' is a form of energy. Researchers remove atmospheric nuclei to create the postulated vacuum. They cannot remove space from anything, anywhere, as space is energy.

Recently scientists discovered that when a liquid crashes into a hard surface within a postulated 'vacuum' there is no splash. There is no such thing as a 'vacuum'-void of anything. In the experiment, air or nuclei were removed, not space. The liquid crashing on the hard surface did not produce a splash, as disintegrating nuclei, and molecules, were not able to convey this disturbance to other nuclei, or molecules and therefore, disintegrated/amalgamated into/as space. The presence of air will always result in a splash, as elements/molecules (as the liquid crashes onto a hard surface) disintegrate, integrating again as other air constituents, or into other forms of energy.

As nuclei constitute molecules of the liquid, they will disintegrate from the molecules (the disintegration per se is the splash) only by colliding with other nuclei, or air around, at certain temperatures and pressure/force. It is wrong to assume that air had no part to play in splashing, as it is less dense and viscous than water.

You cannot contain or remove space, as space, as well as any material of confinement is a form, state, process or structure of transmuting energy. Nothing can contain, or create a vacuum, as no area can be, or is empty in the sense of a true, absolute vacuum. Space and every action/momentum is the transmutation of energy in/as a form, structure, processes or state.

No such thing as the speed of light

A postulated vacuum (void of anything) cannot be created in/as the experiment to test the speed of light. In the experiment, light as a form of energy, transmutes/propagates through/as another form of energy, or space. In the first place, there would be no light if there was no space-a form of energy. Space waves, as it transmutes to/as light that is apparent. In reality, it is not light or space that is 'propagating', but light is apparent as space waves, to/at certain wavelengths and intensities. We postulate light is/as a different existence than energy.

Light as energy does not really propagate from one place to another. As space propagates, or transmutes as light, and space/energy is everything, everywhere, it appears to us that light is propagating. Light/energy/space cannot propagate one-way, as energy is everything, everywhere. Within that propagating pathway, different forms of energy shuttle to-and-fro, explaining why space waves. Therefore, as light is not an entity it is impossible to calculate the speed of light. We have to take into consideration all other forms of energy propagating/transmuting, shuttling to-and-fro, within the same pathways. How can we measure the speed of light, as there are no entities?

Eventually, propagating as space waving through distance, and depending on the intensities produced at the 'source', light transmutes to other forms of radiation, and does not appear as light. Therefore, we cannot determine the speed of light to be constant throughout. Energy is everything everywhere, and does not propagate one way, and does not move from ‘there’ to ‘here’ or vice versa. Light is a form of energy. Speed is a human concept.

Nothing motions in a vacuum, or within space-a void

Motion or momentum is generated as energy transmutes, and is therefore, a form of energy. As there are no 'entities', nothing motions/actions without altering/transmuting another existence, as/or a form, state, process or structure of energy. Bodies in motion should infinitely continue their velocities or momentum within space-as a void. Moving objects (and propagating forms, states, structures and process of energy) alter their characteristics (decreasing speed) through/as space, as space waves, or transmutes to these existences (as space is a form of energy).

Extraction and expulsion, and atmospheric space interactions

Energy continuously transmutes between the sun and the earth, as some forms of energy are expelled from the sun, and concurrently other forms extracted from earth's atmosphere, and oceans. Energy as heat, or light, or temperatures, and force/pressure, propagate as/within the energy transmuting cycle between the sun and the earth's atmosphere. With these interactions the molecules of water in our lakes, and oceans/surface, and atmospheric nuclei and molecules are continuously disintegrating and integrating.

Existences/beings within the troposphere are comprised of atmospheric nuclei configurations, and are not entities, but an integral part of atmospheric space. Living things can function having been engendered, as parts of the atmospheric-surface-ocean cycle, and as long as this cycle continues, living things will continue to endure.

As energy structural processes, we are able to function within/as parts of these cycles. For example, we are able to drink a liquid from a glass, as the liquid is continuously being forced/weighed down by the 'incoming' force/pressure (energy) from the sun within/as atmospheric space. As energy structures, we process energy (from within atmospheric space) generating an 'extracting force' within our mouth. The liquid being emptied in the glass is concurrently being supplanted by atmospheric space, in an opposite reaction taking place in our mouth. We have to concurrently displace atmospheric space in our mouth to suck up the liquid. If space was a void, there would be no action on the liquid, and we would not be able to create a 'force' in our mouth, to transmute one form of energy into another, or to momentum one state, form, structure or process into any one, or the other.

Transmuting/utilizing air and 'space'-a form of energy-for action and locomotion

Creatures on the surface of the earth would not be able to move, or action, if there was space (a void). As creatures are constituted of atmospheric nuclei, their cells have evolved mechanisms to utilize air, and altering atmospheric nuclei (temperatures, pressure/force, and densities), facilitating action or locomotion (revelation 22), and to keep them down on the surface of the earth, and within the troposphere.

Astronauts and objects float aimlessly in upper atmospheric space, as their cell structures/mechanisms are constituted of, from, and within the biosphere. These mechanisms have evolved to function (move and/or action) only with/within certain particle densities, pressures and temperatures within the troposphere's space.

Birds and air planes soar, glide, thrust, forward or lift, 'utilizing'/transmuting air within space-a form of energy. This would not be possible if space was a void. You need energy (air and space) around to 'overcome' energy, or 'displace' it, and concurrently 'replace' it with yet another state, process, structure or form of transmuting energy. ‘Overcome', 'displace', and 'replace' are human concepts. Energy is everywhere, as/in everything.

Kinetic and Potential energy

Kinetic, potential and negative energy are phenomenal descriptions. In reality, as everything is energy, these states are only transmutations of energy. A rock that is about to fall from a cliff is acted upon by energy supporting it, and the energy that is in readiness to action it. A rock that is about to fall down will not fall, or will eventually do so, depending on both the 'incoming' and 'outgoing' pressure/force generated (through air and atmospheric space) within the energy transmuting cycle (revelation 7). We describe this process as kinetic energy, but in reality, space-as a form of energy-together with atmospheric air (nuclei) within it, is continuously influenced by the energy transmuting cycle.

Whether the rock will stay put, or will eventually fall, depends on the density of air, and how space transmutes. Some things fall, and some things can be weighed, as a result of the energy transmuting cycle within star fields (revelation 10).

Keeping bodies apart

If space (as a void) existed, there would be absolute rest. Nothing can be at rest in ‘space’, and or existential energy, as it continuously transmutes, and everything is flux. It is absolutely impossible to remain at rest in relation to any object, or point in space. Everything is always moving in relation to any, and every other thing, as everything is energy, and everything is interconnected/interlinked and interdependent. Every object in 'space'-a form of energy-is constantly moving in relation to everything, everywhere. Energy transmutations engender and terminate existences, as everything is in constant flux.

If 'space' (as a void) existed, planets, satellites, and other bodies would not have orbited the sun. They would just be, roaming aimlessly without any fixed direction, or they would have long been absorbed, and disintegrated within the sun's processes. Astronomical bodies are interconnected and interlinked, within/as the fabric of space, explaining how they are kept at a distant, and also why they orbit the 'dominant' body in a star field. Revelation seven also explains how planets, and other bodies not only orbit the sun, but why and how they rotate around their imaginary axis.

IS THERE SPACE-TIME?

Space as a void is a human concept. Time is a human abstract construction. Nobody can define space, and time other than scientists. With Einstein's theories on the speed of light, time stands still, and mass is infinite. How incredible!

Einstein remodeled the concept universe, replacing it with the bizarre and unexplainable space-time. Einstein's relativity theories explain that Newton's absolute space does not exist, and space and time depends on the relative motion of observers. Further, space was something that could be curved, and warped by matter and energy. Space-time seen as distinct entities was absolute.

Quantum mechanics transformed our understanding of matter and energy (as matter is conceptualized different from energy), bringing in new involvements for space and time. General relativity postulates space-time that is geometrically smooth, whereas quantum mechanics working at the smallest scale sees relativity as a violent agitation, or change, and therefore, these two theories are incompatible with each other. Space and time do not relate together, and as elucidated in this revelation, there is no such thing as space-a void. In physical reality, neither space (as a void or as a form of energy) nor time can provide points in space where the position of an object, or the time it attains that position can be measured.

There is no space-time, and time is independent of space-as a form of energy. Space (as a void) and time are human words/concepts, enabling humans plan, and live their lives.

As existential energy has no 'beginning', it is preposterous to link space with time. Space-time was created considering that the universe originated from the Big Bang. This theory preposterously postulates, that energy, everything (all existences), all dimensions including space and matter originated out from a point. Space-time has a boundary only at the beginning of the Big Bang. There was no Big Bang, and as there is no existence as the conceptualized 'universe' (elucidated in the following revelation), the concept of space-time is nonsensical.

We will learn that it is not light per se that momentums, but light can only momentum, or propagate in the form of, or within/as space. Space waves as it 'transmutes'/ 'propagates' as light. Transmuting/propagating energy waves space. Light as a form of energy does not 'propagate' like a bullet, but transmutes/propagates as space, and light is apparent. This space could be a particle's surrounding space, atmospheric, planetary, inter-stellar or galactic space.

If light as a form of energy does not propagate (as light) over long distances, and there is no such thing as space-as a void-what is relativity and space-time? What is relativity, as both light and space are forms of energy, and time is a human abstract construction?

Time has to begin somewhere. Time should originate somewhere to be effective, so scientists conveniently made time begin at the Big Bang. If there was no Big Bang, then Space-Time is nonsensical.

Humans created time, and commonly ideate/think of time to drive events in their world, and to assist us plan, and live our lives. We create time to begin, and end at our convenience. Time is merely a sequence of events. In reality there is no time, as there are no beginnings or originations, ends or terminations. Scientists assume that time is a dimension like the three spatial dimensions, and that time and space are inextricably linked.

There are no answers why we perceive the now, or the present, and also why time flows in one direction. Physicists cannot explain how time flows from the past to one direction in the future. Time changes occur in the structure of every object or body. Time can be easily understood.

Space as a physical reality is a form of energy, and time is a human construct, and both these cannot be conjoined to corroborate theories explaining reality. The universe is a concept, and does not exist, and so the addition of time to the assumed three spatial dimensions is nonsensical.

When did time begin?

Many centuries ago, people assumed that time began with creation. Aristotle believed that time had no beginning or end, as the world was eternal. It is easy to state that the world is eternal, or infinite, and time has a beginning, but 'infinite', ‘eternal’, 'beginning', and 'end' are human concepts.

Time has to begin when everything was created. If the world was created from nothing, according to the Big bang theory, or from something else, then when does time begin? If our world was generated from any prior state of material existence, that existence would have terminated and therefore, time could not have been eternal. We could go on querying the first existence, and never reach a first beginning. We created ‘beginnings’ and ‘ends’, thinking within our illusory language.

Time could not have begun with creation, as 'creation' is also our concept. Creation is faith, not science. Both faith and science are also concepts from within our illusory language minds. Time and every thing 'began' when humanity dawned, or when primates started thinking, and created 'beginnings'. Time and everything ‘begins’ when we start to think, and ends when the thinker dies.

Special Relativity and Gravity

We are informed that the speed of gravity must be at least twenty billion times faster than the speed of light, but this would contradict the special theory of relativity, which states that nothing can go faster than light. It is argued, that both gravity and light propagate from the sun, and by the time the sun’s ‘pull’ has reached us, the earth would have 'moved on' for another 8 minutes (the time of postulated light travel), but this is not happening. Gravity must then be faster than light, and also be instantaneous. Astronomers inform us that light from the sun comes to us from one direction, and 'pulls' from a slightly different direction, implying different speeds for light, and gravity-if it exists. This shows how much we know about the physical world. Revelation 10 elucidates why there is no such thing as gravity. As space transmutes to all forms of transmuting or 'propagating' energy (including light), and moreover, as all forms of energy dwindle with distance, the above information is incorrect.

Nobody knows the speed of gravity (as it does not exist). All physicists explain that gravity must go very fast, but they also add, that it should not go faster then Einstein's speed limit.

Special relativity has been confirmed to be correct over and over again, but there is no logic to it. According to this theory, if two observers equipped with clocks move in relation to the other, the moving clock runs slower than the non-moving clock. But the relativity principle itself, which is an integral part of the theory states, that if there are two clocks, either one is moving. So if there are two clocks, and you move one of them, one will move, or run slower than the other, and the latter will run slower than the one that was moved, which is ridiculous. Today we have clocks accurate to a billionth of a second a day, facilitating measuring tiny differentials. Experiments have shown that atomic clocks really do slow down when they move. Does that mean that time itself slows down?

Revelation six details that radiated heat and pressure/force from the sun propagates within/as space toward earth, and concurrently other forms of energy are extracted to the sun, as replenishment material, within the same energy transmuting shuttling to-and-fro pathways. All the experiments confirming special relativity are conducted within earth's atmosphere. Revelation 10 details why things weigh less with less momentum, explaining the slowing down of the clocks mechanisms. The clocks slow down, not time. This reproaches Einstein's theory.

Is space curved, and does light bend?

Revelation nine elucidates the characteristics and propagation of light. All forms of radiation, or light, are forms of transmuting energy. Light as a form of energy propagates within/as space-a form of energy-as it waves. It is incorrect to posit, that light 'travels' in rays, or as a beam, or a straight line through space.

In the theory of relativity, space and time consists of three space co-ordinates, and one time co-ordinate. The expansion of the universe, assumed from the cosmic red shift calculations suggests it is inflating four-dimensionally. The measured space and time depends on the motion, and the mass of the bodies. The theory of relativity disregards absolute space and time, showing that space and time alter with the velocity, and the mass of the body, and bodies in motion generate space to be curved with gravity. The theory predicted that a light beam from a distant star would curve in the presence of a massive object in space.

In reality, as energy is everything everywhere, light as a transmuting form of energy propagating within/as space-a form of energy-cannot 'bend'. Space transmutes to light, and propagates (spreads out), and is everywhere, so what bends? Space as a form of energy is not 'curved'. There are no 'entities'.

Revelation 10 explains why there is no such thing as gravity. Gravity as a force-even if it exists-cannot 'bend' light. Light and gravity as a postulated force, are both transmuting forms of energy. Light propagates/transmutes outward from a star toward the outskirts of its star field. Light (if it is still in the form of light) from star fields interact at their boundaries. Light and some other transmuting forms of radiation from a star do not propagate through other star fields. It is preposterous to ideate any light-from outside a star field-passing close to the sun. Light or any form or energy dwindles, propagating with distance. With distance, light will be transmuted to other forms of energy/radiation, some virtual/invisible, or not observable, although some forms can be sensed.

Light is not an 'entity'. Space transmutes to light, and it appears to us that light is propagating from there to here, or vice versa. As space is a form of energy, and transmutes to light, we cannot measure or calculate the speed of light. If space transmutes to light, and both space and light are forms of energy, which further transmute to other forms of energy, and if space is everywhere manifesting light, how can space curve, and light bend.

Einstein noted that light bends, and this was confirmed during an eclipse, when previously seen star light was observed bent around the sun. We see light at the stars, and/or at the sun. We do not receive light from our sun, or light from any other star. Those stars that observers (confirming Einstein's theory) noted were light at other stars, not from other stars.

The sun has atmospheres extending millions of kilometers. The light seen at those stars, through the sun's many layered extreme temperature atmospheres, would obviously appear to be 'bent' (the sun is a sphere). Moreover, how can light from those stars be the same light, apparent around the sun with its extremely high temperatures and pressures, which constitute the corona? 'Bend' is our word and concept. Light does not 'bend' as it propagates, and does not 'travel', as a line or a beam. Space transmutes into light manifesting 'light', and as space is everywhere, light cannot propagate as a single ray, or a beam. Therefore, there is no such thing as curved space, or bent light.

REVELATION THREE

NO SUCH THING AS UNIVERSE

Fallacious concepts/theories within illusory thinking-Rethinking cosmology and theoretical physics

Numerous books and scientific publications, have expounded the Big Bang theory of the origin of the universe, as an indisputable fact. Readers will be astonished reading this revelation, refuting and disputing this theory, or the idea of a single cataclysmic explosion.

We have conceptualized the 'universe' explaining it to mean a whole body of things, observed or assumed (the cosmos). Only existential energy exists, and as it transmutes there are existences (with human understanding), as states, processes, structures or forms of energy. Existences exist/endure, as/within the fabric of space/existential energy. Therefore, in reality, there are no entities, and no 'universe' or 'cosmos', and no space-as a void. We have to rethink cosmology, and theoretical physics.

NO SUCH THING AS UNIVERSE

CONFUSIONS FROM CONCEPTS WITHIN AN ILLUSORY LANGUAGE

Migrating from their millions of year habitat, primates uttered sounds (words) to describe things or objects, existences, or phenomena. This enabled communications between themselves. Only over a few thousand years with the accumulation of sounds or words unfolding into a language, primates 'became' humans. A sound represents/depicts an idea. Within/with ideas we can 'know', or we 'think' we know, or can know reality, or the real world.

Humans are primates, and we arrogate superiority (creativity, ingenuity and intelligence) to ourselves, as we live (think, feel and action) within our own generated illusory language (revelation 33 elucidates). As humans think within their own generated language, everything, our entire lives is an illusion. As we imagine/ideate (within sounds-a form of energy) within this illusory language, we query about existences/beings. Each sound or transmuting energy symbolizes an idea. Within our language, we can therefore, think, imagine and query. We create as we query. However, as each sound-as an idea-is our own creation, we can imagine, and also query, but we cannot expect a correct answer/explanation to our own creation. We also cannot hypothesize or theorize, and expect reality answers.

Revelations 1, 36, 38 and 40, detail how sounds/words/concepts (from within our own generated language or illusory world) describing things or existences, or phenomena of the real world, has produced confusions and complexities. From the beginning of humanity, philosophers/scientists have been unsuccessfully trying to answer profound questions on gods, existences, beings, beginnings, energy, or reality. This explains why their explanations are incomplete, and or inaccurate, with inconsistencies and paradoxes.

Universe is 'Universus', and in Latin means 'the whole thing'-all that is-the whole system of things, or in Greek, the 'cosmos'. The concept ‘universe’ is postulated to contain the heavens, or billions of galaxies, each with billions of star fields, or their systems containing stars, nebulae, planets, asteroids, and astronomical objects.

There is no such thing as space-as a void-but the existence 'space'-as a form of energy. Therefore, in reality, there are no entities. As there is no space (as a void), and there are no entities, and everything is energy everywhere, it is wrong to conceptualize the 'universe'. The concept 'universe' is postulated to exist within space-a void. With the Big Bang theory the concept universe is finite. We cannot finite any existence without ‘entities’. This explains why we also have to postulate causation and creation. Only existential energy exists, everything and everywhere.

We created the concept 'universe,' as we have always assumed every existence results from causation. As the concept 'universe' is a human construct, every hypothesis, speculation, assumption and postulation about such an 'existence', its extent, and contents, is fallacious, and preposterous. The whole body, or system of things, we have conceptualized, and postulated as the 'universe' is in reality, boundless existential energy.

As there is no finite 'universe', there is no need for a bang. Boundless existential energy does not need a creator. Energy is/as everything, everywhere continuously transmutes, as it has been eternally, and ever will it be. The postulated 'universe', and everything else, exists only within our thoughts, or illusory language minds.

THEORIES OR STORIES

Some people argue that everything in science is theories, and they are not scientific laws, or laws of physics. A scientific law must be hundred percent correct. But what is law? Humans create both laws and theories. Who is to judge a law, a theory, or if both are correct or incorrect? What is reality, and how can we describe it? Theories are just what they are-stories. As the theory and/with its concepts 'originate' from within an illusory language, trying to explain the concepts of the theory engenders more questions than answers. More questions generate more theories. Eventually, all these related collective theories/concepts become stories. As we were brainwashed with fairy tales throughout our adolescence, we are now accustomed to believing theories. Inadvertently, the majority have accepted the Big bang universe originating theory or story, as a confirmed indisputable fact.

For some years now the genome project became a reality. We were brainwashed to accept the fact that we would now finally know what life is, and how it came to exist. Today, we are more perplexed, unable to find out what came first, life or DNA.

We grew up being believers in the gods postulated to have created everything, as science cannot prove otherwise. We accepted the Big Bang universe, as other related theories cannot describe the postulated 'beginnings' of the theory 'universe'.

Cosmologists proposed the Big Bang theory, the Steady State theory, and other theories to describe the ‘beginnings’, and existence, of the empirical enormous entity they postulated as the ‘universe’. With the Big Bang theory, cosmologists are continuously trying to demonstrate the origin, formation and existence of their concept ‘universe’ from a point, estimated originating 15 billions years ago. In other words, they are trying to explain their previous concept 'universe'-one unexplainable theory with another.

There are theories that the concept universe is one of an infinite number of universes, all probably popping out of the vacuum, from a point! In the meanwhile, many of them proposed the theory of membranes, multi-dimensional universes, and parallel universes.

Relativity, quantum mechanics and the ideas from the string theories are stories flummoxing and boggling minds (beyond our wildest imaginations) all these years. Theories generated from within our illusory language trying to explain reality, produces more questions than answers, and more confusion, contradictions, complexities and paradoxes. When, and will this madness ever end?

The following are two good examples of scientific madness.

Quote

Quantum mechanics speculates the 'universe' could appear as a tiny object, out of nothing! Normally such a universe would disappear again quickly, but under circumstances it might not. A universe would inflate gaining extreme energies with the original gravitational field, and would not disappear. However, eventually it would slow down, and halt its expansion, begin to contract, and return to its original pin-head size, and extreme temperatures, and then disappear into the nothingness from where it came from. Unquote. What preposterous logic! Why should a universe appear? Why should this universe disappear? Why should it inflate? Why should it stop its expansion? Why should it expand faster like never before? Why should it shrink in size? Why should it eventually disappear into the nothingness it came from? Where do the enormous temperatures come from? These are good stories to keep you thinking, guessing and waiting for explanations, which will never be forthcoming in your lifetime.

Here is an excerpt, explaining the dimensions of the postulated 'universe'. Get your minds flummoxed.

Quote

Cosmologists are now postulating that the universe is 156 billion light-years wide. It seems that the universe is about 13.7 billion years old, as light has been reaching us from the earliest known galaxies, and has been traveling for more than 13 billion years. So one might assume that the radius of the universe is 13.7 billion light years, and the whole thing is double that, or 27.4 billion light years wide. But this distance covered by light in the early universe, gets increased by the expansion of the universe (a batch of light traveling for a year is the equivalent of about 5.9 trillion miles, or 9.5 billion kilometers). At that time, the universe was about 1000 times smaller than it is today. Thus, that one light year has now stretched to become 1000 light years. All that adds up to 78 billion light years away (the radius), and twice that is 156 billion light years or the diameter. The findings rules out any chance of seeing our ancient selves unless we can invent time travel.

If the universe was finite, and had a size of about 4 billion to 5 billion light years, then light would be able to wrap around the universe. With an advanced telescope, we could view the earth just after it solidified, when matter formed, evolving first life forms, but unfortunately the results rule out this teasing or tormenting possibility". Unquote.

This information must have certainly stretched your mind. We are informed with so many theories, that we would have, and/or can find out how life first formed, if cosmologists had a big enough telescope, if the postulated universe was finite, and if light years have not stretched!! We can clearly understand that as science gets deeper into the nature of things, everything gets more and more bizarre, warranting more explanations, and yet further theories/stories. This is the consequence from thinking, imagining, and querying within an illusory language mind. Theories are just stories.

Here are two theories contradicting one another. The Big bang theory states that the universe was once a dense hot body of matter that exploded, and has been expanding since 10 to 30 billions years ago. Einstein's famous formula shows that the material could not exceed the speed of light, as it expanded. Mass then becomes infinite (and time stands still), as the speed approaches the speed of light. The energy increases from the bang increases the speed, beyond the speed of light. Mass will therefore, increase proportionately, as the initial energy is increased.

One cosmologist has suggested that the material forming their postulated 'universe', blasted out from the Big bang at a speed greater than the speed of light. No theory can explain how matter can move at a greater speed than the speed of light, and this also violates Einstein's theory of special relativity. Now wild ideas are being presented to support, and save the Big bang theory

The Big bang theory violates the basic foundation of science, which is Einstein's theory of special relativity. However, even this theory is also incorrect. The famous formula E=MC2 gives the relationship between Energy (E), mass (M), and the speed of light (C). This theory/formula is behind the atomic bombs for both fusion and fission reactions, so it presently stands without challenges. What is the difference between matter and mass? Mass and light are forms of energy (detailed revelation 11), and therefore, as energy is everything, everywhere, it continuously transmutes into different forms, states, processes or structures of energy, explaining different forms of energy in fusion and fission reactions. Matter is not different from energy, but a form of energy. Matter is condensed energy, and as it integrates or disintegrates, different forms, states, processes or structures of energy manifest.

THE BIG BANG THEORY

The Big Bang is the dominant of all theories on the origin of the postulated 'universe'. The big bang theory explains the postulated universe consisting of energy; space, time, and everything else, were created after one enormous explosion-the big bang. All empirical objects/things (existences), and continuing phenomena are part of an enormous entity, that oozed out from a point-whatever and wherever that is.

The Big Bang theory is a familiar story with its contents changing with time, and as more and more problems and paradoxes arise, it continues to change. Speculating and formalizing the Big Bang theory to prove their postulated 'universe', or the so called collections of galaxies, originating from a point somewhere, has generated more and more problems for cosmologists. Untold human hours with vast sums of money (as space probes and other research), are spent to present evidence for, and against this prevailing unproven theory.

Why should there only be a single Big Bang? Are there many universes that originated from many other points? Would there be another Big Bang originating another universe from another point, and amalgamating with ours? What happened before the big bang? What powered the Big Bang? What event started the Bang? What started the ignition? What caused it to explode? Where is the point to be able to find out the centre of the universe? What was before the 'point'? How can the 'universe' ooze out with such extreme conditions, or enormous energies- temperatures/forces, space and densities?

How can the material 'universe', including time, all of space-as a void, all dimensions, all energies and matter, be condensed by gravity into the size of a point, which exploded into the massive universe of stars and galaxies? How can something come out from a point-from inside, terminating everything inside, and then create an outside, facilitating that something to now exist outside of the point? Some cosmology books describe the conceptualized universe from the big bang containing all of space and time, as part of a larger possibly infinite universe.

Space-a void-as explained in revelation two is a human construct. Space-as a void-cannot be formulated from non-existence, and something non-existing, or a void, cannot be created from another void, or non-existence? Where is the space, or time that existed before the Big bang? Why cannot we ask what was before the Big bang.

The singularity, the smoothness question, the low density, the age of the universe, the flatness of galaxies, the horizon problem, the inflation theory, the missing mass, dark energy, dark matter, the initial background radiation, the abundances of light elements, the red shift anomalies, inconsistencies and their interpretation, the accelerating and sudden re-accelerating faster expansions, and a host of other problems associated with this theory is continuously being brought up for debate or discussion.

People still believe in the Big Bang expanding universe theory as real. Astronomers suggest that some type of mysterious antigravity force is pushing the galaxies apart-faster than ever before, and they are more distant to each other than they were. Nobody knows where this force comes from, how it is powered, and how can it be continuous to expand galaxies.

If the concept 'universe' is a human construct, why do we propose theories to prove its origins? Obviously, we will never be able to explain our concepts, thinking from within our own illusory language. As our ideas 'originate' from within our illusory language, we live in a no sense state, and can never understand the states of nature. However, this generally accepted nonsensical story is good, as it tickles curious minds, and fires up human imagination.

The theory of the Big Bang has many weak points, and too many complexities, and unanswered questions, and it is still being modified to being perfect.

As existential energy has no beginnings or ends, the Big bang theory is just what it is-a story. Defuse it before it blows up minds. Without a 'universe' and 'space' (as a void), Newton's dynamics are preposterous, and Einstein's space-time and relativities have no meanings.

Revelation 17 details why, and how curiosity stimulates both physiological and psychological mechanisms of creatures. These natural processes evolved humans, progressively from amphibians, reptiles, and other mammalian ancestors. This innate, or intrinsic characteristic, makes us seek, and like stories.

Why was the steady state universe theory discredited?

The steady state theory states that the 'universe' is the same throughout without changing. Even if there are changes within, existences eventually transform into the same thing as before. Existences spontaneously appear and disappear, from, and to the same thing-energy. In a way, the steady state theory is correct, as energy continuously transmutes creating existences, and existences eventually terminate into the same thing-energy. No existence possesses the same characteristics at any moment of time.

The Big Bang theory explains the universe is expanding, and galaxies are receding away from each other. Eventually, all matter will end in a big rip-off, or come together in a big crunch.

Humans have conceptualized 'universe'. Creating an existence we have to explain it, one theory after another. However, there is no such thing as a 'universe', as we have to finite it to create an existence.

The steady state theory was discredited for two reasons.

Quasars (as a theory) are very distant, bright, star like objects, and we have not yet completely studied them. Quasars reproached the posited 'universe' steady-state theory, as it would have them distributed evenly throughout the 'universe'. It is possible, that a number of stars are exploding, and their material drowning into black holes, and this phenomena being postulated as quasars. Therefore, we cannot expect them to be evenly distributed throughout existential energy's expanse, as galaxies are everywhere. So discrediting the steady state theory with the explanation of Quasar theory is incorrect.

It is postulated, that deuterium and tritium were produced at the Big Bang nuclear-synthesis. We are informed that there are no known natural processes producing deuterium. This is why arguments favored the Big bang theory over the steady state theory. This is also incorrect. Revelation 4 and 8 elucidates how fragments from an exploding neutron star generate nuclei, or deuterium and tritium.

REFUTING AND DISPUTING THE BIG BANG AND OTHER RELATED THEORIES

From many observations, there are some unanswered questions or problems with the Big Bang theory. Some are discussed following:

- The uncertain origin of the postulated universe-where did it come from?

- What created the particle-antiparticle asymmetry in the first moments of the big bang universe?

- The finding of too many very distant galaxies, when the Big bang universe was only about a tenth of its assumed age. This created a great confusion among cosmologists and theorists.

- The calculated age of the oldest stars and galaxies is greater than the calculated age of the 'universe'. This would mean that we live in a 'universe', which is younger than some of its bodies?

- It is uncertain how the expanding early 'universe' managed to produce ordinary galaxies and cluster of galaxies.

- The clumpy large scale galaxy distribution (the clumps of matter) is a major theoretical problem.

- The ejection of material from the galaxies, which present evidence of the existence of an unknown force against the postulated gravitation.

Formation of galaxies

Even if the conceptualized 'universe' 'began', as theorized at the Big bang, logic denies the possibility of the accretion of matter, required for the formation of new galaxies in an expanding Big bang 'universe'. The formation of galaxies and clusters is a big mystery. The Big bang theory indicates the universe started out so smooth, but such gigantic clusters, and clusters of galaxies could not possibly have formed within the time since the Big bang took place. How does an explosion like the Big Bang create such smoothness, which is experienced, and observed today?

We see galaxies distributed throughout and everywhere, some are close, and some far away, and they are every side of us. We have been informed that everything is expanding, but the original mass or energy would have to keep moving out galaxies continuously, over billions and billions of years, to produce the uneven distribution we see in the universe.

The standard theory explains that cold dark matter, or unidentified sub atomic particles provided the base around which matter collected into large uneven concentrations. The cold matter theory does not explain the initial smoothness, and also the large clusters of matter appearing relatively soon after.

As cosmologists look further into their 'universe', they always see galaxies collected into groups, and those groups conjoining into larger groups-the structures look smoother and smoother. The structure seems the same on all scales, and on the very largest scales, galaxies are homogeneous, or they are same in all places, and isotropic, or the same in all directions. It seems that some disturbances in matter distribution condensed to form stars and galaxies. Nobody knows what these disturbances were, and how it was initiated.

It is not known how the postulated universe remained very dark and featureless for over 200 million years. It is assumed that the first stars appeared, which were very huge, as much as a thousand times, as large as our sun. Only after one billion years later that stars and galaxies began to proliferate. However, today new stars continue to be born and die.

The Separated galaxies

Why have galaxies separated? Why have star fields not separated, but clump together to form galaxies?

All galaxies are in momentum, engendering varying densities. Continuous altering densities engender new star fields, as old ones explode, or collapse. Different galaxies are made up of varying quantity of star fields. Some have more or less, or larger or smaller star fields. Star fields engender anywhere within a galaxy, as nascent or old ones collapse, or die with abnormal energy transmutations. If all energy, matter, dimensions and space sprung from a single point, there should have been only one gigantic galaxy occupying the whole of space. This galaxy would be entirely filled with only matter, or only radiation, or star fields.

Galaxies or a collection of star fields are like islands within this boundless stretch of existential energy. Galaxies are not entities, but are interlinked and interconnected, within/as the fabric of this boundless stretch of existential energy, three dimensionally. The separated galaxies are solid proof the concept 'universe' is incorrect, and obviously did not originate from a point in the Big Bang Theory. The next revelation details why galaxies exist as islands within existential energy's expanse.

Stars and Galaxies

The Big Bang states that the original explosion sent out clouds of hydrogen and helium gases, which compressed and coalesced into galaxies and stars. Clouds of gases do not compress, and coalesce into a more dense mass. Gases expand with heat and pressure, 'becoming' less dense and therefore, stars cannot be formed this way (revelation four details the engendering of stars). How can we explain star fields terminating, and engendering again?

The age of the postulated universe

Seventy years ago, Edwin Hubble discovered that the 'universe' is expanding, and assumed that the postulated 'universe' was ballooning from the point in the Big Bang theory, 2 billion years ago. In the 1920s, researchers estimated the 'universe' to exist for 2 billions years. By the 1960s, they improved their measurements, that it existed somewhere between 8 and 20 billion years. Then for more than one third of a century, little progress was made in reducing the uncertainty.

During 1998, astronomers stumbled upon some evidence, that the hypothetical entity 'the universe' is not only expanding, but the expansion is accelerating over time. Now different methods converges the estimates around an age of 14 billion years.

Recently researchers using the Hubble telescope, found the faintest of stars called white dwarfs, and were with full agreement with other estimates, that the age of the 'universe' is just under 14 billion years. An international team of astronomers discovered a galaxy, appearing to be the most remote ever observed, at an estimated distance of 14 billion light years. As distance translates into time-it takes light from a galaxy 14 billion years away to reach us-the distance estimated also measures the age of the object relative to the Big Bang-the origin of the 'universe'.

Many cosmologists speculate that their postulated universe has just enough mass to balance its expansion, and this will slow down until it stops. If this is so, the age of the universe would be only two thirds the original estimates-between 8 and 11 billion years. Globular clusters are calculated to be 12 and 18 billions years old, and it would be interesting, and embarrassing to find 14 billion year old clusters in a universe, which is aged 8 billions years.

Nuclei and everything else are unstable, and will transmute, transform or change, depending on alterations in temperatures and pressure/force. A nucleus can therefore, last for an instant, or for any duration. As everything is energy in continuous transmutation there is constant flux. Radioactivity, decay, or half-lives, cannot accurately determine chronological time existence. Therefore, the age of the hypothetical 'universe' (incorrectly calculated-vicinity 15 billions years) with various observations and mathematical systems cannot be measured by the movement of stars (analyzing light), or from the age of stars, and from the age of nuclei.

The discovered regions that hold the 'edge' of the universe, expanded from a point, can be observed with technology that can take them only so far. In the near future, advanced, and more advanced instruments, will discover more and more galaxies, some further away, defying the expansion, and altering the age of the concept 'universe'.

'Light-years', 'distance', 'travel' are human concepts. We can calculate a light year only if light 'travels', and importantly if it travels, does so, at the same speed through any distance, throughout any form of energy. Light is a form of energy, and all forms of energy dwindles with distance, altering their properties, propagating/transmuting as space-also a form of energy. Light does not propagate as 'light' (at light intensities), but other forms of radiation with/through distance (revelation 9). Now what is a light year? Without a light year we cannot calculate the age of the postulated universe. Calculating the 'beginnings', or the age of the postulated universe considering 'light-years' is therefore, preposterous.

Totality and varying types of matter in the Universe

After the big bang, why should one second pass when the basic building blocks of matter are formed? After the first three minutes of the Big Bang, why should a long time elapse (as temperatures were too high) for nuclei and 'electrons' (to be at low enough temperatures) to coalesce to form atoms? Why should three minutes elapse for helium, and other light nuclei (deuterium) to form? It is posited 400,000 years after the Big Bang (the theoretical beginning of the universe), nuclei and electrons cooled enough to form atoms. From where did neutrons come from, and how were the forming nuclei shielded from radiation?

According to current physics ideas, energy in the early universe should have produced an equal mix of matter, and antimatter, which would later annihilate each other. It is supposed, that some mysterious, and very helpful mechanism tipped the scales in favor of matter, enough to produce galaxies full of stars, but nobody knows what it is.

It is crazy enough understanding something coming out from nothing-from a point-but it is gets crazier still to comprehend why particles, and anti-particles (invisible and hypothetically conceived), or matter and anti-matter, must not only be created, but immediately annihilated. We theorize anti-matter, but as there is no other theory supporting this idea we have to annihilate it. We emphatically state, that energy cannot be created or destroyed, but freely use the word 'annihilate'. What is annihilated if matter is a form of energy, and energy cannot be created or destroyed?

With improved telescopes physicists measured, and calculated the matter in the universe, assuming they reached the first galaxies from the Big bang. To correct their equations, as the calculations did not meet their expectations they introduced antimatter, and Weak Interactive Massive Particles, or WIMPS, and MACHOS.

We are informed that the observed motions of stars and galaxies require the existence of large mass that binds them gravitationally in galaxies, and cluster of galaxies. This unknown invisible dark matter is calculated to account for at least 90% of the mass in the universe. We see only a fraction of the world that emits light or radiation. It is assumed that there is dark matter, as gravitational attraction is necessary to motion stars and galaxies in their systems, and galaxies would motion faster if only visible matter is present. Things become even more confusing, with the discovery that the motion of stars and galaxies, cannot be confined from the gravitational attraction of the light emitting matter. There is a dark matter problem in the Big bang universe, which is very dark-not because of the darkness! Now physicists are trying to solve this problem creating existences for dark matter like neutrinos, axioms and so on.

As this dark mass or matter is conceptualized, and created, it is uncertain. We cannot assume, that stars represent about one percent of the 'universe's mass, and non luminous ordinary matter such as asteroids, nebula, planets and black holes making up another 9 percent, and the other 90 percent is dark matter. How can we account normal matter to be only about 5 percent of the material in the 'universe'?

How do we define mass and/or matter? We have to first define a mass particle, and matter. How much of energy condenses into the smallest invisible energy particle, and how much of it into an observable energy particle? Everything is energy, and any mass or energy particle exists, as it is observable (or according to physics, if it can be measured, to conform to their quantum theories).

Energy is both dark (invisible), and also observable, as it condenses to a state what we call matter. All observable transmuted forms, processes, structures and states of energy are matter. As we cannot observe energy in its static, or invisible state (radiation, temperatures, light, color, sounds, and particles) we cannot speculate its totality. Everything being energy, invisible or observable, we cannot in reality, calculate, individualize, or differentiate energy from mass, to calculate the amount of helium or hydrogen, or anything for that matter, anywhere.

The postulated 'universe' is a human concept, and until we can physically determine (observe 'invisible' energy) and measure its edges or boundaries, this postulation is preposterous.

The background radiation

A faint background radiation coming from all over the sky filling the entire 'universe', left over from the Big Bang, remains the single greatest piece of evidence in support of the Big bang. But the background radiation implied that the early universe was evenly distributed, and evolved at a constant rate. In this case we should have observed evenness of distribution, but we find stars and galaxies.

To continue to support the Big bang theory, cosmologists then introduced the 'inflationary theory', which states the early universe underwent a quick expansion, at a rate much faster than the speed of light, and then slowed down.

For several decades the weak radiation measured the same, but now this radiation is not perfectly uniform, as it is not the same in all directions. Why? But now the background radiation has gone missing.

The back ground radiation is too weak, and if it came from the Big bang it would have a stronger measurement. The radiation is too smooth, or even, comes from all directions, and not from the suspected Big Bang origin.

Within our solar system, there are a number of energy transmuting cycles between the sun and planets/bodies. The microwave background is the consequence of these processes, as each star field will produce this phenomenon.

Varying temperatures at different regions

According to the Big Bang theory, the postulated universe came into existence about 15 billions years ago in a colossal fire burst of energy and matter, and has been expanding ever since. Vast regions of space widely separated, have same temperatures throughout. How can there be hot and cold regions? This would seem that every part of the postulated universe, that came out from the point in the Big Bang was, or was not connected. How could this be possible? Moreover, the temperatures of the background radiation would have had to be extremely high in the past.

The spin of astronomical objects

The Big Bang theory does not account for the spin of astronomical bodies/objects (on their imaginary axis) within star fields, and why they orbit their star, with the star itself rotating around on its imaginary axis. There is also no explanation why entire star fields rotate, and why they revolve around the centre of their galaxy, making the entire galaxy to revolve, and also in continuous momentum. Revelation 7 has all the details.

The expanding universe and moving apart galaxies

The expansion of the ‘universe’-as cosmologists try to prove-is the easiest and almost inevitable explanation, that the postulated universe is the result of an explosion at the beginning-the Big Bang Theory. For some decades now, cosmologists have stated that the postulated 'universe' is expanding, but during this decade they have expressed the expansion of the universe appears not to be slowing down, and it is not only continuing to expand, but accelerating in its expansion.

Accepting the Big Bang theory as real, astronomers from studies of the 'red-shifting' light have suggested, that the postulated universe is expanding, and accelerating faster than ever before. Moreover, they have stipulated some type of mysterious anti-gravity force (which they now assume as dark energy) is pushing the galaxies apart-faster, and faster-the more distant they are. This force has never been observed in any experiment, neither are there any explanation where it can come from, and how its continuous accelerating effect is powered.

During ongoing refutations, cosmologists changed the explanation of the expansion of the material 'universe', explaining that it was space expanding, not the galaxies, and not the universe itself. They state that the Big Bang explosion was an expansion of space. As expanding space has no speed limitation, the speed of light is not a limitation in the expanding space theory. As the speed of light is only a limitation within the boundaries of space, this means that space is finite, or has edges. This concept is similar to those old ages, when sailors were warned not to venture out too far to sea, for fear of falling off the edges of the flat earth.

Theorists suggest space or time began only at the Big Bang, and did not exist earlier, and the matter of the 'universe' is not expanding, but it is the 'space' of the universe that is expanding. This is another theory proposed in an attempt to keep the Big bang story alive to a longer life.

How can we comprehend that such an enormous force from a point will continue this inflation thousand million years later?

The proposition of a faster expanding 'universe' now than before ironically proves that the Big Bang theory is incorrect. The expanding 'universe' (if it came out from a point, and keeps on expanding) would also expand galaxies, and obviously star fields. With carbon dating, and other records, and analysis, our solar system has been calculated to exist for about 5 billion years. If galaxies were expanding, our solar system would not have endured in the same format as it is today. It would have been long disintegrated, as it is localized on one of the spiral arms of our Milky Way Galaxy. Around the region of spiral arms, there is greater density fluctuations, with greater possibilities of star fields terminating and engendering (spiral arms are observed with more star fields), than other star fields closer to the centre of a galaxy.

Measuring expanding galaxies, and the present acceleration

Astronomers measure exploding stars called supernovae, which are far away, and postulating their brightness can tell how distant they are to each other. The distance also tells them how old they are, as it takes time for light to get here. It is also possible to judge the speed with which they are retreating by measuring the spectrum of the light stretched in its travel. This is incorrect, as all stars are not of equal dimensions, distances, densities and luminosities. Over distance, light does not propagate as light, but alters its characteristics to other forms of radiation.

We have been informed, that as astronomers examined the light from more and more distant supernovae (whose light is assumed to arrive from the early universe), they were surprised to see the closer, and more recent they were, the faster they seemed to be rushing away. These stars were rushing away, as they were carried along by the expansion of the universe, and not by their own power. It seems that many billions of years ago, dark energy was weak, and gravity was 'causing' the universe to slow down, as it expanded, but at a crucial point dark energy took over, and began speeding the process. Why did dark energy weaken in the past? Why and how did it strengthen? Why should dark energy take over from gravity?

If the red shifting measurements are incorrect, the whole Big bang theory collapses. Let us turn now to some very important refutations, which will have serious implications, on how these observations have brought about the speculation of the expansion of the hypothetical universe, and also the acceleration of the expansion, faster now than before.

The Hubble value constant

There is a controversy about the value of the Hubble constant that gives the rate of the universe expansion, and hence its age. The calculated ages of some stars are older than the calculated age of the universe. We are also informed that the real universe may have begun, and evolved differently from the Big Bang one. The age problem resulting from the measurement of the Hubble constant suggests, either the standard universe model needs to be revised, or the present theories, and or observations, dealing with the evolution of star fields and galaxies must be looked over again. How can the universe be younger than any of its bodies? This will put the Big Bang theory once again in disgrace.

The Hubble constant is estimated from assumptions, and used for the calculations of cosmic distances. The Hubble telescope studied a very small piece of the sky during consecutive days, collecting light from the observed most distant galaxies. To obtain a detailed 3D distribution of light from these galaxies will take millions of years.

The irregularities of the Red shift

The Big bang theory rests on a key assumption that the postulated universe is expanding. Before the twentieth century, astronomers assumed the universe had existed as it was, 'forever'. In the 1920's theorists uncovered evidence the universe was changing.

Edwin Hubble found that stars exist in huge groups called galaxies, and by tracing the light of those stars, he also found that all matter in space is moving away from each other. Milton Humason who was Hubble's colleague photographed distant galaxies, and determined some galaxies were moving at about one-seventh the speed of light-proof that the universe was expanding. This assumption made the Big bang theory much more likely to be true. This ushered in the era of modern cosmology-the study of the origin, evolution, and structure of the universe.

We are informed that the speed with which a galaxy is moving is determined, by observing the light spectrum emitted by a galaxy. The degree to which the spectrum is shifted toward the red end (known as the red shift) in a spectroscope, indicates the speed at which a galaxy is moving away from us.

The red light shift experiment is the very foundation for the Big bang theory. The concept that the postulated 'universe' expanding from a single source, at a real point of time in the past, was developed as Hubble discovered the red light shift. He found that very distant galaxy clusters were emitting light with a red shift.

The thought was that a distant body was moving away from us, 'because' the light emitted had a shift to a longer wave length, ‘caused’ by the velocity. This is the Doppler effects in sound waves.

As described at revelation 9, sound waves are different from light waves. Sound is produced by activated nuclei, which conveys sound by relaying the disturbance from nuclei to nuclei, by waving space in between. White light from an object moving away from us appears in the red end of the spectrum, and if that light moves toward us, it appears in the blue end of the spectrum. This red and blue shifting has earth at the centre of the postulated 'universe', as nearly all galaxies are moving away from us. This is Galileo's story resurrected again.

What are the red and the blue shift?

A red-shift is one type of Doppler effects. It states, that if a sound or light is moving away from an observer, its wavelengths will be lengthened. Conversely, if an object emitting light or sound is moving toward the observer, its wavelengths will be shortened. In the case of light, longer wavelengths stretch to the red end of the color spectrum, and shorter ones bunch up at the blue end of the spectrum. The shortening of wavelengths of an approaching object is a blue shift.

Edwin Hubble and Milton Humason photographed distant galaxies, and learnt their spectra were all shifted toward the red wavelengths of light. Further studies showed a relationship between the degree of red-shift, and that object's distance from Earth. In other words, the greater an object's red-shift the more distant it is, and the faster it is moving away from us. This finding suggested that due to the large degree to which these galaxies' spectra were red-shifted, they were receding at a very fast rate. Every galaxy is moving away from every other galaxy at a speed proportional to their distance, and therefore, the universe is expanding.

During the 1890s, Percival Lowell gained fame stating he could see a vast network of canals on Mars. Turning his telescope to Venus he also could see peculiar lines crisscrossing the surface of Venus, which were called the spokes of Venus. He speculated the spokes were rocks or sand, weathered by millions of years of exposure to the sun. Recently ophthalmologists pointed out that by severely constricting, or making some adjustments to his telescope, Lowell had converted it into an instrument eye doctors use to see blood vessels at the back of the eye-Lowell was seeing inside his own eye.

It is possible, that our measurements using sophisticated telescopes are incorrect, and refutations corroborating this statement are as follows: Any observed red shift is always accompanied by an equal, and opposite blue shift, with the magnitudes varying widely, and not dependant on distance. Why should this be so?

Red shift data results in the comprehension of extremely great masses, and brilliances of quasars. Variations in the level of radiation from these sources necessitate their size to be extremely small, and their densities to be extremely great. These properties suggest that the present calculations of red shift data as Doppler shift, do not present the correct, and complete story about the speed, and distance of remote bodies in space. Distant supernovae cannot be measured for distance, as they are not of all the same size, and therefore, have different luminosities. So using the red shift to measure distances, brilliances and magnitudes of distant objects is incorrect. Moreover, quasars seem to exist farther away, and are intrinsically brighter, and they are like stars with varying brightness. How do they stay so bright at such a low temperature, and in such a small volume? They must have a different red shift, and should exist nearby to be explained.

In addition to the quasar problems, there is evidence showing that the presently accepted calculations of red shift data are to some degree incorrect. Some galaxies show higher red shifts than their companion galaxies. Radiation from newly formed galaxies is in some manner red shifted by other than Doppler effects.

New evidence found proves a number of galaxies were in existences at even earlier times than previously thought. Red shifts between 4 and 6 mean that they were in existence about 10 billion years ago, when the 'universe' was six times smaller than it is now. This makes the era of galaxy formation earlier in the history of the 'universe' than astronomers previously thought.

Energy in the form of temperatures and pressures/force and plasma, continuously propagate/transmute from stars or the sun, throughout their solar system/star field. There is light propagated from the sun with other forms of radiation, or temperatures and pressures, but light (as is the case with any other form of energy) dwindles with distance. We are incorrectly informed the speed of light is constant, but we cannot measure the dwindling intensities over large distances. Light from a more distant galaxy would dwindle, and alter to other forms of radiation, as a result of its distance from earth. The more distant galaxies are the greater is the red shift in the velocity of light, showing that those galaxies are not only moving away, but moving faster now then ever. This is exactly what the red shift measurements show. Therefore, dwindling intensities of light prove the Big bang theory wrong. Energy propagation from the sun in/as any transmuted form-light, temperatures, pressures or force, sound, color, dwindles with distance. This explains why light from a street lamp, candle, or a fire, can only 'light' up areas to a certain distance.

Temperatures, other forms of radiation with pressures/force from the sun reach earth's atmosphere with dwindling intensities. Light (produced at the sun) as a form of energy dwindles with distance, and does not reach earth as 'light', but other forms of radiation. This explains the darkness at night at both sides of the horizon, and throughout interplanetary space. When the side of the earth is not facing the sun, that part of the atmosphere is not radiated heat interacted, and there is no visible light and therefore, darkness.

We see light (incandescence) at the sun/stars (as it glows), not light 'arriving' from the sun/stars. Light is a form of energy that dwindles, propagating within/as space (a medium through which it propagates-waving space-and not 'travel'-as traveling independently-there are no entities). Radiation/light from stars cannot propagate outside their star fields, as it would confront radiation/light from other star fields. Moreover, light and other forms of energy from a star/sun dwindle with distance, explaining why the outskirts of a star field/our solar system is cold, and or icy. The lines in the spectra of certain stars were displaced toward the red end of the spectrum, not as a result of light propagating to us.

Let us assume light 'travels' at the same speed. During daylight observations, light/radiation from the sun is also propagated on the other side, toward the outskirts of our solar system. Therefore, during the day, light/radiation from astronomical bodies (outside our solar system) would have to first confront light/radiation from the sun. Then, as that light/radiation passes the sun, it would have to propagate together with light/radiation from the sun toward our telescopes. How do we calculate light/radiation from a star outside our solar system, confronted with light/radiation from the sun, and within the same pathway then accompanies light/radiation to our telescopes from the sun? If viewing was at nights, light/radiation from bodies outside our solar system would confront the light/radiation from the sun within our interplanetary space. With night observations, the zooming in to far distances, together with the radiated heat from the sun within interplanetary space, and the confronting light from bodies (if light travels) will give us distorted observations. We cannot see stars during daylight, as the light at the sun is greater than light seen at the stars, behind the sun. In addition, our entire upper atmosphere is filled with atmospheric light. Night calculations would also be incorrect, as radiation from the sun would enhance our telescopic observations.

Telescopes sent above our atmosphere, or areas anywhere in interplanetary space would also present distorted measurements. How can we measure this light/radiation? How does light or radiation from other stars, nearby or distant (if light dwindles with distance), propagate to, and within our interplanetary space?

Our visual systems are comprised of atmospheric nuclei, and we observe light at the sun, and at the stars, through our atmosphere dominated by nitrogen. Nitrogen turns red with radiated heat. Simply, by passing light through gases or plastics, it turns red shifted. Revelation 9 details how atmospheric nuclei with radiated heat manifest colors.

The structure of light (propagating within/as space, as it waves) waves expands with time (and distance), as that of a water wave traveling from its source. Distant starlight (if it travels) passes through vast areas of various plasmas, and gases of star fields, before arriving (if it arrives) at our telescopes. Therefore, it is no surprise that a further a galaxy is, the more red shifted is the light seen.

Our telescopes zoom to a certain point in the sky, and it is from that point or area, from where we actually observe. Focusing on a particular object would obviously bring either the object, or the telescope, closer to each other. This should manifest a blue shift, or a neutral calculation making it nearly impossible to gauge, measure, or calculate distances, intensities, and other phenomena of energy transmutations, or light-if it travels. The further telescopes zoom on to distant objects, the further their point of observation increases. The blue and red shifting calculations are therefore, incorrect.

As we zoom in with our telescopes, we go through only certain areas of the atmosphere. Atmospheric light, or daylight is generated from radiated heat from the sun (interactions at our upper atmosphere), as we do not receive sunlight (or light from the sun-see revelation 9).

Our atmosphere or air consists of the dominating three gases, nitrogen, oxygen and hydrogen (being continuously extracted). With a certain intensity of heat, nitrogen turns red, oxygen green, and hydrogen blue. These gases interacting with the continuously incoming radiated heat produces our visible light. White or visible light is manifested with the combination of these colors. As nitrogen is the most dominant gas in the atmosphere, and hydrogen is continuously departing, there is an equal red and blue shifting. This explains the equal, and opposite, red and blue shifting.

Hydrogen is present in the atmosphere, but is not noticeable, and cannot be measured, as our invented instruments can measure only what is present (continuously transmuting) in the atmosphere, not what is being extracted/departing. Continuously disintegrating atmospheric nuclei and the water molecules in the oceans/rivers generate hydrogen, which is continuously extracted, quickly leaving our atmosphere.

Blue-shifting is normally assumed to explain the approaching of an object. As hydrogen is being swept away, there is an equal generation of atmospheric pressure the opposite way (the incoming pressure/force from the sun)-this is the blue shift. The red-shifting results from the dominant nitrogen in the atmosphere. As our telescopes zoom to areas outside our atmosphere, our atmospheric visible light is not significant in the shifting. Red shifting, and the equal and opposite blue shifting, cancel out each other, as our telescopes operate within the atmosphere. Revelation 9 describes how space transmutes as light, and other forms of radiation, waving at different frequencies. It is waving space generating the red and blue shifting. In reality, therefore, those stars and galaxies being observed are actually stationary. The postulated 'universe' is not expanding, and galaxies are not receding away from each other.

Nothing is a one way phenomenon. There are no entities, but only continuously transmuting energy. Within any star field or our solar system, energy continuously transmutes, shuttling to-and-fro between existences, continuously altering forms, states, processes or structures of energy.

Light or any form of transmuting energy dwindles, propagating with distance, and cannot be measured. This phenomenal reality is obvious, as light is manifested as space waves.

These arguments put the red and blue shift measurements or its calculations in serious anomaly. It will have profound implications on the origins of the postulated 'universe', the expanding 'universe', and measurements of distances of astronomical bodies, and distant galaxies using 'light-years'. Of course, there is no such thing as a 'light-year'.

There are billions of stars in our galaxy, and many galaxies have been discovered. Galaxies have always existed, and more will be discovered. To observe distant galaxies (if the speed of light is constant, and it travels), we have to probe through our solar system, and then confront light from other millions of star fields within our galaxy, and then further probe through dark energy between galaxies, and yet go further through star fields of millions of galaxies.

With our present technologically advanced instruments, we can probe to certain galaxies, which cosmologists speculate to be the earliest ones from the Big Bang. Finding more distant galaxies with more sophisticated telescopes will shatter the Big Bang theory. Five to ten years later with more sophisticated telescopes, and other technology, we will see distant galaxies that were formed even 20 to 30 billion years ago, changing our estimates of the postulated concept 'universe' origin estimates. Future generations will keep on observing more and more galaxies, proving that there is no such concept as ‘universe’, and energy is everywhere and is/as everything, boundless and eternal.

There is no such thing as the universe-period. Close the door to this preposterous Big Bang that is forcing cosmologists to bang their heads for many years now.

Dark Energy and the Expanding, Crunching and Ripping apart 'universe'

Cosmologists have been talking about a mysterious force called dark energy appearing to dominate the 'universe'. It is thought to penetrate their universe, and work in opposition to gravity, and appears to be a constant presence. This postulation produced a theory that the 'universe' will not end soon. As dark energy is a theory, and we have absolutely no information, or clue about it, how can we propose another theory? It is similar to presenting the Big bang theory to explain the postulated concept 'universe'. There is another theory questioning whether the highly accelerated expansion in the early universe, and the current accelerated expansion (due to dark energy), are related.

Dark energy was created in the first place to explain a phenomenal discovery in 1998 that nearly all galaxies in the 'universe' are receding from each other, at an even faster pace. It seems that gravity is loosing some unknown battle, and dark energy is changing over time. If this is right, the universe will continue to expand at an accelerating rate forever. If these theories are wrong, then the universe will not be destroyed by dark energy for another 30 billion years. Cosmologists think that right now dark energy does not seem to be changing fast enough to 'cause' an end to the universe very soon.

Scientists use the Hubble telescope to find stars that exploded when the postulated universe was about half its present age. A certain type of these supernova shine with a known brightness. So examining the light that 'reaches' Hubble, astronomers calculate how far each one is, and the rate at which the universe was expanding, when that star exploded. That rate of expansion has changed over time, other studies have shown. The initial expansion after a theoretical Big Bang was the most rapid, and was called inflation. Then things slowed down before another round of acceleration, which is apparently underway now.

Now there are other theories about dark energy. It seems that it might ooze out from empty space, cannot alter, and of a fixed strength. The other holds that dark energy is associated with a changing energy field called 'quintessence', something like a magnetic field. In this theory, the field 'causes' the present acceleration of the 'universe'.

Another theory circulating is the repulsion from dark energy as it gets stronger, and the 'universe' could expand so dramatically that it would end in a Big Rip, tearing away all matter-galaxies, stars, planets, and everything right down to the atomic level. If dark energy could change, it might also one day reverse course, and put the 'universe' back together in a Big Crunch. Now there are questions relating to the strength of dark energy today, and whether it grows, or decays with time? Presently we are informed that it seems dark energy's repulsive force is changing very rapidly.

Revelation 10 discusses why there is no such thing as gravity. Revelation 4 elucidates dark energy, and why it cannot transmute, but continues to exert its force/influence confining galaxies, and vice versa.

Edges and Boundaries of the postulated 'universe'

Expansion and Contraction of the postulated 'universe'

If there is no 'universe' there is no need of its expansion and contraction theories.

Energy is everything, everywhere. Energy transmuting as existences, as processes, structures or forms have no dimensions being interconnected and interlinked, within/as the fabric of space-a form of energy-where then are the boundaries or edges?

The postulated 'universe' is a concept, and this explains why cosmologists cannot explain, or describe its constitution, dimensions, shape, expansion, contraction and destiny. Revelation 10 describes the non-existence of a force we call ‘gravity’, and therefore, we cannot speculate expansion, contraction, densities, or destiny of the concept ‘universe’.

We cannot suppose sometime in the future the ‘universe’ will stop expanding, and begin contracting, until it eventually collapses itself in a big crunch. Just as the concept universe is preposterous, so also are any speculations, postulations and theories on constitution, expansion or contracting.

Galaxies may be rushing away from each other, but are existential within/as the fabric of energy. The space surrounding each galaxy may be increasing, but everything is moving in relationship to everything else. There is no external finite or infinite void, within, and toward which the postulated 'expansion' is taking place, as everything is energy, everywhere.

We will never see an actual physical edge of the hypothesized Big bang universe, with empty space beyond it. With more and more advanced technology, we will continue to see more and more distant galaxies, which will be posited to have been created even earlier. This will prove the Big bang theory, and the concept 'universe' incorrect.

How are we told that we live in an expanding and not a collapsing universe (although it may collapse at some time in the distant future)? Simply, there is no universe. If there is no universe, we cannot think of expansion, or contraction. Therefore, the concept universe will not end in 2060, as predicted by Isaac Newton.

Reading through revelation five, we will understand that as everything is in constant flux, everything is of certain duration. As things spontaneously appear, they all eventually disappear, or in simple words, exist/endure and terminate, or not exist. Everything does not exist, exists for duration, eventually terminates, and does not exist. This is human thinking or understanding.

Multiple Universes

As we created the concept 'universe', everybody has to get into the act to try and explain this concept. Then there are theories, and questions, and more theories. This is how science operates. Multiple universes are one such theory.

Some physicists warn readers not to believe that there could be more than one universe, while others assert parallel universe exist. We have so many refutations against the Big Bang theory, and we now introduce another to consolidate the others.

Some physicists are talking about very large isolated regions, mutually unknowable by direct observation termed parallel universes. These new universe or universes have many dimensions compared to our three spatial ones. A parallel universe means that this universe contains somewhat different versions of the same entities existing in ours. Therefore, all possible arrangements of energy and matter should be exist somewhere out there, which are missing in ours.

As explained earlier, we generate a theory within our illusory thinking (see revelation 35), and then introduce more theories to try and solve our earlier ones. So some philosophers insist other universes must exist to hide away mysteries. Some of them think all possible universes are real. As explained in this revelation, there is no such thing as a universe, and also no other parallel universe or 'multiverse'. These theories have no empirical content, and hence amount to empty metaphysics.

There is no such thing as 'space'-as a void, and measuring the infinitude of space, and also solving chaotic inflation are exercises in futility. We cannot finite our present theorized 'universe'. There are countless billions upon billions (we could keep on writing the word billion for eternity) of galaxies, and we should close the door to the concept 'universe'.

Reality is existential energy, and existences are processes, states, structures or forms of continuing transmuting energy, with our minds as one of them. Only existential energy exists, three dimensionally. Dimensions could mean front, back, north, south, east, and west, up or down. These words do not mean anything, as they are explainable for human planning, depending only on which side you are facing.

EXISTENCE OF EXISTENTIAL ENERGY

Existence of Existences-No origins-No entities-No gods

We cannot postulate existences collectively as the 'universe', and questions about its existence or constitution, or destiny cannot be explained. Unexplained questions from the Big Bang theory corroborate these self-explanatory facts.

If we agree the Big Bang ‘originated’ the ‘universe’, we have to accept divine intervention. Causation would be the norm; and/as everything would have ‘beginnings’ and ‘ends’. Nothing ‘begins’ or ends’, as using these concepts we create ‘entities’.

We postulate a Big Bang with the same aura, and arcane proposition as a creator, which could engender something, as immense as the postulated ‘universe’ from nothing. We cannot also imagine, or query, or posit the existence of a creator or divinities, as the concept universe is also our creation. The concept 'universe' will never expand or contract, as the 'universe' never 'began', as there is no such thing, or existence. How could it have been created? If it has no ‘beginning’, how can it 'end'? Human sounds or words, and preposterous concepts from illusory minds, generating incomplete and incorrect explanations, dictate human thinking, and lives.

Concluding note:

The story of the evolution of the postulated universe, or the Big Bang theory, is analogous to the changes that occur to a cloud of highly compressed steam when it is released. When steam is released it will expand and cool. When it cools down to 21.4 degrees Fahrenheit, it condenses into droplets of water. If it continues to expand and cool, water will freeze into ice at -32 degrees F. The proponents of the Big Bang story did not realize that these processes take place within our atmosphere of continuing altering temperatures, pressures and densities (revelation 8). With the explosion of energy from the Big Bang into, or as space (as a void) temperatures and pressures would not decrease, but would keep on increasing, with decreasing densities. How can they explain the formation of hydrogen and helium, and also how can these elements later clump together into stars and galaxies, as their present postulated ‘universe’?

PART TWO

REALITY (PHYSICAL EXISTENCES) AS ENERGY CONTINUOUSLY TRANSMUTES

A - Engendering Our World

REVELATION FOUR

NO SUCH THING AS A STAR

The formation and termination of star fields-Energy lights up

ENGENDERING/ENDURANCE OF STAR FIELDS

Existential energy exists, as galaxies and dark energy

Existential energy exists/endures ('energetic'), as existences in/as the structure/processes of star fields and galaxies, and 'static' as dark energy, or energy that cannot transmute.

Why did a single galaxy complete with star fields not form covering the entirety of existential energy's expanse?

Why are galaxies separated from each other, existing/enduring as islands of star fields within existential energy's expanse?

We will learn how free energy in momentum condenses or transmutes into a dense region, concurrently with its containing space-a star field. As energy has this transmuting characteristic, all of existential energy should have transmuted/condensed into a massive body, or a single galaxy, or a massive star field existing together with its confining space, occupying the complete extent of existential energy. But this could not have been, and will never happen.

If existential energy had existed static, there would have been nothingness-no star fields, or no existences, and no humans to think about the states of existential energy. If existential energy would have stayed static, there would be no momentum. In other words, energy is 'energetic' as it transmutes, or transmuting/momentum energy is 'energetic'.

All of existential energy has not transmuted into one single huge galaxy and space, or a single massive star field, as all of existential energy cannot be 'transmuting', as some areas of existential energy must be 'static', enabling existential energy to exist/endure..

Star fields exist/endure, as parts of cycles with other star fields. Star fields conjoin to form a galaxy. A galaxy exists/endures, as part of yet other cycles with dark energy, and with other galaxies, and vice versa. All star fields cannot conjoin to form one massive galaxy occupying the extent of existential energy. Every existence exists/endures concurrently with another existence, as parts of the same cycle, possessing different characteristics, sustaining the existence/endurance of each other. Star fields or galaxies as transmuting energy exist, with/within dark energy, or static energy. If the entire existential energy expanse would be filled with star fields, only transmuting existential energy would exist, and there would be chaos, and star fields, galaxies, or existences would not engender.

Galaxies cannot exist/endure with other galaxies, as both have the same characteristics. Galaxies are confined by their space, and this space is further confined by dark energy, and vice versa. Galaxies and their space exert/force dark energy, and vice versa, as parts of the same cycle. It is the space confining galaxies, which is further confined by dark energy, which keeps galaxies and clusters of galaxies apart-existing as islands-within existential energy's expanse.

Every particle has its confining space. Both particle and space, as parts of the same cycle with different characteristics, sustain the existence/endurance of each other. Therefore, as forms of energy they are integral parts of each other. As particles integrate to form protons, nuclei/elements, molecules, objects or astronomical bodies, concurrently their confining space also amalgamates or conjoins, proportionately increasing in size. Conjoining star fields also increase the amount of accumulating space containing galaxies. As each galaxy has its own surrounding space, all galaxies together with their space confined by dark energy, occupy existential energy's expanse. This also explains, why a solitary or a few star fields cannot engender, just about anywhere within existential energy's expanse.

We may comprehend that energy would/could have stayed static, and there would have been no star fields or existences-only nothingness, so how did the first star field form within existential energy to begin/generate a star field or existences, and transform existential energy into a static and transmuting state full of galaxies?

All existences including star fields and galaxies, engender, exist and endure for duration, and eventually terminate, within/as the fabric of space/energy. Existences randomly 'begin' existing without causation, or creation. Existences are thoughts, imagination, creations, and concepts within human illusory language.

As we can observe, experience, and/or ideate about existences, they exist. Before they were observed, and/or experienced, and after they terminate, there were/are no existences.

As there is no 'beginning' the question when the first star field was created does not arise. The question is preposterous. Existential energy exists, has always existed, and will always exist full of galaxies. We cannot query 'beginnings', or the time when star fields first formed. Star field or galaxies existing as islands within existential energy have 'always' existed.

As all of existential energy is 'energetic', we can postulate every region of existential energy transmuting into a star field, and conjoining into galaxies with their confining space, were/are confined by dark energy, and vice versa. Dark energy remains static, as transmuting energy exists/endures, as star fields/galaxies. This was, is, and will always be the state of existential energy. There is no causation but cycles.

Star fields engender from terminating star fields. No existence can engender, exist and endure without another existence, concurrently engendering/existing, and enduring, sustaining each other, as parts of the same cycle. As an existence exists with another existence of a different characteristic, explains that existential energy has 'always' existed in/as a static, and as a transmuting state. 'Static' and 'transmuting' are human concepts. Static energy as dark energy is as 'energetic' as transmuting energy. It is static, as transmuting energy exists as states, processes, structures and forms or existences. Static energy is dark, as it contains no star fields transmuting light. It is static, as we cannot observe/sense static energy's virtual 'force'.

How galaxies exist

Humans have always postulated existences as a result of causation. We assume existences originate, exist/endure, and terminate. This is why, the Big Bang theory was introduced to try and explain another concept ‘universe’. Cosmology originated, as the Big Bang theory stipulates, that the concept 'universe' originated from a point. Revelation three details, why and how this concept 'universe' is an illusion, concluding the Big Bang theory as preposterous.

Star fields are not sprinkled throughout existential energy's expanse, but are grouped together what we call a galaxy. The entire 'extent' of existential energy contains 'countless' number of galaxies, which have ‘always’ existed.

Star fields exist for duration, eventually terminating and engendering, continuously. Material of terminating star fields transmute generating new star fields, or amalgamate with adjacent star fields, or remain as nebula for some time. This has 'always' been the state of existential energy, and will eternally be.

Each star field exists with bodies within, as parts of energy transmuting cycles. Each star field cycle is inter-dependant and interlinked with other star field cycles. Likewise, a galaxy (constituted of billions of star fields) and its confining space exists/endures, as part of a cycle with dark energy.

A galaxy is a system of many thousands of millions of star fields, together with interstellar gas and dust. Many galaxies are spiral in shape, while others can be elliptical, or irregular. Apart from our own Milky Way galaxy (our solar system is a star field within this galaxy); only three can be clearly seen.

Spiral galaxies are observed, as the entire galaxy rotates with stars, also at the outer edges forming the arms. Elliptical galaxies are oval, and flattened. Spiral galaxies can 'lose' their arms and become elliptical.

Our solar system is located in one of the Milky Way galaxy's four spiraling arms, called the Orion arm. It is calculated, that our solar system travels at a speed of about 220 kilometers a second, and exists millions and millions of miles from the centre, and takes about two hundred million years to complete a single orbit around its galactic centre. The Andromeda galaxy, hundreds of millions of kilometers away, and containing hundreds of billions of star fields is the closest major galaxy to our Milky Way galaxy. It is twice as massive as our own galaxy.

The generation of spin and space

The energy transmuting/propagating shuttling to-and-fro cycle

The engendering of star fields

Free Energy

A star field exists/endures, as energy continuously transmutes, as different forms, states, process or structures between its 'bodies', as parts of a cycle. Within/as a functioning star, energy continuously condenses into particles, particles to quarks, and quarks to plasma, which constitute the atmosphere of the star/dense region. Concurrently, as energy continuously condenses, extreme temperatures, and all forms of radiation, and pressure/force generated, are 'expelled' or propagated/transmuted within/as space, explained following. Therefore, transmuting energy as/at a functioning star field is not 'free'.

With disruptions or alterations in normal energy transmuting cycles externally or internally, a star field terminates. As a terminating star field collapses/amalgamates into space/energy, as a black hole, it continues the rotational processes, as was the norm. Energy as forms, states, processes and structures of terminating star fields disintegrate/transmute with the swirling processes, and energy now becomes 'free'. This 'free' energy exists in/as a black hole, three dimensionally.

How does a dense region form?

As the star, and the bodies/nebula within and space spin, and as energy is now 'free', and in continuous momentum, the amalgamation, or drowning, or disintegration of material increases the rotational speed of the black hole. The increasing rotational speed of a black hole facilitates the formation of a new dense region/nascent star, from/with free energy. Moreover, adjacent functioning, and rotating star fields abets the formation of the new star field.

As particles/plasma form-as free energy condenses-concurrently, star field space engenders-a nascent star field. Condensation processes concurrently engenders spin and momentum, and particles/plasma. In other words, the dense region is particles/plasma.

The increasing rotational speed of the dense region engenders a periphery of particles and/or plasma, or an atmosphere. Eventually the atmosphere is shed off into star field space, and the plasma transforms into a nebula, or clouds, or a body of protons. The dense region now transforms into a star, as it commences an energy transmuting cycle with the nebula of protons-a functioning star field.

Plasma is a naked proton without its confining space. As star field space confines the dense region or the star, and vice versa, a star's atmosphere consists of plasma. A star does not comprise of protons or hydrogen, and if this was the case, there would be no star field space (a form of energy). An existence engenders with another existence with different characteristics, sustaining each other, as parts of the same cycle.

The generation of spin and space

Terminating star field material and space drown/amalgamates into/as a black hole, which continues the rotation of the terminated star field, as was the norm. The rotational speed of the black hole increases, concurrently as energy 'becomes' free, facilitating the engendering of a new dense region, as free energy in momentum whirling at super speeds, condenses/transmutes. Continuous rotation increases the density of the dense region, and a periphery of particles engender. But as the dense region gets denser, and the periphery spins faster, the particles and/or plasma cannot integrate to form protons. The confining space of the nascent plasma amalgamates, concurrently forming star field space-a nascent star field. A nascent star field now exists with the core, or the dense region of particles, the periphery or the outer atmosphere consisting of plasma, and star field space confining the dense region.

The dense region, or its periphery, or the nascent star, and its atmosphere do not contain protons. A naked proton is plasma. Each proton (hydrogen) has it 'individual' confining space. A star's atmosphere contains plasma, not protons, as the dense region of particles and/or plasma is confined by star field space.

As a sphere condenses with increased rotational velocity, eventually a periphery forms, as material are expended from the core. As the core rotates in one direction, the periphery rotates in the opposite direction. As star field space engendered rotating with the core, the periphery of the core now rotates in the opposite direction to the inner core, and its confining space. This opposite direction rotating space confines the periphery, and the dense region, and also facilitates and abets further concentration of the dense region with the increasing rotational speed at the periphery. Pressure/force and temperatures are higher at the periphery, as it is faster rotating than the core, explaining why integrating of particles into plasma is possible.

Free energy in momentum whirls in an area (a black hole) in a two dimensional context. A given volume of energy, or a sphere cannot rotate in a three dimensional context. A single complete rotation is one directional. Therefore, all star fields spin in a two dimensional context. A star's periphery rotates in one direction with its star field space rotating in the opposite direction. As star fields revolve, and conjoin to form a galaxy, the entire galaxy also revolves. Galaxies are disk or flat shaped, as a result of their rotation in a two dimensional context.

As bodies/planets within a star field exist within/as the fabric of space/energy, they orbit the star/sun, with/as the direction of star field space. As the equatorial atmosphere of the star/sun, or spherical bodies, is faster rotating than at other latitudes, planets/bodies orbit their star at the same plane. For the same reason, rings and disks orbit around planets at the equator.

Momentum and spin

If existential energy had stayed static, there would be no transmutations of energy. There is momentum, as energy transmutes, and vice versa. Therefore, momentum is a form of energy.

Every existence exists, concurrently with another existence with different characteristics, sustaining each other. A particle engenders/exists with its space confining it, and vice versa. A star engenders/exists with its star field space confining it, and vice versa. A galaxy engenders/exists with its space confining it, and with dark energy confining this space, and the galaxy, and vice versa. Existential energy exists, as static dark energy confining transmuting energy (as star fields and galaxies), and vice versa. A periphery of the dense region or the atmosphere of a star, a star field's interstellar medium, and a galaxy's spiral arms continuously rotates, with confining space rotating in the opposite direction.

Without spin or momentum, energy would not transmute, or condense forming a dense region. Without continuous rotation energy would not condense, engendering, and integrating particles. Without condensation of energy, space would not engender, and there would be no star fields, and no existences. Momentum/rotation, or spin, and space, are keys to engendering existences. Without momentum/rotation/spin, and space, energy does not transmute. Therefore, spin, momentum, condensation or concentration, particles or plasma, and space are forms, states, processes or structures of transmuting energy. We have to demarcate various states, processes, forms or structures of energy, enabling us to describe, or understand phenomenal reality.

How does a dense region transform into a star?

If you half fill a bucket of water, and swirl it with an oar, a hole forms, and it gets deeper, as the rotational speed is increased. Concurrently, water at the side of the bucket rises. This is how a black hole exists. As material from a terminating star field sinks/drowns (amalgamates into/as space/energy) into/as a black hole, the black hole swirls (as a result of, and continuing the norm of the previously functioning star, and star field rotation).

The functioning star's atmosphere rotates in an opposite direction than the core, and the core rotates in the same direction as star field space. A star disintegrates, amalgamating with star field nebula, or bodies, and space. As the core gradually disintegrates, the entire star field space continues the same directional rotation. The spin/momentum increases, as transmuting energy 'becomes' 'free'. New star fields engender, only from whirling black holes. As a new dense region engenders, the core rotates in the same direction, as the core of the previous functioning terminated star.

The increasing super speed rotation of a black hole facilitates condensation. Free energy condenses into/as a dense region, concurrently engendering space. The dense region and its confining space occupy/supplants the same area/volume of energy from which they engendered-a nascent star field.

Free energy condenses further into a dense region full of particles. The increasing rotational speed forming particles form the periphery of the core. As the dense region further concentrates, the concurrently increasing rotational speed at the periphery integrates particles into plasma. After all of free energy transmutes into a dense region of particles, its atmosphere of plasma, and star field space, the internal cycle terminates, and the outer plasma layer of the dense region is shed off. Plasma propagating through star field space transform into 'individual' protons (plasma with confining space). Still in momentum, and within the rotating star field space (in the opposite direction), protons integrate together becoming nebula, clouds of protons, or an extra solar planet.

A dense region may turn into a nascent star, and this nascent star may become a full fledged star, as/if an energy transmuting cycle (explained following) engenders between the nascent star, and its shed off nebula/extra solar planet-a functioning star field.

The energy transmuting/propagating/shuttling to-and-fro cycle

With the condensation processes at a star, or the continuous engendering of particles and plasma, concurrently, heat or temperatures and pressure/force generated, propagate both inwards and outwards. The force or pressure propagating toward the core increases the integration, or density. The propagating outward force and temperatures extend the outer atmosphere.

Atmospheres of all bodies continuously disintegrate and integrate, and are not fixed bodies. As plasma is seen leaving the sun/star, it appears to us that their atmosphere are constituted of plasma.

At some star fields the dense region forms/is a rotating globule. With increasing pressures and temperatures, the dense region transmutes into a globule central core, which is surrounded by a flattened disk of dust (plasma expended from the globule). The globule forms a flattened disk, as spin can only take place in one direction. The flattened disk forms, as with spheres, rotation is faster at the equator decreasing toward the poles.

Eventually, as free energy within the black hole (star field) is completely utilized (transmutes into/as a dense region and space) the disk splits off. As space is spinning in the opposite direction, the shed off atmosphere, or disk, transforms into an extra solar planet, or a cloud of protons, or a nebula. If an energy transmuting cycle forms (explained following) between the dense region and the extra solar planet, or nebula/body, the dense region (globule) becomes a full fledged star, and a functioning star field is born.

If a dense region does not shed out its outer atmosphere, or the globule its disk, and/or if an energy transmuting cycle cannot take place, the dense region transmutes/transforms in a brown dwarf for some duration, until adjacent variations in energy transmutations within the region, triggers the disintegration of the dormant nascent star field.

With the outer atmosphere or disk discarded, concurrently temperatures and pressure/force decrease at the core, and the core partially disintegrates. As rotation continues and increases, a new periphery forms from the disintegrating core-this is the photosphere. Plasma from the continuous integration of particles at the photosphere is expelled into star field space. This continuous expulsion of plasma generates an upper atmosphere, explaining why atmospheres of stars appear to be constituted of plasma. Now we have a new star field with a new dense region, and also an extra solar planet, or nebula/body. This new star field endures, as an energy transmuting cycle between the new dense region or a star, and its body or bodies is operational.

With processes at a star, temperatures and pressure/force generated propagate through/as star field space, together with plasma, throughout the star field. However, this energy must be replenished to continue processes at the star. Plasma transmutes to protons propagating through star field space. Interacting with other protons, and nuclei at atmospheres and bodies, protons/nuclei disintegrate to elementary particles, which are extracted as replenishment material toward the star, within/through the shuttling to-and fro energy transmuting cycle pathways between the star, and its bodies/nebula. This stabilizes the pressure/force and temperatures, and the star continues to exist. This is the energy transmuting shuttling to-and-fro cycle between the star and its bodies/nebula. Without these continuous energy transmuting cycles, stars, star field bodies, or star fields would not exist.

Forms of energy can only engender/endure, as energy propagates/transmutes within/as star field space, as space concurrently transmutes to these forms, and/as space waves. These forms of energy propagate/transmute, as space is also concurrently generated (and endures) with the continuous condensation of energy.

Energy transmuting cycles within star fields with a nebula, or other bodies

Plasma generated from the integration of particles at the outer core, or photosphere of a star constitute the enduring atmosphere (corona), stretching to millions of kilometers. These processes produce the incandescent at the star, as plasma glows with/as radiated heat.

The energy radiation/propagation/transmuting cycle is not a one-way phenomenon-nothing is. Energy continuously transmutes, as it 'expels' energy from the dominant body, or the 'star', and 'extracts' energy from other bodies within a star field as replenishment. A star continues to function, as lighter particles are extracted from nebula, atmospheres and bodies within a star field, within the same energy transmuting, and or propagating pathway, continuing, or enduring the cycle/processes. Receiving a continuous supply of particles, a star processes these form of energy into plasma/protons.

We have to describe each state, process, structure or form of energy, as different existences to understand phenomenal reality, or transmuting energy.

Energy transmutes/propagates as space waves

Forms of energy within a star field transmutes/propagates, as a shuttling to-and-fro cycle between 'bodies'. Light or heat, or temperatures/all forms of radiation, and/or pressure/force propagate through/as space, as space is not a void, but a form of energy. Space engenders, concurrently with these forms of energy, and therefore, transmutes into light, heat, temperatures or forms of radiation, and pressure/force, or momentum. As space transmutes as/into forms, states, processes or structures of energy it waves, and as it waves, energy transmutes, and/or 'can' 'propagate'. The varying momentum (wavelengths and frequencies) of waving space, manifest momentum, or propagation, or transmutation of different forms, states, processes and structures of energy.

We observe ripples/waves propagating outwards on the surface of the water, from the point where the pebble thrown hit the water. The disturbance, or the ripples/waves (force/energy) generated, propagates from the point, and it is not water that is moving outwards from the point. The water stays at the same place. The water is only a medium, and waving, it transmutes/propagates the force generated. In other words, it is force, as a form of energy propagating in the form of ripples/waves. This is only possible as/within/through space-a form of energy. In the above example, nuclei constituting the water molecules have space containing them, explaining how and why different forms, states, processes and structures of energy transmute/propagate.

As only energy exists everywhere, space as a form of energy does not propagate. As there are no entities, nothing can. Energy transmutations propagate waving the fabric of space, three dimensionally. The propagation and/or the momentum per se are processes of energy. As energy transmutes 'waving' space, or as space waves, there is a concurrent reverse propagation/momentum (as energy transmutes) of other forms of energy, within the same shuttling to-and-fro pathways, explaining why space waves. Nothing is a one way phenomenon, as energy is everything everywhere.

This concurrent shuttling-to-and-fro of energy transmutations can be sensed, standing at the sea shore with waves striking our feet. We can sense and observe waves, or the approaching water molecules. If we are attentive we can also feel, and observe water concurrently leaving the shore, amalgamating with those same incoming water molecules, generating waves. In reality, it is not water coming in, and/or leaving, but energy (in the form of a breeze, or a wind, or a force/pressure) propagating/transmuting in the form of water waves (a structure of energy). As everything is energy, everywhere, it cannot 'travel' from there to here, or vice versa.

Energy transmutes shuttling to-and-fro, as different states, processes, structures or forms of energy. This explains why a paper boat, or a piece of wood floating on water, does not accompany each ripple/wave. The paper boat, or a floating piece of wood, would eventually move from there to here, but not as a result of the water moving it, but with wind/force (another form of energy). As energy transmutes as force/pressure (another form of energy), and as it can do so, as space (another form of energy) waves (in momentum-another form of energy), it appears to us that there is movement. In reality, as energy is everywhere and everything, propagation, or momentum, or pressure/force is manifested, as energy transmutes into states, processes, structures or forms of energy.

Energy as it propagates shuttling-to-and-fro, or transmuting in/as processes, states, structures, or forms, is in reality, only waving the fabric of space. As energy is everything everywhere, energy does not really 'propagate' from here to there, as all existences are structures/forms/processes or states of energy, within/as the fabric of energy. The waving of space/energy per se is the transmutations of energy (momentum), observed, and experienced.

How do protons propagate if space waves?

Plasma is not yet a 'structure' (matter by human understanding) within/as star processes. Therefore, plasma propagates with, or as heat/temperatures and pressure/force, within/as star field space, as space waves. Within star field space, plasma transforms into protons. Protons and nuclei wave within space, analogous to bubbles within water. Protons, nuclei/elements and molecules comprise or constitute matter, or condensed energy.

With temperatures and pressure/force in the form of waving space, protons interact with nuclei at atmospheres of bodies, and nebula within a star field. These interactions disintegrate nuclei into particles, which are extracted within the same energy transmuting shuttling to-and-fro cycle pathways, as replenishment material at the star/sun.

As all forms of energy dwindle with distance, plasma expelled from a star/sun transmutes to protons, or other forms of energy, and may amalgamate into/as heat or pressure force, which is propagating. Energy propagates shuttling-to-and fro, and it is difficult to ascertain what form of energy propagates, between two imaginary points within star field space.

Why star fields exist/endure

An existence will engender/exist/endure, concurrently with another existence, sustaining each other's existence/endurance, as parts of the same cycle, as these existences of different characteristics transmute different forms, states, processes or structures of energy. In other words, a star field will engender, exist and endure, only if the dense region/star forms an energy transmuting cycle with the nebula or bodies, within the star field. This cycle engenders, and/or endures a star, its bodies, and the star field.

Different forms, states, processes or structures of energy propagate through/as star field space, to and from the star, within the star field. As the star rotates at super speeds, energy is both 'expelled', and concurrently 'extracted', to and from interplanetary space, explaining the equal centripetal and centrifugal space, both within the star, and within the star field.

'Stars' or star fields and everything exist, as energy is continuously transmuting in its varying forms, states, processes or structures between its bodies. 'Stars' will continue to burn, as long as it can attract material, replenishing the expended material to continue its processes. As long as the 'expulsion and extraction' radiation, or propagation, or transmuting shuttling-to-and-fro cycle within star field bodies continue a star field will exist/endure. Star fields are duration energy structures, and like every other existence will continue to exist, as long as their internal processes are sustained by the internal/external environments within each cycle, and adjacent cycle(s). Within a star there are cycles between the core, the surface, and the corona-existences with different characteristics (processes).

Planets/bodies, and their atmospheres (if any) are structures of energy, existing within/as the fabric of space/energy, and therefore, as star field space orbits the star, these bodies also orbit the star. If these bodies are parts of an energy transmuting cycle with the star, they also rotate around their imaginary axis.

As particles are disintegrated from nuclei at/from the atmospheres of planets, or bodies they become 'light', and are easily extracted toward the star with/within the energy transmuting cycle. Within star processes, these incoming energy particles integrate to quarks, and quarks to plasma continuing the cycle.

A star requires continuous replenishing material, as it expends energy in the form of temperatures, pressure/force and plasma. If an energy transmuting cycle cannot form, a star field does not engender, and/or a new star field will soon collapse, or explode, or disintegrate. The nascent star/dense region can end up as a brown dwarf, or disintegrate, or transmute into feed for other star fields, or remain as star field nebula, or amalgamate with the interstellar medium to be utilized as star field material some time later.

How star fields exist

Energy is everything, everywhere. As a single star field terminates, it is possible that many adjacent star field cycles may be disrupted, terminating other star fields. A new star field easily engenders from a single terminating star field (black hole). However, with many star fields terminating together it is a different situation. It can be postulated, that free energy from massive black holes (many terminated star fields) may engender many dense regions, each eventually forming their own star fields.

How star fields will be born, how long they endure, what type of star fields will take birth, and how hot, how rare, or how massive they will be, will depend on the quantity/utilization/condensation, and transmuting energy processes, within, and adjacent star fields.

As energy transmutes everything is flux, and all existences including star fields exist for duration. Durations depend on energy transmuting variations or fluctuations, internally and externally. Everything will eventually terminate/transmute instantaneously, or after a few seconds, minutes, days, years, or hundreds, thousands, or millions, or billions of years.

There are no entities, and as everything is part of a cycle, existences cannot survive/exist functioning alone. There are star fields cycles, and not 'individual' 'stars'. 'Individual' or 'entities' are human concepts.

There cannot be more than one dominant body or a star within a star field. Binary stars do not exist. Two star fields orbit each other, and not two stars within the same star field. Two stars cannot exist (with the same characteristics) sustaining the existence/endurance of each other, as parts of the same cycle. Within a cycle bodies must have different characteristics to sustain each other.

The energy transmuting cycle shuttling all forms, states, processes and structures of transmuting energy keep star fields existing. The energy transmuting cycle (s) within a star field keeps bodies 'apart', and orbiting the dominant body or the star. The star or the dominant body also spins around its imaginary axis. All star fields orbit the centre of their galaxy.

We describe the dense region as the 'dominant' region. 'Dominant' is a human concept. These concepts are continued to be used here enabling easier description or explanations. As everything is energy in a static or transmuting form, state, structure or process, we cannot distinguish or compare anything-there are no standards. 'Apart' and 'orbit' are also human concepts.

All star fields have different characteristics, depending on their random formation. Some follow an elliptical orbit around the centre of a galaxy, and some circular. Some orbit slower than others. Any existence is randomly structured, and as these/their characteristics have enabled them to survive, they are therefore, still existing for us to observe them.

Why star fields form?

From what do star fields form? Where do they engender?

Why do star fields conjoin into a galaxy?

Can an entire galaxy terminate?

Star fields engender, as existential energy is in continuous momentum. Three dimensionally energy cannot momentum in one direction, and regions of existential energy swirl, each generating a star field. Star fields engender within the same massive region maelstrom, and this facilitates, and abets them to conjoin as a galaxy. For the same reason, star fields will engender only on the 'termination/transmutation' of another star field or star fields, from previously terminated ones, and within the same region.

Although everything is energy, energy within a galaxy (within existing star fields) must continue to transmute, so that a galaxy confines dark energy, and vice versa. As galaxies have formed/exist within existential energy's expanse, there is no room for the engendering of a new star field outside a galaxy.

Galaxies consists of billions of star fields, and each galaxy is confined by their space, and further by dark energy. We cannot imagine an entire galaxy terminating, sinking/drowning, into/as a massive black hole. We can postulate that an entire galaxy cannot terminate/disintegrate, as this would disrupt vast regions of existential energy.

Why did massive star fields not form larger than the largest ones presently existing? Why did smaller star fields not form, smaller than the smallest ones now existing?

Why could not all the star fields of a galaxy amalgamate/conjoin to form one massive star field as large as their galaxy?

Why did a single massive star field or galaxy not form, occupying the extent of existential energy?

Present star fields exist at their respective sizes, as energy condensing/processes as/at a star, concurrently produces the most extreme temperatures/heat and pressures/force possible anywhere. With human understanding, we would assume that larger star processes would produce increased heat/temperatures and pressures/force. Heat/temperatures and pressures/forces, greater or more intense than those being produced by the presently existing largest stars cannot be generated, having reached their maximum intensities with processes at the largest of stars. Therefore, larger stars or star fields, larger than the largest of stars presently existing cannot engender. This explains, why a single star field as big as a galaxy cannot engender/exist. This also explains why a single massive star field occupying the extent of existential energy could never exist.

Large stars die or disintegrate faster, as they cannot replenish their material easily. They expel and extract more material than other average size stars, explaining why they continuously incite/exact abnormal energy transmutations, or disrupting situations, within its star field cycle bodies, and importantly adjacent fields.

Stars are of certain sizes, as they can commence energy transmuting processes together with planetary nebulae, and or other bodies within their star field.

As extreme temperatures and pressure/force have been reached at the largest of stars or star fields, so also smaller stars or star fields cannot engender, as certain limits of heat/temperatures and force/pressure must prevail. If the heat/temperatures and pressure or force generated would be lower, it would not be possible for a dense region to transform into a star, or a star field transmuting cycle to engender, enduring bodies, and a star field. This explains why there are different sizes of stars, none smaller, or larger than the present ones.

Is there a single star field within existential energy?

Galaxies of millions and millions of stars fields have always existed, within/as existential energy. There were no first stars as we have been incorrectly informed.

There is not a single star field existing alone, anywhere within existential energy. This could never have happened, as collectively for all the galaxies within existential energy, which in other words is transmuting energy, there is equal amount of dark energy, or static energy. Two existences exist/endure (with different characteristics), as parts of the same cycle sustaining each other. Therefore, a single star field cannot engender, exist, or endure outside a galaxy, as dark energy would disintegrate it instantaneously.

A star field engenders from free energy. Dark energy is not free energy, as explained earlier, so a single star field cannot engender outside a galaxy.

Molecular Clouds/Interstellar matter

A molecular cloud is a cool area in the interstellar medium where molecules are formed, mostly molecular hydrogen, although there are other sixty different molecules detected like carbon monoxide, ammonia and water. Nuclei combine chemically to form molecules, as forms of energy from the sun dwindle with distance.

Our solar system's interstellar matter is made up of both gas nuclei, and solid particles. The gas accounting for about 99 percent of interstellar matter is about three-quarters hydrogen, and nearly one quarter helium, plus trace amounts of nitrogen, oxygen, carbon, sulphur, and other elements. The solid particles also called cosmic dust consists primarily of carbon and silicate material (silicon, oxygen and metallic ions), possibly with frozen water, ammonia, and solid carbon dioxide. Plasma/protons expelled from the sun, which are not disintegrated within inter-planetary space, settle on the outskirts (interstellar matter) of our solar system as hydrogen. Much of the materials are remnants of an exploding neutron star, and also of the exploded planet, as detailed in the next revelation. It may be that many other star field interstellar mediums may not contain many of the above mentioned material.

As temperatures/forms of radiation and pressure/force from the sun dwindle with distance, interstellar material is not disintegrated, remaining cold, icy and dark. We can identify the substances in interstellar gas, as they appear in a characteristic imprint.

Within molecular clouds, the concentration of interstellar matter is hundreds to thousand of times greater than in the surrounding interstellar medium. Molecular clouds are enormous, and do not precipitate a dense region or a nascent star, as there is no revolving free energy.

Why are spiral arms of galaxies formed, and how do they last for long periods of time?

Spiral arms of galaxies are not fixed or solid objects, but are patterns of bright stars, within the overall form of the galaxy. All regions of a galaxy are rotating, but it is only those filled more with star fields that are observed, making spiral arms apparent. More importantly, the space containing a galaxy spins the opposite way generating those spiral arms.

Spiral arm galaxies are differentially rotating, that means the time to complete a full rotation increases with distance from the centre. Differential rotation results in disturbances to wind up into a spiral form (dense area), and this explains their shape. But continuous differential rotation would result in spiral features to wind up quickly, and we would not be able to observe those spiral features for any appreciable length of time. This is why differential rotation is not the only reason spirals form. Spiral arms do not wind up quickly, as the whole galaxy moves, and rotates within its 'space'. Dark energy confines this space, and indirectly exerts its force on the galaxy. The centrifugal force of a rotating galaxy and the differential rotation engenders those spiral arms, but an equal reverse differential force, and dark energy prevents those arms from separating from the galaxy.

Each spiral arm has millions of stars firmly set, exerting and extracting (processing/transmuting) energy from within, and external to their star fields. As the galaxy rotates there is an inexorable force trying to pull star fields (dense areas) toward the centre of galaxies, and an equal reverse differential force outward.

Dense areas spiral arms continue to be observed, as they are supported by orbital motion of star fields within them. Density variations travel round the galaxy and therefore, a spiral arm is not simply a concentration of co-orbiting stars. Within the galaxy itself there are set exchange of forces, resulting in varying densities. As spiral arms are the outermost regions of a galaxy, they are subjected to increased influences, and confining effects of dark energy. Most importantly, the periphery of a rotating body spins faster than the interiors, and as spiral arms rotate in a different direction than the interiors, they are apparent.

Star fields aggregating to form spiral arms are more densely 'set', and endure for longer duration, explaining why they are not easily affected with abnormal energy transmutations, which would terminate other star fields. This explains why spiral arms continue to be observed for appreciable lengths of time. This also explains, why our solar system localized in one of those four spiral arms (the Orion) of our Milky Way galaxy continues to exist for a long duration. However, as we will learn in other revelations, there are also other feasible conditions that have endured our solar system for billions of years.

Spiral arms are not only formed at galaxies. Every star rotating at super speeds has its upper atmosphere broken into spiral arms. Every star field also has its interstellar medium designed into spiral arms. We cannot observe the spiral arms of the sun, as there is too much light, and we cannot observe spiral arms of our interstellar medium, as there is no light.

Why are present galaxies not more massive in size?

As star fields orbit their centre of their galaxy, there is an inward centripetal force, and an equal centrifugal force, explaining the spiral arms of some galaxies. Spiral arms form, as the rotation on the periphery of a galaxy is faster than the interiors. Spiral arms of a galaxy form at the outermost layer of a galaxy, explaining the shape/direction of the spiral arms. This explains why spiral arms form in the first place.

The largest of star fields produce the most extremes of temperatures and pressures/force possible. Likewise a galaxy with its centripetal and centrifugal force can contain only a certain number of star fields. These combined forces equal the force exerted by dark energy, explaining why more massive galaxies cannot exist.

Why are galaxies shaped irregularly?

What randomly structures, and is able to exist/endure with those prevailing conditions is observed by us. 'Irregular' or 'regular', and 'shape', are human concepts.

Why are galaxies disk shaped or flat, and not spherical?

A star field will engender only if the periphery of the dense region, and surrounding space within the star field, spin in opposite directions. A body must spin a complete rotation in one direction, explaining why our planets orbit the sun on the same plane (every spherical body has its equatorial atmosphere rotating at faster speeds than at the poles). Star fields rotate in one direction, and conjoined to form a galaxy, which is also in momentum, galaxies are spiral armed or elliptical, but disk shaped, or flat, and cannot be spherical

Why are stars of different luminosities?

We are incorrectly informed that stars obtain their energy from hydrogen fusion at their centers, and continue shining in a rather steady manner, as long as the quantity of hydrogen in their cores remains above a certain amount. The luminosity of a star (dominant body in a star field) does not depend on its hydrogen contents. At star processes, energy is condensed forming particles, and particles integrate forming plasma, which constitutes the star's atmosphere. There is no proton or hydrogen at a star.

There are near, far and bright stars. Nearer stars are not invariably brighter than more distant stars. This would be so if all stars were equally luminous.

As energy continuously transmutes with everything is flux, there is no such thing as an entity. Energy continuously transmutes at/as a star, and cannot 'hold' hydrogen or energy. A larger star field will 'contain' (transmute) more energy, and obviously luminosities will be proportionate to the amount of transmuting energy (material).

As energy is in constant flux, star fields can exist for a moment, or for millions of years, depending on how steadfastly the star field cycle or cycles are set up, the ongoing energy transmutation variations within adjacent star fields, and also within a galaxy. A star field can collapse, and amalgamate with another star field engendering a larger star field, and the dominant body can grow larger, glowing brighter. Sometimes it is just the opposite phenomenon, as a star field can loose some of its material getting dimmer.

Depending on the interstellar mediums between stars, and their orbit (star fields) around each other, we sometimes observe a star shining bright on a particular day, and sometimes dim on other days. Sometimes stars even disappear out of observation completely.

There are stars of different sizes (radius and mass), brightness (luminosity), temperatures, and colors with different movements (rate of spin).

THE TERMINATION OF STAR FIELDS

Why do star fields terminate?

Existences as transmuting states, processes, forms or structures of energy are continuously altering into yet other states, processes, forms or structures of energy. Consequently, all existences eventually terminate some instantaneously, some within hours, or days, weeks, months, years, or hundreds, or thousands or billions of years.

Changes in densities, pressure or force, and/or temperatures alter/disrupt the characteristics of existences. As existences are parts of a cycle with other existences, a cycle may alter, and either adjust continuing the cycle, or may fail to continue terminating existences. There is continuous energy density, temperatures and pressure variations within star fields, and between star fields, explaining why star fields are continuously terminating, and engendering.

Abnormal energy transmutations influence existing star fields, as well as new star field formation. Larger star fields terminate faster, as their internal energy transmuting cycles are delicate. With the same reason, larger star fields continuously influence adjacent star fields.

Although, each star field has its internal transmuting cycle or cycles, between its bodies and their star, each star field is interdependent and interlinked with other star fields (as cycles) within their galaxy. The termination of any star field would result in density fluctuations within a galaxy. This could result in energy transmutation (alterations) within adjacent star fields, which could lead to chain reaction terminations.

Even with billions of star fields, galaxies are in continuous momentum. There are continuous abnormal energy transmutations within a galaxy, disrupting or terminating existing star field cycles. Star fields can continue existing with minor abnormal energy transmutations, being stretched only to a point. Extreme transmuting variations explode, collapse, or terminate existing star fields.

During a star field's lifetime, much of the continuous expulsion of plasma transmuting to protons, propagating through interplanetary space, settle at the outskirts of the star field. Most star fields eventually terminate, as the star cannot incite an energy transmuting cycle with this accumulated forms of energy, as radiated heat and pressure/force from the star dwindle, propagating through distance. As the star cannot replenish its expelled energy, processes cease, terminating the star field.

Our sun has active energy transmuting cycles with the inner planets, their moons, and the asteroids in the asteroid belt, evidence that they have shrunk, and are solidifying. There is plenty of replenishment material still available to be extracted from the large Jovian planets. However, gradually all the planets will shrink and solidify, as our solar systems' boundary would grow very large, accumulating all the expended plasma/protons. Eventually, the solar system will disintegrate.

Note:

We are informed that a star field collapses under its own weight, as gravity pulls the material of terminating stars. This is incorrect. A star field collapses, but not under its own weight. Weight is measured only within the energy transmutation cycle, between the sun/star, and its bodies or planets. We can weigh things on earth, as there is propagation of protons, temperatures with pressure/force from the sun toward the surface of the earth. As we can weigh things, we also postulate there is a force called gravity. Revelation 10 details that there is no such thing as gravity. Outside an energy transmutation cycle between the star/sun and a body, there is no such thing as weight.

As a star field disintegrates or collapses, the material transmutes into other states, processes, structures or forms of energy. Terminating star fields 'collapse', or 'drown', amalgamating into/with its space, and this process/structure is called a black hole. Weight, gravity, collapse, and terminate are human sounds or words or concepts.

What happens as a star field cycle terminates?

As a star field cycle terminates, it may alter the functioning of adjacent star field cycles. Changes/adjustments in cycles must be immediate, otherwise those star fields also terminate. Some star field cycles temporarily shut down and some time later, with internal/external adjustments start functioning again. In some cases, adjacent star fields absorb/transmute energy from terminating star fields getting larger. They may continue with adjustments to their cycles, or if this does not happen, they too can terminate. At times, terminating star fields will set off a chain-reaction, resulting in groups of star fields terminating. If these star fields cannot reset their energy transmuting cycles quickly enough, they could trigger large regions of a galaxy to 'collapse, or transmute', drowning or amalgamating, into/as a massive black hole. Sometimes, groups of terminating star fields engender a massive nebula, enduring for some duration, but eventually disintegrate, drowning into/as a black hole.

New star fields normally engender within, or around the vicinity of terminated star fields. Depending on the mass, density, control, and or influence, and dominance of adjacent star fields, a nascent star may disintegrate before it becomes full fledged.

What happens to the star and the nebula, and/or bodies within a terminating star field?

We are informed that a dying star may turn into a red star, a red giant, a white dwarf, or a neutron star. It is theorized, that stars about the size of our sun will become white dwarfs. Larger stars about 3 times the mass of the sun may become a neutron star, and stars greater than 3 times the mass of our sun will end up in/as a black hole. But what happens to the nebula, and/or bodies and space within the star field?

There are small, medium and large star fields. Whether stars will become a red star, a red giant, a white dwarf, or a neutron star is explained following. Let us learn about particles, plasma, proton plasma, neutron plasma and protons/hydrogen to understand the engendering and terminating processes of red giants, white dwarfs and neutron stars.

Particles

As free energy in momentum (as a black hole) condenses into/as a dense region, concurrently space engenders, confining the dense region, and vice versa-a nascent star field. The dense region is in reality, virtual energy particles. In other words, particles are forming, as energy continuously condenses/transmutes. Within a functional star field, particles are extracted to the sun/star (detailed earlier), continuing the processes. We cannot study energy particles, as we cannot study invisibility, and so searching for the fundamental units of matter are exercises in futility.

Plasma

After a nascent star sheds its outer atmosphere, and as energy continues to concentrate/condense, a periphery (of particles) engenders at/of the dense core (explained elsewhere). With super speed rotation/momentum, particles of/at/as the periphery integrate to form quarks/plasma, or proton plasma. Plasma can be explained as different existences; as condensing energy, particles, and integrating particles as quarks-all forms of condensing or transmuting energy. As plasma is continuously expelled from the periphery, or photosphere of a star/sun, it appears to constitute/form a star/sun's atmosphere.

Protons

A proton is 'individual' plasma contained by 'individual' space. Plasma expelled from a functioning star/sun, transform into protons (hydrogen), propagating through star field, or interplanetary space. Therefore, there are no protons, or nuclei at functioning stars, or neutron stars.

Neutron plasma, Proton plasma and Protons/hydrogen

As some star fields terminate, the dying star sheds off its atmosphere of plasma. This plasma is instantaneously confined by star field space, transforming into a nebula of protons. Concurrently, the cold dense core disintegrates into/its particles, some of which form a new periphery. As density decreases at the core, and without confining space, the peripheral rotational speed increases, generating extreme pressure/temperatures. The new periphery integrates particles to neutron plasma. This new periphery does not integrate particles into proton plasma, as there is no space around, or space cannot concurrently engender.

Proton plasma engenders, as energy condenses into particles. As energy condenses, space concurrently engenders, confining proton plasma/particles, and vice versa. Therefore, if proton plasma is 'individually' confined by 'individual' space, protons engender. This is hydrogen.

Neutron plasma engenders, as condensing energy disintegrates. Neutron plasma engenders (at/as neutron stars) with no space around, and no space concurrently engendering. Therefore, neutron plasma cannot be 'individually' confined by space, explaining why the hypothesized neutron does not exist-there are no such things as neutrons. A hypothesized neutron cannot form (with its confining space), as it engenders at a neutron star with no space around. This explains why hydrogen or protons as 'entities' exist, and the hypothesized neutrons do not.

As there is no such thing as a neutron, a neutron star exists constituted of conjoined neutron plasma and protons, but they are not fused together. If hypothesized neutrons formed (neutron plasma with confining space) at a neutron star, a nucleus would not engender, as neutrons would then exist 'individually'.

You cannot fuse particles, quarks, protons, or the hypothesized neutrons, as space confines every particle, quark or plasma (individually or collectively). These can only be integrated. As plasma neutron at neutron stars are not confined by space and moreover, as space is not around, a nucleus cannot be fused at neutron stars. Every proton, nuclei, molecule, object or astronomical body is confined by space (integral parts of each other), explaining why they are 'entities' separated by space. Therefore, neutron plasma without its confining space, and a proton with its confining space, cannot be fused together, but only integrated. They also cannot be fused together, as there is no space around to confine them, as a nucleus at a neutron star, or within a neutron star field. Temperatures and/or pressures can conjoin these existences, but they cannot be fused into a new functioning whole. This explains why nuclei can be disintegrated by temperatures and/or pressures.

Therefore, there are no nuclei produced at a functioning star, or a neutron star. A neutron star consists of conjoined neutron plasma and protons, but no nuclei.

Nuclei and matter

Eventually a neutron star explodes, as it completes conjoining/concentrating neutron plasma and protons. We will learn that if fragments of this neutron star hurtle through a nascent star field space, this space confines neutron plasma and protons, generating nuclei.

The hypothesized nucleus is space confining neutron plasma and a proton (which has its own 'individual' confining space). A proton cannot fuse with neutron plasma, but can be imprisoned with neutron plasma, with space confining these two existences-the hypothesized nucleus. Therefore, neutron plasma (outside a neutron star) exists integrated with a proton, and confined by space-as a nucleus. A nucleus is an 'entity', consisting of a proton with its 'individual' confining space integrated with neutron plasma. This existence is further contained by (atmospheric) space. Therefore, as a nucleus disintegrates, neutron plasma becomes 'free', eventually (after a few minutes within earth's atmosphere) amalgamating into/as space. Also, as a nucleus disintegrates, protons also become 'free', and as hydrogen continue to endure, unless subjected to certain temperatures and pressures. Within our atmosphere, hydrogen is susceptible to the energy transmuting cycle, and is immediately extracted toward the sun.

If nuclei were fused at a neutron star, it would immediately disintegrate the neutron star, as helium would form turning into a solid at high pressures/temperatures. Our world would have then not engendered/existed. In other words, there would not have been nuclei/elements, or matter.

As a result of a postulated neutron not possessing it own 'individual' confining space, a nucleus can engender from the integration of a proton and neutron plasma. Also, as plasma neutron does not 'possess' its own 'individual' confining space, any nucleus/element decays, or disintegrates into protons and neutron plasma. In other words, elements decay.

As a proton forms from condensing energy, it continues to exist/endure within star field or interplanetary space. As neutron plasma forms from disintegrating condensed energy (generated only at neutron stars), it endures for only few minutes within interplanetary or star field space. It eventually amalgamates or drowns into/as space-also a form of energy. As neutron plasma is not confined by space, or is disintegrating condensed energy, it is easily absorbed by other nuclei, generating radioactive isotopes.

Current descriptions of Red giants

We are informed that after a few billion years, the centre of a star runs out of protons. What is left is a core of central region of alphas (nuclei of helium atoms). It is stated that the outer layers of the star still contain hydrogen, but they are not hot enough to fuse. As it runs out of fuel, the star begins to cool and contract. The outer layers of the star fall inwards under gravity, and as they fall they heat up, a shell surrounding the central core becomes hot enough to fuse protons into alphas. So the star gains a new source of energy. The core of the star is now hotter than it was during its normal life, and this heat causes the outer parts of the star to swell. The star becomes a giant.

The radiation from the fusing shell has grown weak by the time it reaches the surface of the star. Weak radiation is red, so the star becomes a red giant. Meanwhile inside the shell, the core of the star shrinks, and heats up enough to fuse the helium nuclei, together into even heavier ones. Among the commonest nuclei are carbon, nitrogen and oxygen. Heavier and heavier nuclei are created inside a red giant, the heaviest iron nearest the middle/centre. These fusions release only a little more energy, so they keep the red giant burning for a little longer. But they do not produce as much energy as the fusion of protons. Iron nuclei cannot be used as fuel, because they need to be given energy to make them fuse. So iron nuclei collect in the heart of a red giant star. All the above information is incorrect, and following are questions/refutations.

Energy is everything, everywhere. There is no causation. You cannot create or destroy energy, so it is preposterous to theorize a star 'gaining' energy, or heat 'causing' any part to swell, as 'heat' and the 'swelling' are forms/processes of transmuting energy. As also queried earlier, how does a star accumulate protons or hydrogen enabling it to burn for duration? Hydrogen is an explosive material, and at star temperatures would shatter a star before it is full-fledged. Helium is a solid at high pressures, and a liquid when allowed to expand, so the explanation of a star containing hydrogen and burning into helium at the core is preposterous.

How does a star exist with its outer layers still containing hydrogen, and not being hot enough to be able to fuse? What do we mean when we say a star has run out of fuel? Energy is everything, everywhere, and cannot be destroyed or created, so how can a star run out of fuel? How and where is helium and hydrogen stored, with continuous fusion at a star? How does an old star run out of hydrogen and helium? Why should a star cool and contract, and why and how does the outer layer fall inwards? Why do they not fall under gravity (if it exists) when the star is hot? Where does heat come from, if the outer layers fall into the core? How can this heat be hotter than it is during the star's normal life?

Reality

At any functioning star field extreme temperatures have been reached. A star of a certain size cannot get hotter than its normal operating temperatures. How can additional heat 'cause' the outer parts of the star to swell, and how with this increased heat radiation grow weak, and this weak radiation is red, so that the star becomes a red giant?

It is also stated that as the core of the star shrinks, it heats up enough to fuse helium nuclei together into heavier ones, as carbon, nitrogen and oxygen. Heavier nuclei are created inside a red giant-the heaviest iron is in the middle. Iron nuclei cannot be used as fuel as it needs a given energy to make them fuse, so iron nuclei collect in the heart of a red giant.

Reality

No protons or nuclei are produced at a functioning star, or a neutron star. In other explanations of red giant stars, it is stated that increased fusion 'causes' great expansion, and as this expansion takes up energy the surface cools and therefore, the reddened luminosity. As helium fuses to carbon the outer layers begin to expand, cools, and shine brightly.

Engendering and termination of red giants, white dwarfs and neutron stars

Small and medium star fields

Normally, the terminating star, and the star field disintegrate, drowning/transmuting/amalgamating into/within its space, as a black hole. Some stars explode their outer atmosphere of plasma. Propagating through space, this plasma transmutes into protons. What remains is nebula of protons and the core, which eventually disintegrates, drowning/transmuting/amalgamating into/within its space, as a black hole.

Engendering and termination of Red giants, and White dwarfs

Depending on the number of energy transmuting cycles in the star field, and the way they terminate, some stars do not explode, and expand their outer atmosphere. As energy transmuting cycles terminate, the star stops transmuting particles into plasma. As extraction/expulsion processes cease, the star may explode its presently generated plasma or outer atmosphere, or expand its outer atmosphere of plasma. The outer atmosphere expands ten to one thousand times the diameter of the sun, taking on a reddish color (what gives it its name) turning into what we call a red giant. With the decreased temperatures, and the expanded outer core (area is increased) the luminosity is reddened-a red giant star. This red giant endures, but cannot expand continuously, eventually casting off its expanded atmosphere, which floats off as a nebula. The core then becomes into a white dwarf. The remaining core (few thousand kilometers in diameter) is a type of white dwarf, which is glowing (from left over heat) a bit is now left to cool. There are plenty of white dwarfs observed, evidence that most stars end up this way.

Our sun will turn into a red giant, and then into a white dwarf. The term 'white dwarf' is a bit misleading. The core of a white dwarf starts out white, but displays a range of colors, as temperatures gradually decrease (as temperatures are linked to color). White dwarfs are difficult to observe, as when the heat within the core drops, the body gradually stops glowing, and it becomes a black dwarf.

Larger star fields-the engendering of a neutron star

Large star field cycles are very sensitive, and abruptly terminate, even with small alterations in/as energy transmutations cycles. This results in the star to explode its corona or upper atmosphere of plasma. Plasma immediately transmutes to protons propagating through space. What remains of/in/within the terminated star field is the cold dense core of the exploded star, and the nebula or clouds of protons without any star field space (utilized to produce protons).

Without the periphery, and also without space, the core gradually disintegrates into particles, as it is not confined. However, as rotation continues, some of the disintegrating particles now form a new periphery around the decreased dense core. As there is no star field space confining the new core, its rotational speed increases in gradual steps, generating temperatures, and pressure/force similar to/at intensities of functioning stars. Now with opposite processes which engender a normal star, as the condensed energy, or the dense core begin disintegrating, particles at the periphery integrate to neutron plasma instead of proton plasma.

Neutron plasma forms, as there is no space within the terminated star field, with the blasting off plasma being enveloped by space, transforming into protons. If there was space, proton plasma would engender. As neutron plasma form, and the rotational speed increases, the nebula or clouds of protons are extracted toward the core. With great pressures and temperatures, protons and neutron plasma are conjoined together-this is a neutron star.

The neutron star (without confining space) rotates at high velocities, generating great pressure. As protons and neutron plasma condense, the velocity, and proportionately the density increases. A 'star' 1.4 to four times the mass of the sun can be compressed as a neutron star, only about 20 kilometers across. The material is so tightly packed, that a teaspoon of it would have a mass of millions of tons. One cubic centimeter would weigh around one kilogram.

All our planets are fragments (highly compressed energy) of an exploded neutron star. This explains why the inner planets could eventually engender heavier nuclei, or matter (revelation 9). As nuclei gradually disintegrate into protons, and particles, which are extracted back, as replenishment energy to the sun, the sun endures, together with all the bodies of our solar system, explaining why our solar system has lasted for billions of years.

It is postulated, that the neutron star which started out about the size of the earth, is compacted into a sphere less than 100 kilometers across. A neutron star is very hot and dense, and as it has no surface area to radiate energy it is dim, and very hard to see. In other words, as heat and pressure/force do not propagate (without any energy transmuting cycles) there is no light.

The neutron star is a rapidly rotating body-some at 30 spins per second (this powers radio pulsars-powerful radiation), and about once every second-as it spins-gives out radio waves, and x-rays called radio pulsars, and x-ray pulsars, gamma rays, and x-rays. We see beams of light (like a lighthouse beacon) at the poles. If the neutron star's is positioned in such a way, that its on-and-off signal is visible from Earth, it is then called a Pulsar. Simply, pulsars are rotating neutron stars, and pulsars pulse as they rotate. It can be postulated, that under great pressures, integrating protons and plasma neutrons produce those pulsars.

Termination of a neutron star

As all protons and neutron plasma are integrated solidly dense under great pressure, the neutron star explodes. This explosion is a supernova, which is a spectacular explosion that lights up the sky. The surface shines as brilliantly as thousands of suns, while the emitting surface expands at several thousand kilometers per second.

We will learn, that our solar system engendered from a supernova-from fiery proton and neutron plasma, or fragments of an exploding neutron star.

Why are protons and/or nuclei not produced at normal stars or neutron stars?

Free energy condenses into a dense region, concurrently engendering space-a star field. In reality, condensing energy generates particles, and with further concentration particles integrate forming plasma. Particles and plasma are human concepts. However, as particles and plasma concurrently engender with star field space, a particle or plasma are not 'entities' (but condensing energy), as humans understand existences. As human ideate/create existences, a particle and plasma are 'entities' (for explanation purposes) only if 'individually' confined by space.

Proton plasma propagating through space transmutes into protons. Neutron plasma (a form of disintegrating condensed energy) disintegrated from a nucleus immediately transmutes/amalgamates into/as space. As space confines the dense region, protons/nuclei cannot form at functioning star processes. At star processes, condensing energy transmutes to particles and plasma. Plasma generated and expelled at/from star processes, transmute into protons confined by space.

A small medium or large star will engender with highest possible temperatures and pressures, proportionate to the dimensions of the star field and therefore, engendering of different nuclei/elements at a functioning star cannot take place, as different degrees of temperatures and pressure intensities are required. At a neutron star there is no space, as plasma (the cast off outer atmosphere of a dying star) is enveloped by space forming protons.

As a large star sheds off its atmosphere, the remaining core disintegrates, and the particles at the super speed periphery generating great pressures integrate to form neutron plasma, as there is no space around to form protons.

A proton is proton plasma, or condensing energy confined by space. Neutron plasma is disintegrating condensed energy. At a neutron star, without confining space, neutron plasma and protons cannot fuse together as a nucleus. If the hypothesized neutrons (with its confining space) existed, they would exist as separate 'entities'. With tremendous pressures at/as a neutron star, disintegrating condensed energy or neutron plasma is conjoined with protons. However, as space does not contain the neutron star, no nuclei are formed. A nuclei forms only as a proton conjoined together with neutron plasma is confined by space. There are no protons or nuclei produced at neutron stars.

Do stars collide?

Stars are confined by star field space, and also by their interstellar medium. No existence is an entity, as energy continuously transmutes. As massive astronomical bodies, stars do not collide. Energy is transmuting as/into extreme temperatures and pressure/force, enduring stars.

Black holes

As star fields terminate, energy 'becomes' 'free', or in other words, as energy is not transmuting into states, structures, processes or forms, it is now 'free'. As some star fields terminate, all the material from the dying star, and star field material, 'drowns'/amalgamates with star field space into/as a black hole. As a terminating star field continues the rotating norm, free energy in momentum exists, as a swirling black hole.

There is no causation, and black holes begin concurrently existing consequent to terminating star fields. Simply, a black hole is a terminating star field, or an ephemeral end of transmuting energy cycles, as a star field.

One-by-one, as stars collapse into/as black holes, it grows in strength and size. The increasing pressure intensities, and extreme temperatures at/as the black holes can now rip off, and pull in existing star fields 'swallowing' their material. Some times black holes spin existing adjacent star fields many times faster, terminating their cycles. This generates yet further energy transmuting variations enhancing the maelstrom. In some cases, as star fields drown, or are swallowed whole, there would be little or no radiation marking the event.

Bodies, and gas, and dust within a star field continue their spin as the cycle ends. In fact, disintegration increases the velocity of the spin of the entire star field. This spin at great speeds and the emitting radiation are observed, as something 'drowning' into a hole and therefore, the name black hole. Although material is 'sinking' into/as black hole the spin is still maintained, continuing even after all the material is ultimately sunk-amalgamated.

Black holes swirling at great speed facilitate free energy to begin concentration forming dense regions, engendering new star fields. It can be postulated, that with massive black holes free energy would transmute into many dense regions, and each may form a star field cycle-star field.

Black holes are not holes. There is no void, or cavity, or space (as a void) anywhere within a galaxy, or existential energy. Black holes are not black. There is no light once a star field terminates, as star field material transmutes, or drowns, or amalgamates into/with/as its space/energy. We cannot see black holes directly, but we can see their effects from surrounding matter, such as nearby gas, and other star fields.

It is theorized that all the mass of a giant star is squeezed into a single point-the black hole. At this point both space and time stops. We can imagine how preposterous this theory is. This concept is just the opposite to the Big bang theory, postulating the concept 'universe' oozing out from a point of no time, and no space. We are made to imagine a black hole, as a place where mass has no volume, and time does not pass. If the Big bang theory has been accepted by most scientists, why would this black hole (as a void or passing into another universe) concept not work?

We have been informed a black hole is a region of space, created by the total gravitational collapse of matter whose attractive gravitational force is so intense, that no matter, light or communication of any kind can escape. This is incorrect. A star field cycle transmutes energy. There is no such thing as 'gravity'. Matter or light are transmuting forms, states, processes or structures of energy. Energy is everything, everywhere. Collapse means to transmute into yet other forms, states, processes or structures of energy. Light or matter is a transmuting form of energy. Energy cannot 'escape', and only transmutes. There is no such thing as space-a void- as everything is energy everywhere. There is no place to go, as energy cannot be annihilated, created or destroyed. A black hole is a phenomenon of transmuting matter together with its integral space (of a terminating star field). Material or energy drowns/amalgamates, or in reality, transmutes into other states, processes, structures or forms of energy. With energy particles being transmuted at a high velocities (pressure and density are extremes), there is radiation/x-rays, and gamma ray bursts spewing out, glowing from black holes. A disc around a black hole can be seen, as matter is sucked/transmutes into its center, accelerated at high velocities, and heated, giving off x-rays and light.

We are informed that when anything (object or light) gets too close to a black hole it gets pulled in, and forever trapped. There is no void anywhere for anything to be 'trapped'.

Functioning stars (fields) continue to fall (amalgamate/transmute) one by one into a black hole, which sometimes grow very large, some emitting increasing radiation. Quasars may be the consequence of this phenomenon.

This drowning material (black hole) is not cut off from the 'universe', or the fabric of energy, as people assume. Moreover, it is preposterous to assume, that the disappearing matter travels through the black hole into a parallel universe. As discussed earlier, there is no such thing as a universe, and also parallel universes.

Astrophysicists tell us that the star drowning into a black hole sinks, disappearing from the postulated 'universe' into a point of infinite density. This theory is a familiar story (as the Big bang theory) as the postulated 'universe' appearing out of a point of infinite density. This is preposterous, as nothing disappears. We assume that something 'disappears', as it cannot be observed after it has existed.

Black holes are regions within a galaxy, or existential energy transmuting a terminating star field's material into free energy. As star fields and galaxies have already formed within existential energy's expanse, we can be sure another star field will engender from ‘material’ 'drowning' into/as a black hole.

What are gamma rays?

Gamma and cosmic rays can only be seen by advanced sophisticated instruments. Gamma rays are bursts of energy, or energy in high momentum, or the most energetic form of light, and can come from outside the earth from all parts of the sky. Gamma rays are brilliant flashes of light, and are high frequency rays, but there are million times more energetic rays called cosmic rays. Gamma rays can be found propagating from supernova explosions, and neutron stars/pulsars, but gamma rays striking earth propagate only from our sun.

What are x-ray bursts?

X-rays bursts likely result from the flashes in the surface layers of the neutron stars, as neutron plasma form, while rapidly repetitive bursts almost certainly result from a spasmodic accretion inflow of protons onto the neutron star, as neutron plasma are forced to conjoin with protons.

As clouds of protons and neutron plasma conjoin, this produces a very powerful flash-an x-ray burst. The interval between the burst depend on the proton flow onto the neutron star, and on the temperatures of the neutron star.

It is postulated the so called x-ray bursts typically emit ten times more light than our sun does. These x-radiation burst of x-rays are experienced a few times a day. The bursts reach their maximum intensity in only a few seconds, and are sometimes faster lasting for a few minutes, or shorter or longer. At burst maximum, the x-radiation is about ten to one hundred times stronger than the continuous x-radiation between bursts.

Quasars

Quasars (Quasi-Stellar Radio source) are compact objects beyond our galaxy, that shine bright (bright objects-bright x-ray sources-emitting radio waves) more intensely than many galaxies combined, or trillion times brighter than our sun. They look like normal 'stars', and over short times change their appearance.

Quasars have not yet been exhaustively studied. With abnormal energy transmutations star fields explode or collapse. As each star field adjacent to each other is interdependent, and interlinked, the termination of one star field may chain-react terminating other adjacent star fields. Material from many terminating star fields would become incredibly hot, and collectively begin to glow million times more brightly, and with great intensity than an ordinary star field, as they plunge/drown/transmute swirling in/as a massive black hole. This phenomenon can be postulated as quasars.

The mightiest energy jet fountains are probably functioning star fields, violently pulled into a black hole. As each star field explodes/collapses/transmutes into black holes, there are changes in their appearance. The occurrence of violent explosions, which throw energetic particles out from the central source, may be postulated, as a result from the termination of each star field.

Quasars are so bright, drowning out the light from all the other stars in the same galaxy. Despite their brightness due to their great distance from earth, no quasars can be seen with an unaided eye. In addition to radio waves and visible light, quasars also emit ultraviolet rays, infrared waves, x-rays and gamma rays

REVELATION FIVE

A SHATTERING STAR AND PLANET

The formation of our original, and present solar system

The prevailing theories on the formation of our planets, and our solar system

The traditional model of planet formation states, that the dust in a disc of gas around a star gradually clumps together into rocks, which eventually merge to make planetary cores. The cores then accumulate gaseous atmospheres. Gas giants such as Jupiter would have formed in the relatively cooler outskirts of the system. This is incorrect.

Another theory states that planets, or celestial bodies, form with accretion. Planets or extra solar planets are not cocoons of dust and gas, where accretion of particles takes place, as we have been informed. Accretion is explained as small dust gather into grains. These grains gather into large dust grains. These large grains gather into small particulates, and small particulates into large particulates. These large particulates gather into gravel, then into small balls, then into large balls. These large balls conjoin into tiny planets, and these tiny planets conjoin into large planets, until all matter in the region form a terrestrial body. What a preposterous story! The nebula (hypothesis) from which the planets allegedly condensed should have been rather homogeneous in most respects, and planets' masses should have had a smooth radial gradient with solar distance. How can planets form from dust and gas clumping together, or form from the debris of star field material, or even from the interstellar medium filled with nuclei, molecules and elements? How can dust, ice and gobs of gas come together to form the planets without the sun devouring them all is still unclear? How does dust clump into rocks?

The formation of our planets

Billions of years ago, somewhere around the Orion spiral arm within the Milky Way galaxy, a star field collapsed, disintegrating, or drowning, or amalgamating into its space, into/as a black hole. As free energy from/as the black hole began condensing, concurrently space engendered. This was our nascent star field. Concurrently, another adjacent larger star field collapsed. A neutron star engendered, as the dying star discarded its outer atmosphere, and the core began concentrating processes. As is the norm, the neutron star eventually exploded. As our nascent solar system was not full fledged (no energy transmuting cycles), and in close proximity, the fiery fragments of the exploded neutron star blasted through our nascent solar system space.

These fiery fragments of neutron plasma and protons, darting through, were confined by our nascent star field space, engendering nuclei, or deuterium and tritium. Deuterium and tritium with increases of heat and pressure/force, integrated into helium. As helium was heavier, it mostly constituted the interiors. These nuclei were the original atmospheres of planets in our nascent solar system. Concurrently, as atmospheres enveloped the fragments/nascent planets' interiors, the dense region within our nascent star field was activated, concurrently commencing the further integration of particles into plasma. Condensation increased the engendering of particles and plasma, and concurrently temperatures and pressure/force. These expelled forms of energy, interacted with the newly formed atmospheres of every body/planet, as energy transmuting cycles took hold, concurrently steadfastly set the planets, and the sun (the dense region), and our solar system began existing.

As atmospheres continuously disintegrate with interactions, and deplete, the interiors of planet/bodies also continuously disintegrate, replenishing the constituents of the atmospheres.

The key to our planetary formation was the quick engendering of atmospheres, or nuclei enveloping the fiery fragments, and the concurrent activation of the dense region to 'commence' its processes. A dense region will transform into a star only if its expelled plasma, temperatures and pressures (or forms of energy) can be replenished, concurrently within/as energy transmuting cycles, shuttling to-and-fro, with bodies/nebula within its star field.

As soon as energy transmuting cycles were in place, the fragments engendered into what we call planets. The sun also took hold (becoming full-fledged), and continues to process/transmute (propagate/extract) energy, interacting with atmospheres of planets/bodies within the solar system, as/and cycles steadfastly set.

A planet will endure, as long as it disintegrating interiors continue to sustain its continuously depleting atmosphere. After completely degassing its interiors, the atmospheres will deplete, and/as the planet solidifies, it eventually disintegrates.

Fiery fragments blasted within a neutron star field space, disintegrate and collapse into/as a black hole. It can be postulated, that if fiery fragments were to hurl through a functioning star field space, it may disrupt fixed energy transmuting cycles, and terminate the star field.

Our present solar system

Our solar system is a unique steadfastly set star field, as there as many planets, as there are energy transmuting cycles between them and the sun. Over millions of years, our inner planets' interiors, have been continuously degassing, or disintegrating, sustaining their continuous depleting atmospheres. All the planets/bodies and moons of our solar system were initially fiery bodies, not solids, and not with rocky cores.

Atmospheres of these bodies continuously deplete, as disintegrated nuclei are extracted toward the sun, replenishing the expelled material.

The following table detail sizes, distances from the sun, and orbital timings of planets.

The present solar system (including the sun)

	Diameter in Km	Average Distance from the sun- km (million)	Time taken to go around the sun (earth day/years)
Mercury	4878	57.8	88 earth days
Venus	12103	108.2	225
Earth	12756	149.6	365
Mars	6794	227.9	687
Exploded planet	???	Present asteroid belt	---
Jupiter	142800	778	11.9 earth years
Saturn	120000	1427	29.5
Uranus	52500	2870	84
Neptune	49400	4500	164.8
Pluto	3600	5900	247.7

Our solar system consists of planets (the large bodies orbiting the sun), their satellites (moons and various sized objects orbiting the planets), asteroids (small dense objects orbiting the sun, with most of them localized in the asteroid belt), comets, and other smaller bodies. There are now nine discovered planets, and other bodies-asteroid, comets, moons and gas rings, and other bodies in the Oort cloud and the Kuiper belt. Our original solar system consisted of nine planets, Mercury, Venus, Earth, Mars, Planet XYZ, Jupiter, Saturn, Neptune and Uranus.

As there was a number of energy transmuting cycles with as many planets, there were continuous disruptions in extraction and expulsion forms of energy from the sun, and as a result, a planet between Mars and Jupiter exploded. The fragments of this exploded planet darted both toward the sun, and away from it, and were captured by Mercury, Venus, Earth, Mars, Jupiter, Saturn,

Neptune and Uranus, transforming into gas rings, and moons. Some of these fragments settled in the far reaches of the solar system, within the Oort cloud, and the Kuiper belt. Some of the fragments remained around the same location (the asteroid belt) of the exploding planet. Asteroids in the asteroid belts are remnants of this disintegration, explaining why 95% of asteroids are stationed there.

Asteroids were/are small fiery fragments of the exploded planet, immediately enveloping atmospheric nuclei (as energy transmuting cycles were in force) around their interiors. Having commenced energy transmuting cycles with the sun, they transformed into small planets (asteroids), or moons. However, over time, and being small, most of the asteroids' interiors degassed, sustaining their depleting atmospheres, and are now solid bodies.

Some fragments of the exploding planet, captured by existing planets became moons, and parts of their planet's energy transmuting cycle with the sun. Some of the moons of the inner planets, which were degassing over time, sustaining their depleting atmospheres are now nearly solid bodies-our moon is an example.

Mercury and Venus may have had moons, and Earth and Mars may also have had more moons. However, in close proximity to the sun they may have disintegrated over time, or have solidified, as our moon, and Mars' moons. Mercury and Mars have also solidified with little or no atmosphere.

Some fragments of the exploded planet were captured by planets forming into gas rings, becoming parts of their planet's energy transmuting cycle with the sun. However, being in close proximity to the sun, the inner planets may have lost not only their moons, but also their gas rings.

There may have also been other planets, moons, and gas rings, closer to the sun than Mercury, which may have over time disintegrated.

Pluto like all other moons may have been a fragment of the exploded planet. It may have also been part of an energy transmuting cycle with one of the planets, or it may have been a darting away fragment that escaped capture, subsequently transforming into a planet on its own. Pluto is discussed later on.

All the comets and bodies in the Kipper belt, and the Oort cloud may have either been fragments of the exploded neutron star (that were thrown far away), or fragments of the exploded planet XYZ. They could not, and cannot form an energy transmuting cycle with the sun, as all forms of energy dwindle with distance, remaining as icy bodies in the dark far reaches of the solar system.

Why are the inner planets rocky?

Why are planets of different size?

Why are planets not lined up with regards to size in sequence from or toward the sun?

The exploding neutron star blasted away multifarious fiery fragments, some toward the nascent sun, and some away from it.

As we have learnt, the sun expels plasma with temperatures/forms of radiation, and pressures/force throughout the solar system. These forms of energy continuously interact, and disintegrate atmospheres of planets/bodies within our solar system. The interiors also disintegrate, and atmospheres are regenerated again and again. What is expelled must be replenished, as energy is everything, everywhere. Disintegration produces particles extracted toward the sun, within the shuttling-to-and-fro energy transmuting cycle, as replenishment.

In the case of earth, radiated heat and pressures/force from the sun, delves down through our altering atmosphere, disintegrating the water molecule. Our atmosphere and our oceans are continuously being interacted/disintegrated, and depleted. As atmospheric nuclei, and the water molecule, is continuously disintegrated (into hydrogen, oxygen and other nuclei), hydrogen is swept away faster into space than oxygen, explaining our blue skies. Some of the oxygen further disintegrates to be swept away, some to amalgamate with air, and some to produce water vapor, and ozone, constituting our continuously altering atmosphere.

Our oceans and atmosphere continuously alter their characteristics, regenerating again and again, with the continuous inflow of plasma, radiated heat and pressure/force from the sun, and the outflow of hydrogen, and nuclei from the oceans, surfaces and the atmosphere. As atmospheres and our oceans are continuously being depleted, the interiors have to belch out material (disintegrating and degassing) sustaining them, explaining continuous volcanic activity. As the material of the cores deplete, the interiors shrink, and the mantle and the crust sink, explaining earth's plate tectonics, earthquakes with tsunamis, and recent increase in flooding all over the globe. Moreover, as the interiors shrink (with degassing), the body/planet with its atmosphere (if it still has any) have shrunk in size.

At close proximity to the sun, the atmospheres of the inner planets were/are continuously disintegrated, and extracted. As planets/bodies interiors subsequently completely deplete, all that is left is solid material explaining their sizes, and the solid/rocky interiors of the inner planets. The moons of Earth and Mars, and the bodies in the asteroid belt are examples. Jupiter's moon Io next in the active disintegration processes, now has its interiors violently belching out through its volcanoes, sustaining its depleting atmosphere. Presently, Io is our solar system's most volcanically active body.

Looking at the size and distance of planets from the sun, we can now understand that the closer a planet is to the sun, the greater and faster is its disintegration. The closer the planet/body is to the sun the faster its atmosphere is depleted, and proportionately their degassing activity, and solidifying processes. Obviously, the radiation/transmuting cycle is stronger at close proximity. As planets shrink in size, concurrently their orbital distance from the sun also decrease. All the planets may have been the sizes of Uranus or Neptune. Jupiter and Saturn are larger, as they attracted lots of vaporized matter from the exploded planet, which was localized what is now called the asteroid belt.

Mercury being closest to the sun has completely degassed over time, sustaining its atmosphere, and solidifying. As it is closest to the sun more material is/was being extracted, and its diameter is now only 4878 kilometers. Its present wobbling is a sign it will soon be disintegrated.

Venus has engendered an atmosphere which is not easily disintegrated by radiated heat and pressures, explaining why it rotates very slowly around its imaginary axis.

Earth's early atmosphere altered to one that engendered its oceans, which now sustains its continuously depleting atmosphere, so there is less degassing of the interiors, and it has till today maintained some of its molten core.

Mars has completely solidified, as it did not have a unique atmosphere/ocean enduring combination like earth. The lengthy volcanic ranges are evidence that Mars belched out profusely, shrinking and solidifying faster than Venus and Earth. With no more interior material to belch out, Mars presently has a very thin atmosphere mostly of carbon dioxide, which is more than that found on earth. Without oceans, there is no hydrological cycle as it is with earth, and oxygen from the last of little water left has now disintegrated into carbon dioxide. With its energy cycle with the sun nearly terminated (no expulsion and extraction processes), it suddenly stops rotating in its normal direction, rotating in the opposite direction for some months.

As the constituents of their atmospheres engendered, are of a type that prevents easier disintegration than those of Mercury and Mars, Venus and Earth still have their molten interiors. Although closer to the sun than Mars, they have not shrunk in size as Mars and Mercury. They still have considerable amount of inner material, continuing to sustain their continuously depleting atmospheres.

There may have been other planets probably with their moons and gas rings, closer to the sun than Mercury, which over time may have solidified and disintegrated. Mercury and Venus may have had rings around them, together with moons. However, these may have disintegrated over time. Earth also may have had more moons, and also rings, and also Mars, which also may have disintegrated over time. Our moon has taken the same fate as Mercury and Mars. They all have nearly lost their atmospheres, as their degassing internal cores have nearly dried up, and all that remains are the solid material. Mercury, Mars, its moons, Earth's moon, and some inactive asteroids in the asteroid belt are next in the disintegration line. Eventually, after Mercury, Mars, its moon, and Earth's moon, and the asteroids have disintegrated, the next in line would be Venus, Earth, and Jupiter's moon Io with its present violent volcanic activity.

Over time, our moon has completely degassed and solidified. Without a direct energy cycle with the sun, and its close proximity to earth, it is now an impotent part with earth's energy transmuting cycle with the sun. This explains why it does not rotate around its axis, showing the same face to the Earth, as it orbits it. Only planets/bodies with active transmuting cycles with the sun rotate around their imaginary axis. As all existences exist within/as the fabric of space/energy, the moon per se does not orbit the Earth, but does so, as its surrounding space (within which it is stationed) orbits the Earth.

Jupiter Saturn, Uranus and Neptune are still very large, as radiated heat, and pressures/force from the sun does not reach them (to actively disintegrate their atmospheres) with those intensities, as with the inner planets (all propagating forms of energy dwindle with distance). Although all have energy transmuting cycles with the sun, their atmospheres are only partially disintegrated for replenishment material, as the inner planets have active energy transmuting cycles.

The sun as we can note, receives ample replenishment material from the inner planets, and the Jovian planets continue to maintain their sizes. Their interiors do not have to continuously belch of material to sustain their atmospheres, which are only partially depleted. Their fast rotation around their imaginary axis is evidence. This also explains, why they have smaller rocky cores (and less solidification) than the inner planets, as they have not degassed much over the years, and other reasons discussed elsewhere.

Pluto was discovered in 1930, as one hemisphere faced the sun. Its orbit around the sun takes 248 years. When discovered, Pluto did not seem to have any atmosphere. Later on it was observed, that when Pluto approached the sun, its dark areas transformed into light, warming up its nitrogen and methane ices, which when evaporating produces a rarefied atmosphere. Presently Pluto is still close to the sun, but as it moves away from its perihelion, its poles will face the sun. Towards the middle of this century, its gases will freeze again, evidence that Pluto is an energy replenishment reserve, called upon to balance extraction and expulsion material of the sun. Pluto normally has an atmosphere of methane, and other unknown gases. Pluto takes 6 days, 9 hours and 18 minutes to rotate once around its imaginary axis.

We have to decide how small a cosmic body should be for it to be classified as a moon/planet. Pluto is only 2344 kilometers in diameter, much smaller than our moon. Five moons within our solar system are larger than Pluto, and two being larger than Mercury. As it has a direct energy transmuting cycle with the sun, rotates around its axis, and 'independently' orbits the sun, it is considered a planet.

Planets/bodies are of different characteristics, as they were multifarious fragments of the exploding star, and the exploding planet. Moreover, they localized at different distances from the sun, experiencing varying forms of energy at different intensities. Over the years, the interiors of inner planets have been degassing, or disintegrating, sustaining their depleting atmospheres. As a result, their atmospheres are of different constituents, and they are presently of different sizes. Some have partially solidified, with their interiors containing a mix of nuclei/elements.

Some other characteristics of bodies in our solar system

Mercury is the closest planet (58 million kilometers) from the sun. Closest to the sun, it was most subjected to the extraction-expulsion radiation cycle. Without an active energy transmuting cycle, Mercury's present orbit tilts into a long ellipse. Mercury has no material (atmosphere) left-the kind that is required for replenishment at the sun-and as a result, its energy cycle with the sun is nearly inactive. Therefore, being impotent, it takes 59 earth days to complete just one rotation around its imaginary axis. Being closest to the sun, it takes 88 earth days to orbit the sun. Although most bodies (within our solar system) have no energy transmuting cycle with the sun, they exist within/as the fabric of interplanetary space/energy, which is orbiting the sun. This explains why the orbital duration of planets/bodies increases in distance away from the sun. This also explains, why Mercury and other bodies even without an active energy transmuting cycle continue to orbit the sun. Mercury's interior has nearly completely degassed, and is now solid metal composed of iron and nickel, and therefore, has a very thin atmosphere. Mercury has temperatures reaching 427 degrees Celsius during its long day, and -173 degrees C during its long night. It now has the densest core than any other planet in our solar system. As it is closest to the sun, and traverses the energy cycles between the sun and other planets, Mercury has a strange perturbed motion. Without an active energy transmuting cycle with the sun, it could disintegrate any time in the near future.

Venus has an atmosphere mainly of carbon dioxide, preventing the radiated heat from the sun to penetrate its atmosphere, and concurrently heat from escaping its interiors. As a result its surface temperature is a fierce 482 degrees Celsius, even hotter than Mercury, its neighbor closer to the sun. As the radiated heat interacts with Venus's atmosphere, it shines as/with a bright light. As Venus has its unique atmosphere preventing most of the radiated heat from the sun interacting (integrating/disintegrating) it, there is minimum material extraction from Venus. As a result, Venus rotates very slowly, taking 243 earth days to complete a single rotation around its imaginary axis. Moreover, a day on Venus lasts even longer than its year, taking 225 earth days to orbit the sun.

We will learn that craters on the surface of our solar system bodies are not the result of asteroids/meteorites striking them, as we have been incorrectly informed. Those craters are vestiges of volcanic eruptions of the degassing processes. The relatively few craters appearing on Venus is evidence its interiors have not degassed, as much as the other inner planets, our moon and the asteroids. More craters can be counted when viewing a section of the moon through a small telescope, than those occurring on the entire surface of Venus. Venus has large craters, and no small craters, evidence of less, or uncommon volcanic activity.

As a planet exploded, earth's interiors profusely degassed, forming a unique new atmosphere producing water vapor, clouds and precipitation. Continuous rain solidified parts of the fiery interiors into the mantle and crust, which held the rains into/as rivers, lakes and the oceans. As a result, radiation delves through the atmosphere, and plunges into the oceans disintegrating the water molecule. Earth developed a many layered atmosphere, engendering a troposphere where life could flourish. Energy is absorbed, processed and extracted through the processes of the biosphere from living things. With these processes, both Venus (with its unique atmosphere) and Earth have not yet drastically shrunk, as Mercury and Mars. The Earth, its moon, its atmosphere, oceans and body are discussed in another revelation.

The interiors of Mars belched out material engendering a certain type of atmosphere, easily interacted, integrated, and disintegrated by/within its radiation/transmuting cycle with the sun. As a result, Mars has shrunk faster than Earth and Venus, and is now about half the size of Earth. Mars has a 24 kilometer high Volcano called Olympus Mons, which is larger than any other in the solar system. There is a 3220 kilometer long canyon called Valle Marineris, twenty-six times as long, and three times as deep as the Grand Canyon, evidence that Mars at some time in the near past degassed profusely, and shrank very quickly. Mars two moons Phobos and Deimos have also shrunk, and are now very small. As Mars is further away than Earth, it takes 687 earth days to orbit the sun, and has its seasons twice as long as ours.

As the Jovian planets' atmospheres are not easily interacted by the radiated heat from the sun (propagating forms of energy from the sun dwindle in intensities with distance), they have maintained their large sizes. As soon as the inner planets solidify (and eventually disintegrate) the sun will source its replenishment material from the Jovian planets. They will then move closer to the sun, and with their atmospheres slowly being depleted, their interiors will increase the belching out of disintegrated/degassing nuclei/elements. Eventually, they will transform, solidifying like our moon, Mercury and Mars.

Jupiter consists mostly of hydrogen and helium, and Saturn with a less than a third the mass of Jupiter follows it fairly closely. Planets rotate (their speed) on their imaginary axis, depending on their closeness to the sun (the radiation intensity), extraction/expulsion requirements, and the type of atmospheres that can, or cannot be easily disintegrated.

If Jupiter and the other Jovian planets spin faster than the earth, it would logically explain that the radiation intensity from the sun is greater at these planets. This is not the case, as the atmospheres of these planets are not being interacted, as intensely as the inner planets. They have maintained their gaseous interiors, and their huge sizes. With their gaseous bodies, they spin faster than the earth and the inner planets, which are more solid bodies.

As the Jovian planets do not have very active energy transmuting cycles, as the inner planets, their fast rotation results with their gaseous interiors. Their different rotational speed also depends on their constituted material, the amount of material, and/or their size.

Over millions of years, the sun replenished its material from the inner planets, and possibly together from some moons, and some gas rings around inner planets, that may have long been disintegrated. Jupiter still has its 16 moons (21 more were discovered this year), evidence that the extraction power of the sun is still not as intense reaching the Jovian Planets. Although Jupiter's diameter is about 142800 kilometers compared to the Earth's 12,700 kilometers, Jupiter's rotation around its imaginary axis is about 10 earth hours compared to the earth's 24 hours (one day).

Saturn is very dense, as its surface is covered by a sea of liquid hydrogen and helium (at the beginning all planets were comprised of deuterium, tritium and helium). As a Jovian planet further away from the sun, it spins very rapidly, and this has flattened its poles. It takes 10 hours and 30 minutes-less than half the time it takes Earth-to complete one turn. Saturn holds as many as 18 moons, and three distinct rings around it.

There are faster winds within Neptune's atmosphere generating densities (greatest mean density that other Jovian planets), producing heat which is ejected outwards. As temperatures, pressures/force and densities are all forms/states of energy, the transmutation of decreased temperatures (as with the Jovian Planets and other bodies further distant from the sun) into densities, generates Neptune's heat radiated outward, more than it receives. Uranus, Neptune are also comprised mostly of gas, mainly hydrogen and helium.

Bodies of the Kipper belt and the Oort cloud are fragments of the exploded neutron star, and the exploded planet. Radiated heat and pressures from the sun dwindle with distance and therefore, the outskirts of our solar system have very low temperatures, explaining why icy comets are distributed at extremely great distances from the sun. This also explains, why at the outskirts of a star field or our solar system there is cold dark matter-nuclei/elements and molecules form bodies-containing primarily of frozen hydrogen, helium, methane and ammonia. In the deep freezes of space, or outskirts of our solar system, the ices could remain unchanged over millions of years, or until the star field terminates.

Astronomers have found new astronomical bodies recently calling them planetoids. Quaoar smaller than Pluto was recently discovered, followed by Sedna. Sedna takes 10500 years to orbit the sun being many million kilometers away from it. Researchers calculate the size of astronomical bodies by brightness, making assumptions based on how much light they 'reflect'. These measurements can be incorrect, as those bodies 'light' up, proportionately to the amount of atmosphere enveloping them. Moreover, it is not only the amount of atmosphere, but the type of atmosphere that 'light' up with radiated heat. For example, Venus' brighter atmosphere is dependant on the type of its atmosphere. Again, we must also take into consideration the reflectivity of the surface material of those bodies. Our moon does not 'light' up with any radiated heat (as it has no atmosphere to be interacted), or even by sunlight (which does not reach it). The moon's surface comprises of a kind of basalt shiny material, that 'lights' up 'seeing' the light at our upper atmosphere (revelation 9), or possibly some radiated heat produced at our upper atmosphere reaches its surface. There may be other planets/bodies within our solar system, which cannot be observed, as they possess no atmospheres to be interacted by radiated heat to produce light, and be apparent.

Concluding:-

The terrestrial or rocky planets Mercury, Venus, Earth and Mars are composed primarily of rock and metal, and have relatively high densities, slow rotation, solid surfaces, no rings (that may have been), and few satellites (that may have been). It can be postulated that earlier in our solar system history, there may have been more planets closer to Mercury, which may have now disintegrated.

Before Mercury and Mars shrunk in size, there could have been less radiated heat, or the intensity of the energy transmuting cycle with/on Earth would have been less intense. In this case, Earth would have been rotating slower on its axis, and our days would have been shorter. It can be postulated, that over thousands of years, Earth once had a 16 hours day (and was not tilted) with equal day and night hours. Could this be the reason why most mammals including humans (evolving mechanisms) need about 8 hours of sleep?

The inner planets are rocky, having depleted much of their gaseous interiors sustaining their atmospheres. Without an abundant atmosphere (for material replenishment), and as they are rocky they are slow rotating, and have a partially solid surface.

They have no rings, as they must have been disintegrated by now. Earth has a moon which has gradually shrunk in size. Mars has two small ones, and has also decreased in size together with their host.

Over time, Mercury not only lost its atmosphere, but has nearly completely solidified. Being closer to the sun, there was more extraction of its material.

The Jovian or gas planets Jupiter Saturn, Uranus and Neptune, are composed primarily of hydrogen and helium, and generally have low densities, rapid rotation, deep atmospheres, rings, and lots of satellites. They possess a number of moons, as radiated heat/pressure from the sun, has not yet actively begun sourcing replenishment material from the Jovian planets.

The small inner planets Mercury, Venus, Earth, and Mars have diameters less than 13000 Km, whereas the giant outer planets each have diameters greater than 48000 Km. As we will learn elsewhere, as the inner planets transformed their atmospheres and their interiors into heavier nuclei/elements, they partly solidified. Over the years as their atmospheres were being depleted, the interiors which belched out material have now dried up. Our moon, Mercury and Mars are examples.

Different types of elements are found in different layers, and at different places on all moons and inner planets, as their molten core belched out diversified material, at different times. Atmospheres have also changed their characteristics (densities, composition, temperatures and pressures) over the years.

Venus and Earth possessed unique atmospheres, which were not actively disintegrated, and the sun sourced its replenishment material from Mars and the asteroids, explaining the present small sized Mars, and the solidifying of most of the asteroids.

Why are Jupiter and Saturn massive planets?

As Jupiter and Saturn were adjacent to the exploded planet XYZ, the solar wind, or radiated heat, and pressure/force from the sun would have ensured that all the nuclei/nebula of this planet would be driven toward Jupiter and Saturn. These planets swept up this material increasing in size. This also explains why Jupiter has so many moons.

As Jupiter was more aggressively involved in these processes, Saturn was secondary in the running (being further away), attracting the left over material, which drifted away passed Jupiter. This explains why even though Saturn has many moons, they have rather unevenly spaced orbits, with several huge gaps interspersed with rings of material. This also explains why out of Saturn's eighteen moons only Titan is large, the others being relatively small.

Fragments not captured by Jupiter and Saturn were attracted to/by Uranus and Neptune, as moons and gas rings. Other fragments that were not captured were thrown to far distances, settling down in the Oort cloud, and the Kuiper belts. As radiated heat and pressure/force dwindle in intensities through distance, these bodies did not, and could not form an energy transmuting cycle with the sun, remaining as comets, and other icy bodies.

As radiated heat and pressure/force from the sun dwindle in intensities through distance, the outer planets are still large, and still hold on to their moons and gas rings, as there is plenty of replenishment energy available, extracted from the inner planets, over all these millions of years.

Why are planets observed as different colors?

Planets/bodies have atmospheres composed of different nuclei configurations, from their multifarious elements degassing interiors. Forms of energy from the sun interact with atmospheres, producing varying kinds of lights.

Mercury looks grey, as it has practically no atmosphere, and has solidified, and so we see its rocky surface. Venus looks white-yellow, as a result of its sulphuric and carbon dioxide atmosphere. We can only see the thick layer of colorless, featureless sulphuric acid clouds.

With radiated heat hydrogen looks blue, oxygen green and nitrogen red-these are our atmospheric colors (revelation 9). From outer space, Earth looks blue/green (more on the blue side), as a result of hydrogen continuously leaving its atmospheres, oceans and surfaces. The green is atmospheric oxygen, and green plants. Earth's oceans look blue/green from above, as the water molecules are continuously disintegrated. As we see our earth from outer space (the area facing the sun), and as its upper atmosphere transmutes to light, this light exhibits the brown, yellow and green continents.

Mars looks red/orange, as a result of metal oxides (rusty rocks) at its surface, as it has a very thin atmosphere. Jupiter has orange clouds from ammonium hydrogen sulphide and white bands from ammonia clouds. Saturn looks pale yellow, as a result of the ammonia haze, but it has redder clouds below. Uranus and Neptune looks light blue from methane clouds.

What are moons and gas rings?

Why do planets have moons? Why do Jovian planets have rings around them, and also moons?

Planets are fragments of the exploded neutron star. All asteroids, moons, gas rings, comets, and bodies in the Kipper belt and the Oort cloud are fragments of the exploded planet, which was localized between Mars and Jupiter-the asteroid belt.

Fragments of a planet that exploded after our solar system formed would comprise of deuterium, tritium and helium. These fragments would enhance the already existing energy transmuting cycles of existing planets with the sun. Fragments from the exploded planet darting towards the sun were captured by the inner planets, some engendering into moons, and others, into gas rings. Some darting away from the sun was captured by the Jovian planets, as moons, and gas rings, with some settling down in regions called the Oort cloud, and the Kipper belt.

Fragments formed into moons, as the fiery interiors were enveloped with nuclei/atmospheres. Fragments that did not engender into moons formed gas rings.

All planets may have possessed moons and gas rings. However, over time, at close proximity to the sun with intense radiation, the four inner planets may have lost their rings, and Mercury and Venus their moons. These planets together with Earth and Mars may have also had many moons like the Jovian planets.

The rings contain dust and a large quantity of water, frozen in various forms such as snowflakes, snowballs, hailstones and icebergs, ranging from 8 centimeters to 9 meters in diameter.

Why are the Jovian planets or outer planets gaseous? Why are they large?

All the planets initially engendered as gaseous fragments. It can be postulated all the planets engendered, at more or less, the size as Uranus and Neptune. Pluto is small, and was a fragment of the exploded planet, and was possibly a moon to Uranus, and/or Neptune, later breaking away from their cycles orbiting the sun on its own.

The inner planets have shrunk over time, as their atmospheres and interiors continuously deplete, as the sun replenishes its expended material from their atmospheres. There is plenty of material still to be disintegrated, and extracted from our solar system outer bodies, as replenishment material to continue the sun's processes. The large gaseous interiors, and smaller rocky cores of the Jovian planets, tell us that there was/is very little internal degassing to sustain their atmospheres, which are not easily disintegrated and depleted. As they continue to remain in their far away orbits, they are still gaseous, maintaining their large size.

Jupiter and Saturn are the largest of the outer planets, as they captured the vaporized material from an exploding planet, which was localized between Mars and Jupiter. It can be postulated that both Uranus and Neptune may have also captured the departing vaporized material, and are larger in size than the inner planets. With this scenario we can speculate, that all the original planets were sized slightly smaller than Neptune and Uranus.

Why is Pluto rocky if it is the most distant planet from the sun?

Pluto was/is a fragment of the exploded planet. As the inner planets, and their moons, and the asteroids in the asteroid belt have now all nearly turned solid, the sun has now involved Pluto in its replenishment processes.

Pluto was perhaps a moon to Uranus or Neptune, later breaking away from their cycle, as it was not steadfastly set, orbiting the sun on its own. It can also be postulated that some time in the recent past, as there was a expulsion-extraction energy balancing requirement, Pluto being a fully gaseous body (with the right type of required extraction material), and moreover a much smaller one than the other Jovian planets, was easily perturbed out of its planetary hold, as a moon to one of the outer planets, and now orbits the sun.

We can also postulate that Pluto was once a comet, jostled away some time ago from the Oort cloud. As and when more replenishment material is required, comets are pulled away from their localizations extracted toward the sun. Normally, they are disintegrated approaching the sun, but some of them orbit the sun, and are thrown away with the energy propagating cycle. Pluto may have been a comet that was perturbed from the far reaches of the solar system (the Oort cloud), but before it approached the sun, replenishment material may have been sourced elsewhere, and Pluto remained in its present orbit. Charon as a comet may have been perturbed together with Pluto, sometime before, or after. Like Pluto, as replenishment material was sourced elsewhere-as at that time it was at close proximity-it was attracted to Pluto as its moon.

Pluto is composed of rock and ice with surface temperatures averaging around -220 degrees C. Pluto's atmosphere is probably made up of nitrogen, carbon monoxide and methane. As Pluto comes closer to the sun, its atmosphere turns into a gaseous state, but otherwise during most of its orbit it is frozen. This may explain a comet with an eccentric orbit. As it orbits around the sun, material from its atmosphere (transforming into a gaseous state) is being extracted. To support/maintain its atmosphere the interiors have over time degassed, and Pluto now possesses a mostly solid interior.

Jupiter's moon Io has active volcanic activity, explaining the fact that energy radiation from the sun is now extending further with more intensity.

What is the distinction between a planet, moon, asteroid and comet?

This question is evidence that we create existences giving them names, without understanding phenomenal reality.

A planet will only engender (a body with its atmosphere) ‘commencing’ an energy transmuting cycle with the sun/star. Without an energy transmuting cycle, continuously interacting, and disintegrating its atmosphere, this body cannot be called a planet. Bodies rotate around their imaginary axis, as energy is being extracted from their continuously interacting, and disintegrating atmospheres. Moons and asteroids, with an energy transmuting cycle with the sun, should be called mini planets.

The moons (previously mini planets) of Earth and Mars used to rotate around their imaginary axis. Earth and Mar's moon(s) have now solidified, as their atmospheres and interiors depleted of extracting nuclei and therefore, do not rotate around their imaginary axis.

Pluto is much smaller than our moon, and some other moons, but having an independent energy transmuting cycle with the sun it not only orbits the sun, but also rotates around its imaginary axis, explaining why it is called a planet.

Every asteroid in the asteroid belt engendered with energy transmuting cycles with the sun, each orbiting around the sun at the asteroid belt. Over the years, having completely belched out their interiors to sustain their depleting atmosphere, some are now rock solid, with/as their cycles terminated, and are now impotent bodies. As they were/are localized in the asteroid belt, and/or exist within/as interplanetary space orbiting the sun, they still continue orbiting the sun, as was the norm. At this belt there are other asteroids that still have energy transmuting cycles with the sun, and they too orbit the sun, and also rotate around their imaginary axis-these can be called planets.

Some of the outer planets moons may be rotating around their axis, evidence they have an energy transmuting cycle with the sun, but as parts of their host energy cycle.

Comets are fragments of the exploded planet that were not captured by the outer planets, settling down as bodies in the far reaches of our solar system. As radiated heat and pressure/force dwindle with distance, comets cannot form an energy transmuting cycle with the sun.

During July 2005 astronomers discovered Sedna, a new body around a ring of rocky bodies beyond Neptune, called the Kuiper belt. Sedna is much larger than Pluto, but can it be called a planet? Sedna and other smaller, and larger undiscovered bodies, and all bodies/planets orbit the sun, as they exist within/as interplanetary space, which orbits the sun. However, if they are not rotating around their imaginary axis, they do not have independent energy transmuting cycles with the sun, and cannot be called planets.

Why are the solar system bodies constituted of nuclei/elements?

The exploded neutron star fragments were constituted of conjoined neutron plasma and protons. As the fragments of the neutron star blasted through our nascent solar system, they transformed into nuclei, as neutron plasma and protons were integrated, confined by space. Deuterium, Tritium and Helium constituted atmospheres, enveloping the fiery interiors engendered into planets.

The fragments of the exploded planet transformed into asteroids, moons, gas rings and other bodies, as energy transmuting cycles were already in force. As the interiors of planets/bodies, especially those nearer the sun, have degassed over the years sustaining their depleting atmospheres, the disintegration has produced a diversified mix of nuclei/elements and molecules at their atmospheres, surface, and their interiors. Revelation 8 and 11 details the formation of nuclei and elements.

Why have the inner planets shrunk in size? Why has the atmosphere of the moon depleted?

Why do atmospheres deplete?

The sun extracts lighter elements and particles/plasma, from atmospheres of bodies/planets in the solar system, within the energy transmuting cycle, replenishing the material (energy as temperatures/force and plasma) expelled. With the depletion of atmospheres, the interiors of bodies/planets continuously belch out material sustaining the depletion.

However, energy is everywhere and everything, so why should the inner bodies shrink in size? Plasma from the sun are propagated throughout interplanetary space, and over millions of years, are thrown away from energy transmuting shuttling to-and-fro cycle pathways, between the sun and the planets/bodies, settling down at the outskirts of our solar system as protons/hydrogen. Here they conjoin as nuclei forming molecules and compounds, and the interstellar medium grows larger. This explains why atmospheres deplete, interiors belch out to sustain them, and planets/bodies shrink/solidify, and eventually disintegrate.

Over the years, the atmospheres of the inner planets, and the moon were intensely interacted, as they were in close proximity to the sun, explaining why the inner planets have shrunk in size, and presently our moon is without an atmosphere.

Asteroids

As the exploded planet was localized, what is now a band of space between Mars and Jupiter, 95% of all fragments-turned-asteroids are located here-called the asteroid belt. Asteroid belt bodies are not the remnants of a planet that never formed, as we have been incorrectly informed.

These fragments formed quick atmospheres enveloping their interiors forming asteroids, avoiding disintegration from the sun's radiation. Fragments darting toward, and away from the sun engendered into moons and gas rings (captured by planets), comets and other bodies in the Oort clouds, and the Kuiper belt.

Each asteroid, or a group of asteroids, was once part of their own independent energy transmuting cycles with the sun. Some have lost their atmospheres, and without their interiors, and the energy transmuting cycles having terminated, they are now impotent bodies. Most of the asteroids in the asteroid belt still orbit the sun as was the norm, but without steadfast energy cycles, some impotent ones are often perturbed out of their localization roaming out of the asteroid belt.

It is not the same scenario with the Oort cloud comets. Comets are also small fragments of the exploded planet, which settled down at the outskirts of the solar system. As radiated heat and pressures dwindle with distance, they did not engender atmospheres enveloping their interiors. As they still possess easy disintegrating material (without atmospheres), and as they are small, they are perturbed from the localizations, and attracted when more replenishment material is required at the sun. Bodies like Pluto and the comets, compensate for the replenishment material decreases at the sun. This explains Pluto's eccentric orbit, and the seldom observation of comets.

The sizes of the asteroids range from dust particles to nearly one thousand kilometers across. The largest known asteroid is Ceres, 940 kilometers in diameter. In the observable range there are more than ten thousand known asteroids. The orbit of the earth and some planets are crossed by more than one hundred of them having eccentric orbits (explained above). Some of them miss hitting our moon, and the Earth.

Most asteroids cannot be seen, as they have no atmospheres. If asteroids possessed atmospheres, radiated heat interacting their nuclei, would manifest light, enabling us to observe them. Some of the asteroids are observed when they are in close proximity to earth. Some of them appear therefore, only to be points of light. Some are bright enough to be photographed from earth, as these obviously, are still holding on to their atmospheres.

Some time in the near future, Earth and Mars will loose their moons. Mercury, Venus, Earth and Mars may have already lost their gas rings. After that, all the asteroids in the asteroid belt will be disintegrated.

Why are some asteroids (those in the asteroid belt and those that roam the solar system) not spheres?

We can still observe some asteroids in the asteroid belt having pin-point lights. Obviously, they still have atmospheres, and are rotating around their axis, and are spheres. Some asteroids have completely solidified with continuous degassing over time.

As bodies have their atmospheres gradually being depleted, they rotate around their axis. This rotation, and the centrifugal, and centripetal force, endure them as spheres. These spherical bodies begin disintegrating, as soon as the rotations around their imaginary axis stop, and they loose their spherical shape.

Our Moon

Like other bodies, excluding the planets, our moon also was/is a gaseous fragment of the exploding planet.

Earlier with an atmosphere, our moon may have been rotating around its imaginary axis orbiting the earth. Localized within/as interplanetary space, it also orbits the sun with earth. It may have been rotating around its axis, as its atmosphere was being depleted (for replenishment material) by the energy radiation cycle. The interior continuously degassed to endure its continuously depleting atmosphere, and having solidified, now possesses a thin atmosphere. Without an atmosphere the energy transmuting cycle has terminated. Therefore, it does not rotate around its axis, but continues orbiting the earth, with/as the same side facing us. It does so, as it is not the moon that is orbiting the earth, but as it exists as/within the fabric of space orbiting the earth.

Mercury and Venus presently have no moons. Earth has one, Mars 2, Jupiter 47, Saturn 18, Uranus 15 Neptune 8 and Pluto 1. Five of the newly found 11 moons of Jupiter follow retrograde rotation, with their paths going in the opposite directions of Jupiter's spin.

The five prevailing theories of formation for our moon and our refutation

The Fission theory

The moon was once part of the earth, and somehow separated early in the history of the solar system. The present pacific-ocean basin is the most popular site for the part of the earth from which the moon came from.

A-The moon was/is a fragment of the exploded planet.

The Capture theory

The moon was formed somewhere else, and was later captured by the gravitational field of the earth.

A-There is no such thing as gravity. The moon (as a fragment of the exploded planet) was ‘captured’ darting towards the sun, becoming part of the energy transmuting cycle between the sun and the earth.

The Condensation theory

The moon and the earth condensed together from the original nebula forming the solar system.

A-The Earth was/is a fragment of an exploding neutron star. The moon was/is a fragment of an exploded planet. They did not condense to the state they are at present.

The Colliding planetisimal theory

The interaction of earth-orbiting and sun-colliding with orbiting planetisimals, early in the history of the solar system led to their breaking up. The moon condensed from this debris.

A-As explained earlier, the moon or other bodies/planets cannot condense from disintegrating gaseous nuclei. The moon and most of the inner planets are not what they initially were. Some may have disintegrated, some have solidified, some partly solidifying, but most have shrunk in size, as a result of belching out of their gaseous interiors to sustain their continuously depleting atmospheres.

The Ejected ring theory

A planetisimal struck the earth ejecting large volumes of matter, and a disk orbiting material was formed, and this matter eventually condensed to form the moon in orbit around the earth.

A-The earth has a many layered atmosphere. Every existence exists, as/within its space or existential energy. It would be difficult for a large object to strike the surface of the earth ejecting large volumes of matter without disintegrating it. Why and how should this matter condense?

Why does our moon have so many craters?

The moon's craters are not evidence of the maelstrom, which followed the engendering of planets and other bodies. Asteroids and meteorites did not bombard the moon, or the earth, as we have been incorrectly informed. It is also incorrectly postulated that the surface of the moon seems ridden with craters from impacts of meteorites, which somehow mysteriously missed the earth, preferring the moon instead.

The moon is one of the gaseous fragments of the exploding planet captured by earth, becoming part of its energy transmuting cycle with the sun. The moon engendered, concurrently with its own energy transmuting cycle with the sun. Over the years, both Earth, and the moon have been continuously belching (degassing) out their interiors, sustaining their depleting atmospheres. At present, the moon's thin atmosphere is evidence that its interiors have completely solidified, as it is much smaller. This will eventually be the fate of Venus and Earth, as it is now with Mercury and Mars.

Planets and bodies concurrently formed with atmospheres containing them, solidifying with gradual degassing of their interiors. Over the years, the moon's many volcanoes belched out, sustaining its depleting atmosphere. Eventually, as the atmosphere depletes completely, there is an increase of volcanic activity with more volcanoes. The moon's surface is now full of craters, with the cooling/shutting down of volcanic activity, as each crater was once a volcano.

Jupiter's moon Io is presently belching out, as volcanic activity is observed. These volcanic eruptions will also leave its surface full of craters. Callisto one of Jupiter's moon is the most cratered celestial body in the solar system. This is evidence that Callisto, over the recent years, has been actively involved transmuting energy, as replenishment material for the sun.

Why is there volcanic activity on Io-a moon of Jupiter?

Over the years, the sun extracted much of its replenishment material from the inner planets. Mercury and Venus may have had moons. Mercury, Venus, Earth and Mars may have also had gas rings. As the radiation is more intense closer to the sun, these postulated moons, and gas rings, over time, have disintegrated. The atmosphere of Mars, Mercury and our moon has nearly completely depleted, and as a result, these bodies are completely solid.

With Venus and Earth's unique atmospheres, and Earth its oceans, their interiors have to belch out less, explaining earth's molten core. Next in line sourcing replenishment material were the asteroids, which have also depleted their atmosphere, and some are rock solid. Presently, as replenishment material deplete from these solidifying bodies, the sun's radiation activity has extended its interactions to Io, explaining its present volcanic activity. There is also active volcanic activity at the south pole of one of Saturn's moon, and one of the gas rings of Saturn has started to disintegrate, and deplete.

Extra solar planets

We are informed that a body/planet randomly engenders from terminating star field debris, and also randomly form from dusty discs around stars that became unstable. We are also informed that a planet engenders clumping dust from a disc of gas into rocks, and a planet's core forms from the merging of rocks, and furthermore, the core randomly accumulates gaseous atmospheres. All this is incorrect. Only fiery fragments from a neutron star passing through, and confined by a nascent star field space engender into planets.

At some star fields the dense region forms a globule with a disk. Eventually, the outer atmosphere/disk is shed off. As the star field space is rotating this shed off atmosphere becomes an extra solar planet, if it can form an energy transmuting cycle with the dense region. The dense region or a star, and the extra solar planet continue to function, or exist/endure, as the star expels, and concurrently extracts material from the planet, and continues functioning. These types of star fields do not endure long. Extra solar planets have not been detected directly, and are assumed that they exist.

Why are extra solar planets larger than Jupiter?

Our solar system has many planets, and this could be the reason extra solar planets are larger than our largest planet. At some larger star fields the shed off atmosphere transforms into a single nebula of protons or cloud. If an energy transmuting cycle engenders, this nebula transforms into a single large planet.

Why do extra solar planets have elliptical orbits?

Planets orbit their star, as their material (forms of energy) is continuously being extracted, as energy is continuously being 'replenished' at the star. The planet would plunge into a star if there was only extraction. The equal expulsion material/force keeps it at an appropriate distance. There are normally only one extra solar planet in other solar systems and therefore, a single energy transmuting cycle. In a single extraction-expulsion energy transmuting cycle, between the star and the other body, the planet/body must take on more of an elliptical orbit.

In a cycle 'expulsion' and 'extraction' are different forms/states of transmuting energy. As energy is continuously transmuting, and as everything is flux, and every existence is interdependent and intertwined, there are energy transmuting variations everywhere. Therefore, at times expulsion is greater than extraction, and at times expulsion is less than extraction. When extraction is greater, the planet is pulled coming closer to the star. But in/as a cycle many things happen. As the distance of the body from the star decreases, the expulsion increases. As extraction increases and the planet comes closer to the star, expulsion increases pushing the planet away. So the planet moves away from the star, making an ellipse. Again, as expulsion decreases, extraction increases, and the planet again is pulled toward the star. With expulsion and extraction increasing and decreasing, the planet takes on an elliptical orbit. In reality, this extraction/expulsion processes are the reasons why bodies keep their orbit around the star.

Sometimes planets avoid elliptical orbits, by spinning faster or slower, as they are forced toward the star (increased extraction). At times, they also avoid elliptical orbits by moving faster in their orbits. This is only possible if the star can expel or extract its energy from/through other cycles. This may not be possible with a single body (as an extra solar planet) orbiting the star. As each star field cycle is interdependent, and interlinked, with other adjacent star field cycles, energy transmuting variations within the region dictate the expulsion and extraction increases, or decreases. A body cannot spin faster or slower by itself, neither can it move faster or slower in its orbit. The star and its body are parts of a cycle, and alter their characteristics sustaining the existence of one another.

Within our solar system, there are as many cycles as there are planets, so there is continuous energy transmuting cycle variations. Energy or material from the sun is expelled, and extracted to/from many planets that have formed energy cycles. This explains why most of our planets have elliptical orbits.

Pluto acts as a balancing reserve adjusting the flow of energy from, and to the sun (within the solar system) explaining its eccentric orbit. This situation is presently ephemeral, and Pluto's continuous orbit around the sun will eventually end, sometime soon in the near future, as it will not continue to possess the required replenishment material.

Comets are perturbed from their localizations for the same reason, and work the same way. However, as they are comprised of easily disintegrating material without firm atmospheres, and or insufficient degassing interior material they disintegrate, sometimes on their way to the sun, or later on orbiting around it.

Why do extra solar planets closely orbit their star?

There is normally only a single planet in these types of star fields. As all forms of energy dwindle with distance, the close proximity of a planet to its star, steadfastly sets the energy transmuting cycle. This also prevents an extra solar planet's star from wobbling.

B - Enduring Our World

REVELATION SIX

EXTRACTION AND EXPULSION

The energy transmuting cycles within star fields. Flux and the continuous structuring or integration, and disintegration of everything.

This revelation details the existence/endurance, and behavior of bodies within our solar system, as energy transmutes. As energy transmutes, or radiates, or propagates, star fields or our world exists/endures.

The concept Radiation

The concept 'radiation' is incorrectly explained. Radiation is described as rays, shine or glow from the sun sent out as light, or other forms of energy. Radiation is also explained, as something that is spread out from a centre.

As everything everywhere is energy, forms of energy propagate, or transmute, shuttling to-and-fro, but in/as different states, processes, structures or forms of transmuting energy.

Space as a form of energy transmutes to temperatures/all forms of radiation and pressure/force, as space waves. It appears to us that these existences or forms of energy are 'propagating', but in reality, they are not. As we have named each transmutation of energy, as different types of existences, we cannot really explain energy propagation.

The light at the sun-as a form of energy-dwindles with distance, and does not reach any of our planets as light, explaining the darkness throughout interplanetary space. However, other forms of energy as temperatures, forms of 'radiation', and pressure/force also dwindle with distance, but with extreme intensities generated at the sun reach the planets with lesser intensities, transmuting or altering to different forms of energy. Light at the sun transmutes to other forms of radiation en route.

.

The composition of the Sun

The sun has well defined layers of different forms/states/process and structures of transmuting energy. At a functioning star's photosphere, energy particles extracted from bodies within the star field are integrated again into plasma. The super speed star rotation expels plasma with/as temperatures and pressure/force, propagating throughout star field space.

We are informed that the sun is about 1.3 million kilometers in diameter, which is about 109 times the diameter of the earth, and it is so large that it is 332,946 times more massive than the earth, and 1.3 million Earths could fit inside it.

Our earth is presently a partially solid body, still possessing some of its molten core, with its atmosphere and oceans. The sun has a dense core, and as the photosphere expels plasma with great pressures, its atmosphere extends outwards to millions of kilometers.

The sun can be divided into four layers. The innermost is the core followed by the photosphere, the chromosphere, and the corona (the outermost layer).

The Core

The core 502,000 kilometers below the sun's surface, accounts for only about 3 percent of the sun's volume. It is so dense (compacted as frozen particles) that it contains about 60 percent of the sun's mass. We are informed that the sun's core consists of hydrogen. This is incorrect as energy condensed, concurrently generates space confining it. A proton or hydrogen exists with its 'individually' confining space, and this is impossible at/with the sun's processes.

The Photosphere

The core's periphery is external to the core's surface. The photosphere a few hundred kilometers thick, is made up of granules (the bright surface with millions of granules each several hundred meters in diameter), which constantly change in size The photosphere is the region where energy is continuously condensing forming particles, and particles integrating into plasma.

The Chromosphere

The interior of the corona (the outermost atmosphere) is called the chromosphere, punctuated with flares, or bright hot jets of bursting plasma. What we see is not the 'sun', but clouds of plasma bursting out of the chromosphere, which is about few hundred miles thick, and through it the sun's intense heat is given off into interplanetary space.

The Corona

The corona or the extended atmosphere starts at 10000 kilometers from the core, extending for about 12-16 millions kilometers out into inter-planetary space. We postulate that the chromosphere and the corona of the sun are fixed existences, as there is a continuous flow of plasma from the photosphere. Atmospheres around a star/sun and other bodies are not fixed bodies, as they are disintegrated, and integrated, over and over again.

The corona is millions times less dense than the atmosphere of the earth at sea level. At its farthest reaches, the corona is observed as the solar wind-a stream of plasma (transforming into protons) flowing through the solar system. The corona is the thinnest part of the sun's atmosphere (where prominences appear), viewed only during a total eclipse of the sun. It is the only layer hot enough to emit x-rays.

It is incorrect to postulate, that the solar atmosphere extends into interplanetary space, and ends at the outskirts of the solar system. The corona (the uppermost part of the sun's atmosphere) rotates at super speed, and ends much before Mercury's orbital path around the sun. All the planets and bodies exist within/as interplanetary space, confining the sun's atmosphere. Interplanetary space rotates the opposite way to the corona, but in the same direction of the core, explaining why all the bodies within interplanetary space orbit the sun. If the solar atmosphere extended throughout interplanetary space, interplanetary space would be rotating in the same direction as the corona. All the bodies would then be subjected to the centrifugal force, and would not orbit the sun in/at their orbital distance. Moreover, there would be no energy transmuting cycles between the sun and the bodies/planets. Planets exist within/as interplanetary space, orbit the sun as they exist within/as the fabric of interplanetary space, but rotate 'independently', on their imaginary axis, as part of an energy transmuting cycle with the sun.

Present theories on the sun's composition

We have been informed that fusion takes place at the sun. Without permanent fusion in the sun we would not exist, and if that solar fusion was not highly efficient and 'resource conserving', that is utilizing a myriad of different sequences (many hundreds) for achieving nucleo-synthesis, or the production of heavier atomic nuclei from light ones, there would be no matter constituting this earth, and the sun would have burned out long ago. It is assumed, that 98% constituents of the sun are hydrogen and helium, and determine the operational characteristics of the permanent fusion occurring in the sun.

It is stated that the sun's core has a tiny silicon-iron core. It is speculated that the various atmospheric layers generate, or favor various types of nucleo-synthesis, and also conserve matter through limiting neutrino loss, the 'ageing' of stars being primarily by escape of neutrinos. Most astrophysicists believe the sun can continue with its current 'stock' of matter for at least 5 billion years. All the above is totally incorrect.

The facts:

At a functioning star, particles integrate to form plasma not protons. A proton or hydrogen is plasma having its 'individually' confined space, and this is not possible at processes at/of a star. The dense region of particles or plasma, or the star, is confined by star field space, and vice versa. Plasma transform into protons only after expulsion from a star/sun, propagating within star field space. Therefore, a star is not hydrogen, nor does it contain hydrogen. As plasma is continuously produced at the photosphere, and concurrently continuously expelled, it is apparent that the chromosphere and the corona are comprised of plasma.

At a functioning star, the dense core's outer layer or photosphere rotates at super speeds, and it is here that particles produced (as energy in momentum/concentration, generating temperatures and pressure/force), and particles 'received' (from interactions within star field space bodies) are further integrated into plasma. This explains, why granules, which are continuously altering sizes are observed at the 'surface' of the sun. As energy condenses, engendering and or enduring a star, concurrently engendering and enduring star field space, the core is a frozen dense region full of particles. The frozen core is kept continuously compressed with/by processes (force/pressure) at the outer core, or photosphere.

The processes of a star/sun

Understanding temperatures and pressures/force

Spin, momentum, condensation, heat/temperatures, force/pressure, space, all forms of radiation, energy particles, plasma, protons, nuclei/elements and objects are state, processes, structures or forms of transmuting energy.

Temperatures and other forms/states of energy, propagate within/as space-a form of energy, explaining why they dwindle through distance. Energy does not 'propagate' one-way, as concurrently energy is shuttling in the opposite direction, in/as forms, states, process or structures, explaining why space waves. As some forms of energy are invisible, it appears to humans (as we create existences) that energy is 'propagating' one way.

Energy in momentum (as condensing forming a dense region), concurrently generates temperatures and force/pressure. This explains why temperatures/radiated heat, propagating from a star/sun are intertwined with force/pressure, and vice versa. Therefore, temperatures/heat and pressure/force are integral parts of each other. With high or low degrees of temperatures, there are high or lesser intensities of force/pressure, and vice versa. You cannot increase temperatures, and decrease the intensities of force/pressure, and vice versa.

As energy is everything everywhere, all forms/states/processes or structure of energy dwindle, propagating/transmuting through an area/distance. In reality, as energy is everything everywhere, energy does not 'dwindle', but as forms/states/processes or structures of energy transmute, it appears to us that these existences 'dwindle' (as we have incorrectly created existences for different energy transmutations). For example, propagating temperatures and pressure/force will dwindle in intensities, or decrease in degrees. As temperatures and pressure/force propagate through/as space (as space transmutes to these forms of energy and waves, and is also a form of energy), these forms of energy dwindle with/through distance. These forms of energy 'propagate' (spread out) within/as space, explaining the dwindling in degrees and intensities.

Light is a form of energy, and is constituted of certain degrees of temperatures and pressure/force. As temperatures and pressure/force decrease in intensities propagating through distance, light is not light anymore, altering to different forms of radiation.

Within our troposphere, a fire (a form of energy) will continue to burn, as long as the burning material (structure of energy) continues to exist. There will be light (a form of energy) from a bulb, as long as electricity (a form of energy in momentum) continues to supply the force (a process or form of energy) to light (heat) the element. Temperatures/heat (energy) from a fire (energy), or light (energy) from a bulb will continue to be felt and observed, as energy continues to transmute into those states/forms/processes or structures.

Temperatures/heat from a fire, or light from a bulb dwindle in intensities through area/distance, explaining why heat cannot be felt, or light (as temperatures and pressure/force) cannot be sensed at far distances from their 'source', at those intensities. A huge fire can be felt, or a bright light (for example light at the sun) can be seen at further distances from their source. The temperatures and light felt or observed, will be proportionate/dependant to the distance, and also the intensities at source.

These forms/states of energy dwindle in intensity, propagating within/as atmospheric space filled with nuclei/elements and molecules. Atmospheric space and particles alter/transmute with the propagation of these forms/states of energy, explaining the dwindling phenomena.

We would assume, that energy concentration processes generate temperatures and pressure/force. This in incorrect, as there is no causation, and every existence is part of a cycle with another existence. Condensation/transmuting of energy (into particles and plasma/protons) in one state/form/structure/process, concurrently engenders other forms/states/processes or structures of energy (as temperatures/pressure/force/spin/momentum).

Free energy condenses concurrently engendering both heat/temperatures and pressure/force, and space-a star field. At a functioning star, as energy is expelled in the form of plasma, temperatures and pressure/force, lighter particles and other forms of energy are extracted back through/as the same shuttling pathway, as replenishment material. These particles are integrated into plasma, and the cycle continues. With these continuously energy transmuting processes temperatures and pressure/force are maintained, and plasma continuously produced.

At the periphery of the dense region, or the photosphere, the super speed rotation generates particles, and particles are integrated into plasma, and concurrently temperatures and pressure/force are generated, facilitating and abetting these processes.

Particles or plasma are invisible, but as they integrate together, and can be observed, we call this matter. In other words, matter is condensed energy, and or condensed pressure or force (as pressure or force is a form of energy). As particles and matter disintegrates there is a reverse process. Disintegrating particles/matter manifests the confining/constituted pressure/force. The pressure/force 'released' from disintegration, and propagating within/as space also manifests temperatures/heat, as this form of energy is an integral part of pressure/force. This explains why with nuclei disintegration there is both temperatures/heat and pressure/force.

The processes at the sun/star

The sun's core is frozen solid

At a functioning star the periphery of the dense region (the photosphere) rotates at super speeds, condensing energy producing particles, and integrating particles into plasma, concurrently generating extreme temperatures and pressure/force. These forms of energy propagate both inwardly and outwardly. These forms of energy confine the core, explaining why the core's surface has higher temperatures than the photosphere. Outwardly, plasma with/as temperatures and pressure/force generated at the periphery, or photosphere of a functioning star are expelled outwards, via the inner atmosphere or chromosphere, and the corona or outer atmosphere.

As heat/temperatures, and pressure force, and plasma are continuously expelled, the adjacent atmospheric layer or chromosphere has higher temperatures and force/pressure, than the photosphere rotating at super speeds. The corona or outermost layer has yet higher temperatures and pressure/force than the chromosphere, as it is faster rotating than the photosphere, as it is tenuous and spread out (less dense) (explained following).

As plasma propagates with/as temperatures and pressure/force via the corona, we postulate a star's corona being constituted of plasma. If plasma is not expelled from the photosphere, the chromo sphere or corona will not exist.

Plasma with/as temperatures and pressure/force expelled from a star propagate through/as space, as it waves. However, plasma leaving the star immediately transform into protons. Protons do not propagate as space, but through space, as a bullet darting through space.

Space waves, as propagating forms of energy are not a one way phenomenon, nothing is. As temperatures and pressure/force and plasma propagate from a star, other forms of energy are also concurrently propagating, shuttling to-and-fro, to the star, within the same pathways, explaining waving space-an energy transmuting cycle.

Plasma/protons, temperatures/radiated heat/all forms of radiation and pressure/force, interact with nebula, or atmospheres of planets/bodies, disintegrating nuclei into lighter particles, extracted back to the star as replenishment energy. Particles extracted back via the corona into the periphery, once again integrate into plasma, continuing processes at the star, enduring the star field.

The surface of the core, which is adjacent to the photosphere, is confined by pressure/force propagating from processes at the photosphere, explaining its higher temperatures than the photosphere. As the core exists extremely dense with continuously propagating high pressure/force, it cannot transmute, and as particles here are at a standstill, the sun's core is frozen solid.

As the surface of the core, adjacent inwardly to the photosphere, has propagating high temperatures (from processes at/from the photosphere), likewise the chromosphere outwardly adjacent to the photosphere also has high temperatures. As the density decreases from the inner to the outer reaches of the chromosphere, the temperatures increase tremendously, from about 5000 degrees near the outer photosphere to about 10000 degrees (extreme heated corona) near the beginnings of the outer atmosphere. The temperatures increase through the chromosphere, to millions of degrees at the corona to the outermost zone. The continuous propagation of plasma, temperatures, and pressure/force propagating outwardly, generates/endures the upper atmosphere or the corona, stretching for millions of kilometers. As plasma is expelled through the tenuous corona, the force with expansion enhances the temperatures and pressure/force.

Some scientists have theorized that the core of the sun has temperatures of 10-22 million degrees, whereas its outer layer or 'surface' has temperatures of 5500 degrees. This is incorrect. Temperatures and pressure/force engender, as energy momentums. The periphery of a dense core (as with all rotating spheres) is obviously faster rotating, and/as it is less dense, explains why temperatures and pressure/force are generated at the periphery. Temperatures and pressure/force propagate both inwards and outwards from the photosphere. The dense core confined with continuous pressure/force generated at the periphery, or the photosphere is frozen solid of energy particles.

Why does the chromosphere rotate faster than the photosphere, and why does the corona rotate much faster than the chromosphere? Why do particles not integrate into plasma at the chromosphere and/or the corona?

As the corona is the outermost layered atmosphere it is most tenuous, as it spreads out from the 'surface' of the core, explaining why it is faster rotating than the adjacent chromosphere, and also why the chromosphere is faster rotating than the photosphere.

Approaching the core density increases. As spheres rotate, the centrifugal force propagates (spreading) outwards (exteriors are less dense), whereas the centripetal force propagating inwardly generates density. There is a centripetal and centrifugal force, as with rotation/momentum, energy condenses into a dense region/the core. A rotating sphere therefore, engenders these forces.

As plasma with/as temperatures and pressure/force propagate through the chromosphere, and the corona, they appear to be fixed existences. The chromosphere and the corona are in reality, only pathways of propagating plasma with/as temperatures and pressure/force generated at the photosphere. Only the star/sun's core is a solid/fixed existence.

The corona is millions times less dense than the atmosphere of earth at sea level, explaining why particles can only form, and transmute into plasma at the photosphere. As free energy condenses as/in momentum, or as/in rotation/spin, particles engender, and with yet further momentum/spin, concurrently as temperatures and pressure/force increase, particles form into plasma. These processes take place at the periphery of the core or the photosphere.

Why does the corona of a star/sun have extreme temperatures? Why is the corona hotter than the surface or the photosphere?

What increases its temperatures to millions of degrees?

Temperatures and pressure/force, and plasma generated at the photosphere propagate both inwardly and outwardly. As a result, the surface of the core exists with higher temperatures and pressure/force. As plasma with/as temperatures and pressure/force propagate through the chromosphere and the corona, these layers of atmosphere should have had lower temperatures (as forms of energy dwindle with distance) than the surface of the core, and also the photosphere. However, as the chromosphere and the corona are less dense than the photosphere, these layers rotate faster (generating higher temperatures and pressure/force) enhancing/increasing the temperatures and pressure/force, which would have normally dwindled through distance.

With the extreme orbital motion of the corona, temperatures soar to millions of degrees, with proportionate intensities of force or pressure. However, this is not the only reason the corona has high temperatures, as those extreme temperatures and pressure/force propagate outwardly.

If the coronal super speed orbital motion produces extreme temperatures and force/pressure, how does the corona maintain them, if these forms of energy are propagated both inwardly and outwardly?

The photosphere also produces high temperatures, and pressure or force, yet its adjacent layers the 'surface' of the core, and the chromosphere have higher temperatures. We can understand that the chromosphere has temperatures and pressure force, propagating from both the photosphere and the corona, but how are those extreme temperatures maintained at the corona?

The core/surface rotates in one direction, and the adjacent photosphere orbits the core in the other direction. The chromosphere and the corona orbit in a direction opposite to the photosphere. Interplanetary or star field space orbits the corona in the opposite direction of the corona. If interplanetary space orbited the same way as the corona, temperatures, pressure/force, and other forms/states of energy would not propagate, shuttling to-and-fro, through/in between adjacent 'atmospheric' layers.

The orbiting coronal atmosphere is separated as spiral arms, analogous to a ceiling fan with separated blades, concurrently pulling in air, and expelling it with the same rotation. Each rotating sphere will have its periphery rotating in the opposite direction than the core. Every periphery of a rotating sphere (the atmosphere of a planet, star, star field and galaxy) is broken up into spiral arms, as a result of the centrifugal and centripetal force.

It is consequent to the corona (of/as spiral arms) orbiting at high speeds, opposite to the rotation of interplanetary space, that energy is both propagated outwardly, and replenishment material/energy extracted inwardly, within the same energy transmuting shuttling to-and-fro pathway. With these processes, particles (replenishment material) extracted from disintegrating atmospheres/nebula/bodies, from within interplanetary space, shuttle back through/via the corona to the photosphere. The coronal spiral arms form with the high speed orbital motion, broken apart by the centrifugal force. The gaps in between the coronal spiral arms creates a pathway for both, forms of energy propagating outwards, and also for the incoming replenishment material for the photosphere.

The increased heat with the outwardly pressure/force, and

plasma/protons propagate, interacting with nebula, atmosphere and bodies within interplanetary or star field space. These interactions disintegrate nuclei/protons generating particles. The high speed orbiting corona generating the outwardly pressure/force, concurrently creates an equal inward or extraction force. The particles are concurrently extracted by the equal inwardly pressure/force (explained earlier) within the same energy propagating shuttling to-and-fro pathways, via the corona to the photosphere, as replenishment material. What is expelled must be replenished.

As particles are lighter than protons, their velocities are greater, and shuttling back through radiated heated pathways, increase their velocity through distance, and concurrently raise the temperatures and pressure/force intensities. The lighter material (of high temperatures and pressure/force) speed through the coronal atmosphere, maintaining the high speed rotating corona with its temperatures.

Why does the corona with extreme temperatures and pressure/force glow million times less bright than the photosphere?

The photosphere condenses energy producing particles, and also transmutes particles to plasma. These processes concurrently generate extreme temperatures and pressures/force, which are propagated both inwardly and outwardly. With condensation and transmuting processes at/as the photosphere, particles produced transmute to plasma generating the incandescent.

The chromosphere and the corona are apparent, but not solid/fixed structures. Plasma with/as temperatures and pressure/force only pass through engendering the chromosphere and corona, and therefore there is no activity.

We do not receive light produced at the photosphere, as explained later (revelation 9). Radiated heat from the sun interact our upper atmosphere generating our visible light. We see the incandescent being produced at the photosphere through this radiated heated pathway. The corona can be seen only during eclipses.

If the corona is so hot, why does it not heat up the photosphere until it has an equally high temperature?

Nothing is a one way phenomenon. All forms of energy propagate shuttling to-and-fro, as there is no 'source', or 'origin', or 'targets', and as everything is energy, everywhere.

The chromosphere and the corona rotate much faster than the photosphere, explaining why dwindling temperatures and force/pressure are enhanced/increased/maintained at these layers. The chromosphere and the corona are tenuous, and/or spread out, explaining why the photosphere cannot be heated to equal temperatures.

All activity/processes at the star take place at the photosphere, producing those temperatures and pressures, necessary to produce particles, and transmute particles into plasma. Plasma, with/as temperatures and pressures propagate both inwards and outwards. The corona has much higher temperatures for reasons explained earlier.

Does a star/sun produce atomic nuclei?

The conventional view of how stars form is that a cloud of interstellar hydrogen collapses under gravity, and under enormous pressure the atoms of hydrogen become so hot they fuse to form helium. Once ignited, the core of the newly formed star burns continuously, transmuting hydrogen to helium to carbon and so on, until the fuel is exhausted, and the star's life is over. However, when gases are compressed, they also heat up, and this makes them expand. As temperature increases, the outward force due to expansion will become greater than the force compressing the gas, and the gas will simply dissipate in space again. This is incorrect.

How does a star accumulate hydrogen to begin its processes in the first place? If hydrogen engendered after the big bang, how is hydrogen produced after a star terminates? From where do new engendering stars obtain their hydrogen? How could the condensing hydrogen cloud ever ignite spontaneously to begin a star? Why should gravity exist only within the cloud of hydrogen facilitating its condensation to eventually form a star? Condensing energy concurrently generates temperatures and pressure/force. This would explode a nascent star without reactions of/from other forms of energy.

Here is another theory.

It is theorized that under the influence of gravity there is condensation which slowly contracts to form a star. In this process, energy released leads to the heating of the newly formed star. Finally, the temperature is high enough to set off nuclear reactions in the interior of the star. As a result, the hydrogen, forming the major part, is burnt to helium. This creates a pressure which stops the contraction and stabilizes the star so that it can exist for millions of years. When the hydrogen has been consumed, other nuclear reactions take over, particularly in the more massive stars, and increasingly heavy elements, up to iron, are formed. At this stage, the star can no longer resist gravity, and may either collapse (depending on the mass of the star) or explodes resulting in a supernova. This brings about a brief but extremely intense flow of neutrons, which leads to the formation of the very heaviest elements. Smaller stars like our sun will turn into a white dwarf. This is also incorrect.

The facts

Only plasma engenders at functioning stars. Free energy in/as a whirling black hole condenses into a dense region, and with further continuous concentration particles engender. Yet further, these particles integrate into plasma. Although plasma contains particles and quarks, these cannot be taken as 'individual' existences, as their containing space amalgamates to form star field space, confining the dense region of virtual particles/plasma. In other words, as the dense region forms, concurrently space forms, confining it, and vice versa. Both the dense region and space, now occupying/supplanting the area/volume of energy from which they engendered. The dense region (of particles/plasma) is an integral part of space, and vice versa, so in reality, particles/plasma cannot be separated 'entities'.

Particles and plasma cannot be observed, so in some ways they are not real, but hypothesized. In reality, as energy condenses, and as space concurrently forms, condensed energy (as particles or plasma) is confined. As plasma integrates with other plasma, we have a hypothesized proton that is observable, and this is what we call matter. We then state that particles have a mass.

As energy condenses into particles and plasma, concurrently heat and pressure/force is generated. Therefore, as heat and pressure/force propagate, we can postulate that particles and plasma also propagate with, or as these forms of energy. Particles, plasma, heat, pressure/force, momentum or propagation are all forms of energy.

The progressive condensation generates proportionate increases of intensities in temperatures and pressures. As plasma is produced we can conclude that extreme limits of temperatures and pressures/force have concurrently been reached, or being produced. Plasma propagating through, and as, constitutes the atmosphere of a star.

If the star contained gases (as we have been informed) it would heat up and expand. As temperatures increase, the outward force due to expansion will become greater than the force compressing the gas, and explode. If nuclei are produced at functioning stars, helium would engender. As helium turns into a solid at high pressures, it would immediately shatter the dense region, terminating the nascent star processes, or a star would never be born.

Plasma transmutes to protons (hydrogen) propagating through space. There is no hydrogen or a proton produced at a functioning star. Therefore, there is no nucleus or nuclei produced at a functioning or dying star, or a neutron star (explained elsewhere).

Why does the enormous gravitational pull of a star like the sun simply force it to collapse into a ball of crushed atoms?

Revelation ten details there is no such thing as gravity, or a gravitational force. Nothing is a one way phenomenon. Nothing is an entity.

It is incorrect to postulate that a star contains, or has accumulated stores of hydrogen, or nuclei to keep it burning for certain duration. Nuclei are not produced, or processed at a functional, a dying, or even a neutron star.

Why are x-rays emitted only from the corona and not from the surface of the core or the photosphere?

The high speed rotating coronal atmosphere (faster rotating than the photosphere) reinforces the incoming pressure/force and heat (from both the photosphere and interplanetary space) raising temperatures to millions of degrees, and force/pressure to extreme intensities, explaining the hot corona. Some of the plasma propagating through the corona of high speed orbital velocities, extreme temperatures, and extreme pressure intensities, transmuting to protons, alter as cosmic/x-rays, propagating throughout interplanetary space.

The differential rotation of the sun

The centrifugal and centripetal forces of a rotating body expelling out material, and concentrating it inwardly eventually transform it into a sphere/orb. A sphere can only spin in one direction.

As a body is a sphere, and as it spins, it becomes protuberant at the equator, and/as the atmosphere around (over) the core at the equator will be less dense, and faster rotating than at the poles. Approaching the poles (a sphere) the atmosphere is more dense and/or thicker. As the sphere is so shaped, and is spinning in a two dimensional context at the equator, the atmosphere around approaching the poles is of a greater volume, its spin slower and there is an increase in density.

As the equatorial atmosphere of the sun is tenuous, it orbits its core faster at the equator than at the latitudes and the poles. This also holds true for a spherical body like earth. While the sun many layered atmospheres (not fixed) rotates about twenty five days near its equator, a complete revolution takes more than thirty days near the sun's poles.

However, the sun is not a gaseous, or even a plasma filled body. Its core as well as its surface cannot be described correctly. We cannot really say that its core is dense, as it does not contain matter or visible energy, as nuclei/elements or molecules. The core is frozen solid of condensed energy.

Pressures/force, temperatures and densities are different within the core's surface, the photosphere, the chromosphere and the corona. Each atmospheric layer has different directional rotation, different rate of rotation, and variations in rotational periods. Moreover, each layer has different characteristics at the equator, at the latitudes, and the poles. The different processes at the various layers, generating and propagating temperatures and pressure or force, the density within each layer, and the differences of the many layers is why the sun has differential rotation. As each layer orbits differently, and in opposite direction, temperatures and pressures alter, propagating and changing densities in the process.

The sun comprises of many layers of atmospheres, each rotating opposite to adjacent layers, its uppermost atmosphere, or the corona orbits in an opposite direction to planetary space. Planets and bodies existing as/within this fabric of interplanetary space therefore, orbit opposite to the rotating direction of the corona.

Why do planets have tilts?

Tilts change during the life of a planet. Mercury has no tilt as its energy transmuting cycle with the sun has terminated. It is now an impotent body awaiting disintegration. Venus has a 177.3 degree tilt (nearly upside down); Earth has a 23.5 tilt, with Mars 25.2, Jupiter 3.1, Saturn 26.7, Uranus 97.9, Neptune 29 and Pluto 118. Venus, Uranus and Pluto have more than a 90 degrees tilt.

A planet tilts, as some surface areas material is being extracted more than from other areas. As there are more oceans in its southern hemisphere where disintegration of the water molecule, and extraction is greater than the northern hemisphere, earth tilts facilitating these processes. However, as the northern hemisphere has less water surfaces, earth comes closer to the sun during the northern hemisphere's summer, explaining why earth's tilt does not change during its entire orbit around the sun.

What is a meteorite, meteoroid and a meteor?

Meteoroids are debris found in interplanetary space. There are millions of minute particles called meteoroids, some about the size of grains of sand, and cannot be easily seen. Meteoroids as the size of a car hit earth speeding at several kilometers per second, more than a dozen times a year, but are fragmented into small pieces, as earth possesses a dense and a multi layered different directional orbiting atmosphere. Meteoroids entering earth's atmosphere are heated by friction, and the air glow we see are called a meteor or shooting stars. Over 40 million meteoroids enter our atmosphere every day.

Meteorites

Larger bodies are called meteorites comprised of large chunks of rock, metal or both, that break off an asteroid or a comet coming crashing through the atmosphere landing on the surface.

The meteoroid is an actual object in space, and a meteor is the streak of light in the sky, as a result of the incineration of a meteoroid. A bright meteor that appears to break up is called a bolide, but if any part reaches the ground it is called a meteorite.

If it is big and bright the meteor is defined as a fireball. 85 fireballs were sighted over the U.S during the first 6 months of 2004. If a fireball explodes it is called a bolide. If chunks of a meteor, fireball or bolide fall to the ground they are called meteorites.

Meteor

Meteors also known as shooting stars are small particles of dust left behind by a comet's tail. We encounter meteors every time our planet crosses that path of a comet, or the debris left behind by a disintegrating comet. They appear as sparks, vaporizing and disintegrate in the sky, but never reach the ground.

Space is full of dust and chunks of rocks, and the earth is being hit by this stuff all the time. An estimated 25 million meteors fly through the atmosphere every day. Most of the pieces are literally the size of small pebbles, and burn up within seconds of hitting the atmosphere.

95 % of all meteors big or small are made up of the debris spewed out by the tails of comets. These materials, dust and ice and rocks are too fragile to survive their fiery flight, and never survive to become meteorites, and never land. We do not see most of them, as most occur over the oceans during the day, and at nights when we are asleep. All the meteors in meteor showers are comets tail dust. Meteor showers occur when the orbit of the earth around the sun traverses with the stream of debris, spreading out for millions

of miles behind a comet in its orbit around the sun. Some nights you can see as many as 60 or more shooting stars and on an average night about eight. Meteors flash different colors-white, blue, red, green, orange and yellow-mostly depending on what they are made up of. Different minerals emit different wavelengths of light when they burn. That is the same reason, why different minerals are used to give fireworks their colors. Shooting star colors also depend on the speed of the meteor, and the angle at which it hits the atmosphere. While meteors are mostly comet dust, the pieces that land are made of the denser studier stuff of asteroids.

10 and 50 meteorites crash down to the surface of the earth every day. Mostly small rocks, 70% of them land in the oceans, which cover 70% of the surface of the earth. Most others land in inhabited areas where nobody notices.

As meteors incinerate we see flashes of light. This interaction or friction also slows most of them down, and they do not survive to hit the ground, disintegrating at 10-15 kilometers above the surface

Comets

As energy continuously transmutes, and as there is flux internally and externally, and with as many cycles as there are planets with different characteristics and orbits, there are continuous energy expulsion and extraction differences.

Comets and asteroids are perturbed from their localization (orbits), adjusting the increases or decreases of energy expelled from the sun, and extracted toward it. Comets are extracted toward the sun, as/when replenishment decreases. Sometimes they disintegrate before reaching the sun, or after orbiting it, depending on the replenishing/extraction requirements, and whether in the meanwhile, replenishment material has been sourced elsewhere.

We have been informed that comets are left over blocks of the solar system, which formed when a giant cloud of gas and dust collapsed to create the sun and planets. It is stated, that comets were born in the solar system's outer fringes, their core still possessing some of the primordial ingredients. Studying these ingredients could yield clues to how the solar system formed 4.5

billion years ago. This is incorrect. As discussed earlier, clouds of dust and gas just cannot coalesce to create the sun and planets.

Comets are small fragments from the exploding planet, which never formed energy transmuting cycles with the sun, settling down at the outskirts of our solar system, as bodies in the Kuiper belt, and the Oort cloud. These areas are occupied by millions and millions of inactive comets, and remain there until they are perturbed out. As and when their material is required to by the sun, they are forced to go into an orbit closely around it.

Some comets are bright whereas others fizzle out after a bright start. This explains that their material was quickly extracted, and immediately disintegrated. Some of them are noticeable only when they are in close proximity to earth, or after their brief sojourn around the sun. When comets are first observed leaving their areas, it is difficult to gauge their brightness, and judge whether the initial activity is the sign of a truly great comet, or just a temporary flare-up of an ordinary one. Brightness or temporary flare-up is dependent on the immediate replenishment requirements of the sun.

Comets are made of dust and rocky material mixed with frozen methane, ammonia and water. A comet speeds within an elongated orbit around the sun. It consists of a nucleus, a head and a gaseous tail. When a comet passes near the sun, its material such as frozen carbon dioxide and carbon monoxide is vaporized. This obviously is the reason why it was attracted toward the sun. But these ices tend to vaporize far from the sun, giving us the early surge in brightness, which can trigger unrealistic expectations, and subsequent over-hyping like comet Kahoutex in 1973.

It is postulated that sometimes comets complete their orbit around the sun, and settle down back in the Oort cloud. This is incorrect. It does not seem possible that a comet once seen (orbited around the sun) having its material disintegrated, would resettle back to/from where it originated. Even if this was the case, it would need to gather material to be once again chosen to be perturbed for an orbit once more around the sun. Why should this particular body 'return' to the far depths of our solar system, and how does a particular once observed comet gather material to be perturbed out of its orbit, and attracted once again when there are other millions of them? There may be similar types of comets that have similar characteristics. Familiar comets like Hale-Bop and Kahoutex can only be observed once. The look-alike comets were mere coincidences to have appeared on the postulated likely dates. Predicted events that happen generate a great spectacle.

Why comets' tails are always pointed away from the sun?

Comets are clumps of rocky material, dust and frozen methane, ammonia and water. A comet is star like in appearance, and consists of a nucleus, head and a gaseous tail.

The extraction-expulsion energy cycle disintegrates, and extracts material from the comet on its way to the sun, explaining why it is being attracted toward the sun. Furthermore, as it possessed appropriate replenishment material it was chosen, and perturbed from its localization far away from the outskirts of the solar system.

There are several energy transmuting cycles between the sun and the planets, and radiated heat, pressure/force and protons continuously propagate toward interplanetary space. As the comet approaches the sun, parts of the comet are gradually disintegrated, and particles extracted. Some nuclei that are not required or disintegrating are pushed away by the pressure/force, accompanying the radiating heat and plasma/protons from the sun. This produces the comets tail. We cannot see the lighter material extracted from the comet. However, we can see the comet (structure), and its tail (disintegrating departing nuclei).

Some comets are completely disintegrated on their way to the sun. Others have their disintegrated material extracted orbiting near the sun, as they go by the sun. We see comets pass by the sun, as they would still contain materials that were not disintegrated, or required, or utilized. The disintegrating tail, or an exploding comet near earth sometimes produces meteor showers.

Why is Venus extremely hot and dry?

Venus has an atmosphere made primarily of carbon dioxide, water vapor, some nitrogen acids and heavy metals. Its clouds are laced with sulphur dioxide. Earth has large amounts of carbon dioxide and water, but most of it is not in the atmosphere. The water endures in/as the oceans, and the carbon dioxide is mostly bound chemically in rocks in compounds called carbonates.

As the inner planets experienced different temperatures and pressure/force intensities, their degassing produced a different mix of elements. As a result, Venus being in closer proximity to the sun did not generate earth's type of atmosphere, and could not maintain its oceans-if there were any. Radiated heat from the sun kept oceans from forming at Venus, or kept them from staying around, if they did form.

Presently, Venus does not possess oceans, rivers or lakes as earth, where radiation continuously interacts and disintegrates the water molecule into hydrogen and oxygen, extracting these gases into/as air. This process on earth continuously generates a type of atmosphere where water vapor exists, and both hydrogen and parts of oxygen pass through it into space (revelation 8).

As Venus does not have oceans of water to be disintegrated, there are less atmospheric nuclei interactions by radiated heat and pressure/force from the sun, and its atmosphere does not easily alter as that of earth, explaining why Venus suffers from a greenhouse effect run amok, and why it is slow rotating.

It can also be postulated, that Venus does not possess a multi layered atmosphere as the earth. Multi layered atmospheres (each rotating at different velocities and opposite directions) facilitates easier propagation, and continuous alterations of temperatures, pressure/force and densities. Without much depletion of the atmosphere, Venus' interiors do not have to belch out continuously. Without interactions as those within earth's atmosphere, Venus' atmosphere it less interacted and therefore, less depleted. As a result, radiated heat is 'trapped' within its atmosphere with fierce 482 degree C temperatures, hotter than even Mercury, which is closer to the sun. Earth continuously belches out its inner material, as the oceans and atmosphere gradually depletes. The atmosphere of Venus confines the heat generated from its interiors, explaining its fierce temperatures.

The dominant carbon dioxide atmosphere of Venus is hot, and acidic. This type of atmosphere absorbs the radiated heat, but without much interactions/disintegrations of/at its atmosphere, heat does not easily 'escape'.

Earth evolved an atmosphere wherein its energy transmuting cycle is more active than other planets. Earth's molten core continuously belches out disintegrated elements to sustain its atmosphere and its oceans. This type of atmosphere with our oceans facilitated and facilitates the evolution of living things. 'Hot', 'dry', and 'escape' are human concepts. We postulate that the type of temperatures we live by is the norm.

The energy cycle between the sun and the earth

Continuous integration/disintegration of everything

Every existence anywhere engenders/exists, by/from certain temperatures and/or pressures acting on it. If these temperatures and/or pressures continue, the existence endures. If they alter, the existence alters, or terminates. As energy continuously transmutes, everything is flux. There are therefore, no existences maintaining their same characteristics for any moment of time.

Within our interplanetary space the temperatures/radiation with force/pressure, shuttling to-and-fro, between the sun and the bodies, continuously interact, and disintegrate everything.

Atmospheres anywhere are not the same at any given moment. Atmospheric interactions/disintegrations and integrations result in the formation/alterations of climates/weather, seasons, tides, currents, breeze/winds, and other phenomena, within/as our atmosphere, and other phenomena at other bodies/planets. In other words, atmospheric interactions/disintegrations and integrations, generates altering temperatures, pressure/force and densities.

As our solar system is inter-dependant, and interlinked with adjacent, and other star fields within our galaxy, and as energy continuously transmutes, and everything is flux, there are energy transmuting variations, both internally and externally. Every galaxy, star fields and bodies/objects molecules, nuclei, and particles are continuously altering and/or terminating.

Radiated heat and pressures/force from the sun, continuously interacts our atmosphere, the surface, our rivers, lakes and our oceans. Our oceans (disintegration of the water molecule into hydrogen and oxygen) and atmosphere (disintegrating gaseous nuclei) continuously alter their characteristics, regenerating again with the continuous inflow of protons, radiation and pressure/force from the sun, and the outflow of hydrogen, and parts of oxygen from the disintegrating oceans, surfaces and the atmosphere. Hydrogen is swept away faster into space than oxygen explaining our blue skies, and the continuous existence of oxygen, and traces of water vapor in our atmosphere. Moreover, our interiors are continuously degassing to sustain these depletions. This explains our volcanic activity, ocean currents, earthquakes, tsunamis, and plate shifting.

The radiated heat bakes the upper atmosphere to high heat, forcing and abetting hydrogen outwards at supersonic speed (disintegrating enroute as particles)-the same force or pressure/temperatures from incoming radiated heat, as-everything is part of a cycle. Temperatures/all forms of radiation and the solar wind interact with our upper atmospheric nuclei, producing our atmospheric or visible light.

Coronal activity and the concurrent earth' atmosphere alterations

The continuous expulsion of plasma (which immediately turn into protons within interplanetary space) with, or as radiated heat and pressure/force, propagated outwardly from the sun is called the solar wind.

As our atmosphere and our oceans are continuously interacted, disintegrated and integrated, we experience our climates/weather, seasons, tides, currents, wind/breeze, depending on the earth's rotation, and its orbit around the sun.

As energy is continuously transmuting everywhere, everything is flux. Moreover, with the replenishment material dwindling from the inner planets, all the cycles with as many planets are gradually being altered, occasionally more than normal. As a result, there are increases and decreases in expulsion and replenishment material, altering the processes of/at the sun. In other words, there is abnormal coronal activity.

During the past thousands of years, earth's atmosphere and oceans have been playing a very significant role in the sun's replenishment material. As a result, if there is abnormal coronal activity there are concurrent weather, climate, and other abnormal activities within earth's atmosphere, oceans and the interiors.

Sunspots

Within the photosphere there are some lesser hotter regions which appear to be darker, called sunspots. Although sunspots look dark they are actually very bright-they only appear dark by contrast with the surrounding brighter areas. Sun spots on the sun are giant explosions, blasting material hundreds of thousands of kilometers into space.

Forms of radiation, light and matter are different forms of energy. Energy in the form of plasma, temperatures and pressure/force blast away from the sun, but nothing is a one way phenomenon. Within the same energy propagating pathways, energy in other form/states/processes or structures, propagate the other way. As normal incoming temperatures and/or outgoing temperatures alter, sun spots are apparent.

The number of sunspots varies dramatically in a 22-year cycle. At sunspot maximum (every 11 years) the sun can have many spots on it while at sunspot minimum, weeks can go by with no spots visible at all. This has been going on for many centuries. At the peak of the sunspot cycle with many sunspots, the solar wind is the strongest. We can postulate that sunspots occur with an abnormal increase in replenishment material, cooling some photosphere areas (sunspots), and thereby increasing expulsion material. If the 'expulsion' and 'incoming' energy is balanced, there are no sunspots. Sunspots vary as replenishment energy is being extracted more from within the inner planets, with very active energy transmuting shuttling to-and-fro cycles.

Jupiter orbits the sun once in 11 years, and with its abundant supply of replenishment material, acts as a reserve to supply replenishment material shortages from the inner planets, and bodies of the asteroid belt. Depending on its orbital location around the sun, and the different distances from the inner planets, replenishment material for the sun may decrease or increase explaining the 11 (maximum sunspots) year sunspots cycle.

As there is a decrease in replenishment material from the inner planets, presently there is continuous volcanic activity on Io (degassing to support its depleting atmosphere), Jupiter's moon. With this new source of replenishment material Jupiter's atmosphere is less depleted, explaining why recently there was an alteration in the 11 year sunspot cycle.

Coronal Holes

As sunspots appear at the photosphere, coronal holes appear in the corona. As with sunspots, the normal alterations of expulsion and incoming energy produce coronal holes. Coronal holes are areas within the corona where plasma bursts out abnormally. This would explain that coronal holes form, with an increase in expulsion material. Coronal holes literally are gaps in the sun's outer atmosphere, through which the stream of hot supersonic particles (plasma) pours out into space throughout the entire planetary system.

At earth's atmosphere, the abnormal increases of hot flood of particles producing the auroras-the northern and southern lights, interferes with electrical and radio transmission, and may threaten passengers aboard high-flying airliners, or astronauts aboard unshielded spacecraft.

Prominences, Sunbursts, and violent explosions at the sun

What is a solar flare? Why does the sun sometimes have violent explosions?

Energy as plasma, heat/temperatures, and pressure/force is continuously expelled and extracted (into) within all layers of the sun's atmosphere. Increase or decreases of ejecting, or incoming replenishment material disrupts normal cycles, altering heat or temperatures, and pressure or force propagation/processes. Temperature fluctuations in the chromosphere-suddenly heating up-produces what are called solar flares. Solar flares sometimes belch out extra clouds of plasma directly into interplanetary space, via the corona. A solar storm throws out high speed plasma in a shower, preceded by a solar flare. Flares are sudden temporary outbursts of light, which extend from the outer edge of the chromosphere into the corona.

Masses of glowing plasma called prominences are sometimes ejected from the chromosphere. These penetrate the outermost part of the sun's atmosphere. Some prominences are violent, and short lived phenomena while others are much calmer, and may persist for several weeks. Prominences get violent depending on the sudden increases, or decreases, in outgoing, and incoming energy.

Solar flares are another category of brilliant sunbursts in the solar atmosphere. They are associated with sunspot groups, reaching maximum brightness in just a few minutes. They send large amounts of particles and radiation. Violent and short lived prominences are also the result of this phenomenon.

Cosmic rays/x-rays

Cosmic rays bombarding the earth from the sun's coronal activity are high-speed invisible plasma/protons and x-particles, along with x-rays, or short wavelength ultraviolet rays. Cosmic rays are extremely energetic. The process of cosmic rays colliding with particles in our atmosphere, and disintegrating nuclei into lighter particles is called a cosmic ray shower.

Cosmic rays are sometimes disruptive, as these highly energized particles expelled from the sun, generate power surges, fluctuations in electrical systems, disturbances in communications, satellite systems and broadcasts. Solar cosmic rays events correlate relatively well with solar flares.

Climate changes on earth

Abnormal expulsion and extraction energy processes, alter the sun's 'smooth/normal' functioning. This results in coronal holes, solar storms, solar flares, sunspots, prominences, and bursts of plasma in the forms of cosmic rays, and or high speed particles, and other abnormal phenomena.

Depending on the increases or decreases in replenishment energy at the sun, the solar wind is accelerated, disrupting normal weather conditions. There is a clear decrease in terrestrial atmospheric temperature, after the sun's coronal activity is most intense. At this point, there is a decrease of activity, and an enlargement of the coronal holes. Solar activity is therefore, directly linked to our climate changes. The percentage of the sun's surface covered by coronal holes seems to be a fairly accurate indicator of temperature in the earth's troposphere, over months or years.

The Northern and Southern lights

Sunspots erupt as solar flares, ejecting streams of particles, cosmic rays/x-rays, and gamma rays into the solar system. Sometimes, masses of glowing plasma called prominences are ejected from the chromosphere sending off high speed plasma, and x-particles through the corona into interplanetary space.

The strength of the solar wind is not constant, but changes with activity of the sun. Near the peak of the sunspot cycle the solar wind is the strongest.

Strong solar wind particles interact with oxygen, and nitrogen nuclei of our atmosphere at about 100 km producing the Northern (Aurora borealis) lights, and Southern (aurora austral is) lights-the stupendous show in the sky filled with brilliant red and green colors. The most spectacular displays occur about 24 hours after major solar-flare activity. The brightest ones occur around the maximum of the 11-year sunspot cycle, but sometimes they peak earlier as they did recently.

The flow of solar wind is generally continuous, although it occasionally bursts out of holes in the corona. We are most likely to witness auroras, when there is greater ejection of solar wind, and it is during these periods that we notice solar flares, prominences and sunspots.

Why are these lights observed only near the North and South poles?

A rotating sphere's equatorial areas have less density, higher temperatures, and a faster rotating atmosphere, or layers of atmospheres. At the poles there is more density, more atmospheric volume, lower temperatures, and the atmosphere rotates slowly.

Earth has a many layered atmosphere, and as temperatures, densities and pressure/force vary at different places, between the poles and the latitudes, and also between the atmospheric layers, there are different climates, weather, currents, tides and breeze/winds at various areas. There are differences in the Northern and Southern hemisphere at the same time.

The slower rotation at the poles, and a greater density of atmosphere, explains why we see those lights at the north and south Pole regions. At the equator the atmosphere is tenuous, and the atmosphere is moving very quickly to generate these lights, or for us to notice the lights, if any are produced.

Solving the Bermuda Triangle mystery

We are all aware of the mysterious area in the Atlantic Ocean bounded roughly at its points by Miami, Bermuda, and Puerto Rico called the Bermuda Triangle. Within the last century, many people, scores of ship and 20 planes have mysteriously vanished in this area. In 1973 alone, the US Coast Guard answered more than 8000 distress calls in the area.

On December 5th 1945, six navy planes and the crew disappeared. It was a sunny calm day with ideal flying conditions. Prior to disappearance, many air plane crews through radio contact stated nothing was amiss. Cruising above and through this triangle, air plane crew reported sudden power failures, instrument failures, and violent movement of aircraft. At times aircraft suddenly drop, rising up again. On some flights passengers have been ejected from their seats. Some years ago a passenger jet had ice build up on the aircraft.

There are many theories like aliens and time, evil extra terrestrials, residue crystals from Atlantis, evil humans with anti-gravity devices, or other weird technologies, and vile vortices from the fourth dimensions, and human error. All these are incorrect. There are no such things as gravity, aliens, extra-terrestrials, and space time.

All the disappearances can be attributed to the areas' unique environmental features. This triangle is noted for hurricanes and high waves. It is extremely swift and turbulent, and although it can quickly erase any evidence of a disaster there was none. In other words, ships and planes instantaneously disintegrated, and there was no evidence of a disaster, and they disappeared without any trace.

There were no water spouts, or thunder storms, or strong currents over the many reefs generating navigational hazards. There was no human error, as a result of insufficient knowledge of the area's hazards, or lack of good seamanship. There was no reduction of the density of water due to methane that would sink ships. Also, there was no highly combustible gas igniting aircraft engines (a certain percentage of flammable gas to air is required for the atmosphere to become combustible). Boats/ships as well as air planes, disappeared without a trace. Any of the reasons mentioned above would leave some evidence.

As there are many cycles as there are planets, and as energy continuously transmutes and everything is flux, there are alterations generating decreases in replenishment material. As this happens, the sun immediately sources alternative sources. During the past century, areas within the earth surface, and oceans having the right type abundant replenishment material were also tapped. The Gulf Stream or the Bermuda triangle, and also Jupiter's Red spot are one of those areas, where there were/are abundant supplies of stand-by easily disintegrating replenishment material.

There are methane hydrates locked below the sea sediments in the Bermuda Triangle. As the interiors shrink, and the mantle and core sink, subterranean landslides unlock, exposing the vast beds of methane hydrate reducing the density of water, which would make any ship floating above sink like a rock. But ships did not sink; they just disappeared without evidence (nothing has been found) so this theory cannot hold. Furthermore, what about the air planes? How and why did they disappear?

As mentioned above, with the decrease of replenishment material the sun seeks alternate sources. Methane hydrates are loaded with hydrogen or protons. With the decrease in replenishment material, the outgoing force/pressure suddenly increases in intensity, and the energy within the sun-earth's atmosphere/surface ocean cycle brute forces right down to these methane hydrates. Disintegrated hydrate material are quickly extracted, with intensities equal to the incoming brutal force. This sudden incoming and outgoing brutal force is not continuous, but instantaneous, lasting for short durations (until the decrease of replenishment material is compensated).

Anything within the incoming/outgoing pathway of this brutal force would be shattered, or disintegrated into particles or protons instantly. Ships and planes within this brutal force pathway would have been instantaneously disintegrated. In a reaction, as the incoming force suddenly increased, the extraction force concurrently increases, explaining the complete disappearances without any evidence.

People on air planes that have survived to report the violent movements, power failures, and instrument malfunctioning were fortunate to have witnessed either the beginnings, or the ends of this sudden inwardly, and outwardly energy abnormal transmuting processes. They may have also been fortunate not to be in close proximity, but at a safe distance to bear witness to this brutal force shuttling pathway, generating those abnormalities.

During April 2005, a cruise ship experienced 50 foot waves cruising near the eastern coast of the U.S. This resulted from similar events that take place within the Bermuda triangle.

Why does a magnetic compass within the Bermuda Triangle point toward the true North?

North, south, east and west are human constructs, and it depends on which side you are facing. The Bermuda triangle is one of the two places on earth a magnetic compass point toward true north. Normally it points toward the magnetic north-this is called compass variation (an error that must be compensated by a navigator).

During the past centuries the magnetic needle has pointed to different directions. In the northern hemisphere it points toward the sky, and in the southern hemisphere it points toward the ground. Materials at some sea beds, as those within the Bermuda triangle, have been altered from their original direction, and as the earth now has a tilt, as extraction is greater at some surface areas, the magnetic needle at some areas would continue to point at the true north.

What keeps people, buildings and other objects from flying off the rotating earth?

The earth rotating on its imaginary axis at 1670 kilometers per hour at its equator should have had people, buildings, and other objects hurtled out of the biosphere.

The sun expends/propagates/transmutes plasma, together with all forms of radiation/temperatures, pressures/force. Large amounts of material in the form of hydrogen, and disintegrated particles, are vented out of the surface and our oceans, equal to the amount of energy 'received' from the sun.

All organisms and objects have evolved, and/or are structured, constituted of materials from within our atmosphere, and the surface. Organisms and objects are also designed and structured from, by, and of the continuing energy radiation, and/or the environments that are continuously interacted, disintegrated and integrated, over and over again. The expulsion-extraction energy transmuting cycle between the sun and the earth, and the cycle between earth's atmospheres, and its oceans/surface/land have engendered organisms, and other existences, comprised of certain particle and nuclei configurations. These configurations conform with/as the environments, which generated them (revelation 20 details).

We are not aware, and cannot feel the incoming forms of energy, as radiation, pressure/force to the planets, and to the surface of the earth. We are also not aware of material, which is being extracted toward our upper atmosphere, where it transforms into appropriate material, as replenishment at the sun. Organisms have evolved mechanisms, manifesting to them only things that are significant, and necessary for subsistence or survival. Imagine surviving for a few seconds with the amount of noise, which is generated as our atmospheric nuclei collide, integrate and disintegrate continuously. If you wish to experience this, tune in your TV to a non-programmed channel exhibiting 'snow', and raise the volume. Ask tinnitus sufferers about their discomfort.

As living things and objects are comprised of the atmosphere, and the surface/oceans, they exist, as the atmosphere, or as the environments exist. Living things have evolved mechanisms complying to, or conforming with/as the rotating speed of the earth, and the atmospheric constituents and conditions (existing or enduring with/as, or within the expulsion/extraction energy cycle), explaining why they exist/endure, as their environments do. As we are structures of/within, and by the biosphere/atmosphere, we are not hurtled away with the earth rotating at that speed. There are no entities, and/as all existences exist within/as the fabric of space/energy. Everything everywhere is energy.

REVELATION SEVEN

CIRCLES AND CYCLES

The energy transmuting cycle within our solar system-Revolutions and rotations
This revelation explains rotation/spin, and the orbits of our solar system bodies, as energy transmutes.
No existence exists without spin, and/or space.

The Energy Radiation/transmuting cycle

A 'star' engenders/endures, or becomes full-fledged, concurrently with another body or bodies, as parts of the same cycle forming a star field. Within our solar system, the sun is part of several energy transmuting cycles, as there are planets.

THE ORIGIN OF SPIN/ROTATION

We are bombarded with some bizarre explanations, which keep our head spinning. We have been given to understand that within instants after the theorized Big Bang the initial fireball of energy expanded at an incredible speed, then later cooled, and solidified into all the matter in the universe. This fireball was not perfectly uniform, as some regions were denser than others, and some expanded more rapidly. The denser regions in the universe began to collapse under their own gravitational force forming clumps of matter becoming giant cosmic structures.

The spin of such cosmic objects is described by a conserved quantity called angular momentum, which accounts for both, the speed of the rotating mass, and its distance from the spin axis. The largest clumps of matter in the universe had an initial angular momentum, and these clumps broke up into smaller clumps forming smaller cluster of galaxies, groups of galaxies, individual galaxies, solar systems within galaxies and ultimately individual stars and planets. The smaller clumps each got their own share of the original angular momentum, continuing down in scale so that everything today has some spin.

In an earlier revelation we pointed out, that the Big Bang theory is wrong, and bizarre, and therefore, this above explanation is preposterous. If there was a big bang, why and how did the primordial fireball of energy expand first at an incredible speed? We may accept that it later cooled, but all of it should have solidified into matter. There should have been no radiation. If there was radiation, it should have 'annihilated' both anti matter, and matter back to invisible energy. Why were some regions denser, and why did some regions expand more rapidly? If the fireball of energy expanded why should it spin? An object spins if there is continuous force, or another confronting object/force initiating the spin. The Big Bang, if it happened, initially generated a force, and this one time explosion does not, and cannot continue that force. Therefore, we need another story or theory, describing the continuous force driving the expansion of the postulated 'universe'.

Space was 'invented' in the Big Bang theory, and 'space' in this theory was/is empty, or a void. Why, and how should the largest clump of matter in the universe have an initial angular momentum, so that all smaller, and subsequently formed clusters get their share of angular momentum, continuing down in scale, so that everything today has some spin?

The fact that all the planets orbit the sun in the same direction, is 'thought' to be a relic from the rotation of the original solar nebula. Our solar system is assumed to have been formed from the postulated solar nebula, parts of which collapsed due to 'gravity', forming the sun and the planets.

The Big Bang theory does not account for the spin of astronomical objects (on their imaginary axis) within star fields. Why does the corona of the sun rotate in one direction, and interplanetary space and the planets orbit the sun in the opposite direction? Where did this mechanism come from? There is also no explanation why entire star fields rotate, and why they orbit around the centre of their galaxy, and why the entire galaxy also rotates around its imaginary axis.

Spin/rotation and orbits are not the result of angular momentum. Angular momentum cannot explain the following: Why Venus and Uranus tilt on their imaginary axis at more than 90 degrees? Why Mars stops, and rotates for some duration in the opposite direction, rather than at the same West to East direction, continuously like other planets? Why does Venus stop to rotate in an opposite direction once every 243 days? Why do some moons of the Jovian planets, rotate on their imaginary axis opposite than other moons of the same planet?

Explanations

Energy transmutation per se generates/is a force. Force therefore, is also a form of energy. This also explains why we explain energy as 'energetic'.

Every existence exists, concurrently with another existence, as parts of the same cycle. As energy condenses (engendering a force), another opposite force, or a different form of energy of a different characteristic, concurrently engenders (as space engenders). Energy condenses, and space engenders, concurrently generating two equal and opposite forces. Condensing energy spins in one direction, and its confining space (possessing a different characteristic) spins in another direction. This is the only way that energy in momentum condenses, concurrently engendering space. Revelation four also elucidates spin.

Spin within our solar system

Spin facilitates the formation of energy transmuting cycles between bodies; engendering star fields, and existences to exist/endure.

We learnt how within our solar system, energy transmutes, shuttling to-and-fro, between the sun and planets/bodies. As a result, all the planets/bodies with energy transmuting cycles with the sun rotate around their imaginary axis. As they exist within/as the fabric of interplanetary space/energy (which orbits the sun) they also orbit the sun. It is consequent to the energy cycles between the sun and bodies, and their existence within/as interplanetary space, that bodies rotate around their imaginary axis, keep their distance from the sun, keep their distance from each other, as well as orbit the sun.

Bodies not having energy transmuting cycles with the sun, orbit it, as they exist as/within the fabric of interplanetary space, which orbits the sun. However, without an energy cycle with the sun, those bodies do not rotate around their imaginary axis.

Most of the planets rotate the same direction-counter clockwise-when viewed from above, and clockwise from the south. As Mars interiors have nearly stopped degassing (which over the years sustained its depleting atmosphere), it is now a complete solid body, as its energy transmuting cycle with the sun is near termination. As it is slowly disintegrating, and material extracted at times, Mars stops rotating, and rotates for some time in the other direction.

Venus does not have an atmosphere consisting of inappropriate replenishment material for the sun. As its atmosphere does not easily disintegrate and integrate again, its interiors do not belch out, as often as earth, and therefore, does not have an active energy transmuting cycle with the sun. It takes Venus 243 earth days to rotate once around its imaginary axis, even longer (225 earth days) than it takes to orbit around the sun. Moreover, without an active energy transmuting cycle with the sun, Venus stops momentarily after its single rotation (243 days), and then starts rotating again.

Venus and Pluto did not commence spinning hit around by huge chunks, and neither did any hit Uranus to knock it on its side, as has been speculated. Existences structure randomly, and if these/their characteristics facilitate their continued existences, they endure, and are here for us to observe, and experience them.

The sun has many atmospheric layers. The outermost layer or the corona orbits the sun's interiors clockwise. Interplanetary space orbits the sun in the opposite direction. As all the planets, and bodies, exist within/as the fabric of energy, or interplanetary space, most of the planets (as seen from above) orbit the sun's corona anti-clockwise. They also rotate around their imaginary axis in the same direction (explained following).

There are some moons of planets rotating the opposite direction, than other moons of the same planets. These are captured fragments, added later on. They have no atmospheres, and without energy cycles, they rotate in opposite directions to moons of the same planet, as they exist within the planet's many layered atmospheres rotating in that direction.

Why do some bodies rotate around their imaginary axis?

Within energy transmuting cycles between the sun and the planets/bodies, the extraction processes generate the rotation. It is analogous to pulling a thread from a ball, which makes the ball spin.

There is more extraction than expulsion energy, as some of the plasma produced, and propagated from the sun settle on the outskirts of our solar system. This also explains why atmospheres of planets deplete, and planets/bodies shrink, and solidify.

Planets and other bodies having their atmospheres directly interacted and disintegrated, and material extracted within the expulsion/extraction energy transmuting cycle with the sun, are forced to rotate around their imaginary axis, facilitating easier functioning. As energy/material is extracted to the sun to replenish its expelled energy, the sun also continues to rotate.

Moons/satellites and gas rings are parts of the planets, such as atmospheres and oceans (if any) around their interiors, and therefore, these existences orbit the planets, and not the sun individually. However, some as Io (Jupiter's moon) have interiors with appropriate replenishment material, and as they have a direct energy transmuting cycle with the sun, they not only orbit their planet, but also rotate around their imaginary axis, and together with their planet also orbit the sun.

Our moon has completed its degassing explaining why it has little or no atmosphere. Without any active energy transmuting cycle with the sun, it does not rotate around its imaginary axis, but orbits around earth, and together with earth (existing within interplanetary space) orbits the sun.

Pluto was perhaps a moon to one of the Jovian planets, and possessing replenishment material for the sun, was disturbed, and/or released from its captivity by a planet, and now orbits the sun individually.

All asteroids had energy transmuting cycles with the sun. Over the years, most have completed degassing, and are now rock solid with terminated energy cycles. As they exist, within/as the fabric of interplanetary space, they orbit the sun at that same belt. Some asteroids are still degassing, and obviously would be rotating around their imaginary axis.

Other objects/bodies further distant than Pluto still possess their gaseous interiors. However, they do not have energy transmuting cycles with the sun; as they are not in close proximity, as all forms of energy from the sun dwindle intensities with/through distance. Comets from this region are sometimes extracted from their locations to meet replenishment requirements.

As star field rotate, an entire galaxy also rotates. Star fields orbit the galaxy's centre.

Why do astronomical bodies rotate at different speeds?

The different rotational speeds of bodies/planets within a star field, is dependant on the amount of extraction, and expulsion of energy between the body, and the star/sun, the forms/processes of energy extracted, the distance from the sun, their size, and their types of atmospheres.

The different orbital speeds of star fields around their galactic centre, depends on their distance from the centre, and the centrifugal and centripetal force that is generated within them, the size of each star field, and other forces acting upon each of them externally. Each galaxy within existential energy's expanse will revolve depending on their size (the amount and types of star fields).

Why do star fields rotate around, and orbit the centre of their galaxy? Why do star fields and galaxies rotate?

Without spin and space, existences cannot engender, and endure. The high speed rotation of a black hole facilitates the engendering of both the star, and/or a star field. The dense region or the star, and the bodies within the star field space (or interplanetary space), rotate in opposite directions. The different directional rotation not only engenders the star field, but endures its existence. Star fields rotate in the direction its space (or interplanetary space) rotates, with the corona of a star/sun rotating in the opposite direction. However, the core rotates in the same direction as star field space.

All star fields are interdependent and interlinked with other star fields and therefore, rotate. They exist/endure together spinning as a chain reaction, and collectively together as a whole, or a galaxy. As each star field rotates around their imaginary axis, and revolves/orbits around the centre of their galaxy, the whole galaxy revolves. Billions of star fields rotating around their imaginary axis are interdependent and interlinked, as parts of cycles forming their galaxy, explaining why they orbit their centre of their galaxy, and also why the entire galaxy rotates. There is a centrifugal and the centripetal force, as a result of the rotation of any body. With the centripetal force, there is a greater density of star fields around the centre of a galaxy, forcing all star fields to orbit their galactic centre.

As energy transmutes, star fields engender with their space confining them. Existing within this maelstrom region of transmuting existential energy, star fields and their confining interstellar space collectively form galaxies, further confining dark energy, which cannot transmute, as a cycle forms between them. As star fields continuously transmute energy, there is force momentum/spin, explaining the revolution.

SPIN AND SPHERES

Why planets and most bodies are spheres? Why are some asteroids/meteorites not?

It is assumed planets are round, because their gravitational field acts as though it originates from the centre of the body, pulling everything toward it. With its large body and internal heating from radioactive elements, a planet behaves like a fluid, and over long periods of time, succumbs to the gravitational pull from its centre of gravity. The only way to get all the mass, as close to the planet's centre of gravity as possible is to form a sphere. The technical name for this process is 'isostatic adjustment'. With smaller bodies such as the 20 Kilometer asteroids, the gravitational pull is too weak to overcome the asteroid's mechanical strength. As a result, these bodies do not form spheres. Rather they maintain irregular fragmentary shapes. These explanations are incorrect. Revelation 10 details why there is no such thing as gravity and therefore, no gravitational field.

As explained elsewhere, fragments from the exploding neutron star formed planets, and gaseous fragments from the exploding planet transformed into moons, gas rings, asteroids, and other bodies in the far reaches of our solar system. As parts of an energy transmuting cycle with the sun, bodies/planets are forced to spin, as material from their atmospheres is continuously extracted toward the sun, as replenishment material.

As an object spins, its nuclei configurations expand, and get lighter, as nuclei become susceptible to the extraction part of the energy transmuting cycle-this is the centrifugal force. However, there is an equal centripetal force, as spinning objects are in momentum, and as they expel force (centrifugal). As bodies rotate/spin, there is a centrifugal force throwing out unwanted/disintegrating material, eventually producing spheres. In reality, bodies will not form, and endure as spheres, if they are not dense. As these bodies spin the centripetal force enhances the concentration, or the density. All rotating bodies therefore, eventually form spheres. Spin/rotation is what generates the centripetal and centrifugal force, and vice versa. Spin/rotation, the centripetal, and centrifugal force are different types of force and are therefore, forms of energy.

If bodies (solid or gaseous) spin fast enough, and are not confined by atmospheres, they would eventually disintegrate. Without concurrently generating/enduring an atmosphere, cycles would not form, and bodies or planets would not engender in the first place. Therefore, enveloping atmospheres did not only concurrently engender bodies/planets (interiors), but also prevented/prevents radiated heat, and force/pressures from the sun from disintegrating them (the interiors)..

Some asteroids and meteorites, and other objects roaming within interplanetary space, are fragments of bodies (which were once spheres, and parts of an energy transmuting cycle with the sun) within the asteroid belt, and other areas within interplanetary space. Some of these bodies have lost their atmospheres, as their interiors were degassing over time, and have now solidified awaiting complete disintegration. Without rotation around their imaginary axis they partially disintegrate, loosing their spherical shape. As impotent bodies without an energy transmuting cycle, some of them still orbit the sun at the asteroid belt (existing within/as interplanetary space), as was/is the norm. Some of them are at times disintegrated into fragments (depending on energy expulsion/extraction requirements elsewhere within the solar system), and disturbed out of their localizations roaming interplanetary space.

Ceres the largest discovered asteroid in the asteroid belt is not a perfect sphere, as both its ends are squashed. Obviously, it has terminated its energy transmuting cycle with the sun, and is now rock solid. However, it shows planet like features, evidence that it has only recently completely depleted its atmosphere, and turned solid, and impotent.

Why is there a centripetal and a centrifugal force with dense rotating bodies?

As an object spins, its nuclei configurations expand, and get lighter, as nuclei become susceptible to the extraction part of the energy transmuting cycle-this is the centrifugal force. However, there is an equal centripetal force, as spinning objects are in momentum, and as they expel force (centrifugal). Energy transmuting is not a one way phenomenon, and therefore, there is also a centripetal force equal to the centrifugal force. Spin/rotation (energy in momentum) is what generates the centripetal and centrifugal force, and vice versa.

Are all planets spheres?

All planets/ bodies with an energy transmuting cycle with the sun (with their atmospheres continuously being depleted) rotate around their imaginary axis, and are therefore, spheres. Not all planets are perfect spheres. A sphere can only spin in one direction at any given time, as discussed earlier. As bodies rotate they bulge a little (protuberance) at their equator, depending on the receipt and extraction of material/energy, generated by the shuttling to-and-fro energy radiation/transmutation cycle.

ORBITING THEIR STAR

Why do planets and other bodies orbit the sun?

We are informed that planets are caught by the sun's gravity, or a centripetal force, and do not stay in the same place, as they are always moving about the sun. However, they stay in the same orbits. This is incorrect, as there is no such thing as gravity. If there is such a thing as gravity, and according to Newton's dynamics, planets should have been spaced out at certain distances from the sun, according to their size. If there was gravity, why do they orbit the sun? They should have crashed into the sun. According to Newton's' dynamics planets/bodies should have orbited around each other. They should have perfect circular orbits around the sun, and around each other. They should not have rotated around their imaginary axis, and orbit the sun at the same speeds, proportionate to their distances from each other, and their orbital distances from the sun. Why are they on the same plane?

As star field or interplanetary space engenders, concurrently with the dense region, or the sun/star, they are integral parts of each other. Space confines the dense region, or the star, and vice versa. As space, and the dense region, or the star, concurrently engendered, and are parts of the same cycle (with different characteristics), there is both, an inward, or centripetal force, toward the dense region, and an outward, or centrifugal force from the dense region. As a result, star field space or interplanetary space is continuously 'forced' on to the dense region. All planets and bodies with, or without energy transmuting cycles with the sun exist, within/as interplanetary space (which orbits the sun) and therefore, orbit the sun.

Why do bodies in the asteroid belt orbit the sun at that location?

Asteroids are bodies (smaller planets) that were fragments of an exploded planet, which was localized at their present orbital location. All the asteroids collectively, or individually, initially engendered into bodies having formed energy transmuting cycles with the sun. Some still continue their cycles, while others having belched out their interiors, and sustaining their depleting atmospheres have now solidified, and are impotent. Some are being disintegrated, and as asteroids, and other smaller bodies roam interplanetary space, with some settling down at the outskirts of our solar system.

Those with active energy transmuting cycles, and those that have not yet been disintegrated, or disturbed, still orbit the sun at the asteroid belt, as was the norm for many years. Obviously, these asteroids (with energy cycles) would be found rotating on their imaginary axis. Bodies are not 'entities', but as denser structures exist as/within interplanetary space, which orbits the sun.

Why do planets orbit the sun at different speeds at different orbital locations?

Revelation 5 detailed why and how planets formed with their different characteristics. The orbiting speeds of planets set, depending on the receipt and extraction of material, that were/is required, as their cycles formed/continue. This is explained within revelation 5.

As everything is interdependent and interlinked within/as the fabric of energy, speeds were/are randomly structured, taking all the cycles within the star field into consideration. If any process/structure of the cycles alter, both the orbiting speed, as well as their distance from the sun would concurrently alter. Obviously, there will also be a change in their characteristics.

Why do bodies orbit the sun in the same direction?

All the planets and bodies, exist within/as the fabric of interplanetary space, and therefore orbit the sun in the same direction as interplanetary or star field space.

Why are all planets more or less on the same plane?

A star field can only engender, if the dense region, and the space confining it, spin in opposite directions. A star cannot form, as its many layered atmosphere, and also star field or interplanetary space would be spinning in different directions. Moreover, if interplanetary space rotates in the same direction of the corona, the photospshere would not rotate at super speeds. The corona would not be extended for millions of miles, as it is presently, and without being tenuous, it would be slower rotating, and extraction and expulsion of energy would not be possible.

Spin cannot take place at different directions around/as a sphere. One complete rotation must be in one direction. As the sun's atmosphere, or the corona, and the chromosphere, orbit the photosphere in one direction, interplanetary space orbits it in the opposite direction. As all the bodies and planets are existences within/as the fabric of interplanetary space, they all orbit the sun in the same direction.

All spheres spin faster at their equator (as they are so shaped), explaining why equatorial atmospheres are tenuous. It is easier for energy to propagate/transmute from regions around the sun's equator, than at the poles where the atmosphere is denser, and the rotation much slower. This explains why all the bodies/planets are more or less on the same plane. As the equatorial periphery rotates faster than other latitudes of a sphere, all planets orbit the sun at that same plane. This explains why disks or gas rings of planets orbit around their equator. This also explains why disks around planets are flat, and vertical to a planet's atmosphere. If disks were horizontal to a planets atmosphere they would disintegrate (rotational speeds decrease approaching the poles). This also explains why star fields and entire galaxies are flat, not spheres.

As there are much energy transmuting cycles as there are planets, also with some bodies, and as all the planets orbit the sun at the same plane, facilitating transmuting of energy, our solar system is steadfastly set. Energy from the sun is easily 'propagated' or transmuted to bodies/planets in the same plane than if they were in different planes.

Our solar system has endured for millions of years, as a result of the smooth functioning of all cycles. All the cycles have endured or continue to exist with this format of planets being spaced at a certain distance, orbiting at certain speeds, and orbiting on the same plane. The energy radiation cycles with its to-and-fro extraction, and expulsion of energy (transmuting), connects all these bodies in the planetary system keeping them at a certain distance, at certain orbits, and at certain spins. This collective smooth operations, and conditions, have enabled our solar system (steadfastly set) to confront/conform energy transmutations variations, both within and outside our solar system.

The orbits of the planets are all more or less in the same plane called the ecliptic. The ecliptic is inclined only 7 degrees from the plane of the sun's equator. Depending on the amount of extraction and expulsion energy by the sun, bodies/planets may deviate few degrees from their plane.

As Pluto is a smaller planet independently orbiting the sun, unlike moons of other planets (some bigger and orbiting the sun together with their host planet), it is strongly affected within the extraction/extracting cycle, deviating the most from the plane of the ecliptic with an inclination of 17 degrees. Presently, Pluto changes its atmospheric characteristics, as it approaches and departs from the sun.

Why do some planets move faster when they orbit nearer to the sun?

At times, as more material is being extracted, a planet gets closer to the sun, pulled from their normal orbit. In the same way, if expulsion of energy is more than normal, the planet moves away (with the ejected force). This explains why most of our planets have an elliptical orbit around the sun.

As there are many cycles, and other cycles alter (but are firmly in place), the planet sometimes spin/rotate faster, and also moves away slower. At times the planet spins/rotates slower, and moves away faster (see extra solar planets). If there was a single planet in our solar system, its orbit would have been erratic. The star would also wobble like those with extra solar planets. There are many planets and cycles in our solar system, interdependent and interlinked, explaining why all bodies, and/or our solar system, have endured for long durations. This also explains why they get closer, or further away from the sun, and also why they rotate slower or faster.

Most importantly, we must note that it is the atmosphere of bodies/planets that are disintegrated and extracted. Planets' interiors are contained within their atmospheres (if we wish to understand that atmospheres are separate existences). Moreover, atmospheres are continuously engendered, being continuously integrated, disintegrated and depleted.

Cycles have formed wherein at a certain distance from the sun, atmospheres would not regenerate as the norm. Therefore, the elliptical orbit ensures that planets have their atmospheres continuously integrated, and disintegrated, as normal. What structured, and engenders as part of a cycle (which also engenders), will exist/endure. Things are as they are, as they could exist, and continue to endure.

Mercury's atmosphere is presently completely depleted, and has now solidified. It is now an impotent body awaiting disintegration, explaining its peculiar type of revolution/orbit.

Pluto's atmosphere of nitrogen expands, as it moves away from the sun (this process take decades long). Pluto is so far away that it takes 248 years to complete one revolution. It has easy extractable material for replenishment at the sun. Moreover, its atmosphere easily alters its characteristics, approaching, and receding from the sun. It acts as a material balancing reserve, and utilized, as and when there are energy transmuting variations altering other cycles, explaining why it has a peculiar eccentric orbit.

WHY NEWTON'S DYNAMICS ARE INCORRECT

Why do planets not orbit each other?

According to Newton's dynamics, all planets of different sizes should align themselves from the centre (sun) out, corresponding to their size and weight, with the bigger planets spiraling into the sun. No planet or its satellite pulls at each other with a certain force, and neither does the moon pull on the earth, and vice versa. Density and size has nothing to do with why the planets are being controlled by the sun. There is no relation between the mass, and the distance of two bodies acting upon each other. There are also no significant interactions between two planets, or objects, if they are not parts of the same cycle, sustaining each others existence. With Newton's dynamics the planets should not only be orbiting the sun, but should also orbit each other. With energy transmuting cycles interdependent, and interlinked, with one another, some planets have elliptical, not circulatory orbits disapproving Newton's dynamics.

Energy 'propagates'/transmutes between the sun, and a body/planet, as a cycle, explaining why planets rotate around their axis. Planets/bodies exist as/within the fabric of interplanetary space/energy (which orbits the sun) and therefore, as they have randomly engendered at their localizations, orbit the sun at a fixed distance from it. As their atmospheres are depleted, planets/bodies shrink in size, and are gradually pulled closer to the sun, altering their orbits and characteristics-the smaller inner planets are evidence of this.

Planets have formed energy transmuting cycles with the sun, but there are no planets which have formed cycles between themselves. Bodies/planets have to be of different characteristics, sustaining each other's existence and therefore, are parts of the same cycle. This explains why there can be no two dominant (extracting) bodies, or two stars, or two dense regions within a star field. This also explains why planets do not orbit each other.

Do binary stars orbit each other? Do Star fields orbit one another?

Bodies are of different characteristics, enabling them to sustain the existence of each other, as parts of the same cycle. There is no such thing as binary stars. There cannot be two 'dominant' bodies, or 'stars' within a star field, forming parts of the same cycle. All stars exist within their star field space. Although it appears that binary stars orbit one another, in reality, their star fields do. All star fields are interdependent, and interlinked, with other star field cycles, and together form a galaxy.

KEEPING DISTANCES

How and why do bodies keep their distances from the sun?

We are informed that all planets fall (orbit) to the sun being pulled by its gravity. This is incorrect, as there is no such thing as gravity.

All planets/bodies are fragments of the exploded star, and exploded planet, and have localized where they engendered. Densities, material compositions, size, size of member bodies, both within a cycle, and other adjacent cycles, the amount of material being received from, and extracted by the sun, the changing atmospheres, and many other conditions, initially decided/decide, and dictated/dictate (as/and cycles formed) the distance of planets/bodies from the sun, the spacing of planets from each other, the rotational speeds, and their directional orbits. All bodies within the solar system exist as/within the fabric of interplanetary space/energy, which orbits the sun, and therefore, keep their distances.

Why and how do planets keep apart from one another?

We are informed that there is no energy available to change a planet's orbit, as a rocket carries on board fuel to boost its way. This is preposterous.

Everything (as energy) is interconnected and interdependent. Nothing is an entity, and planets are structures, or processes, of transmuting energy, within/as the fabric of energy. The energy transmuting cycle, or the incoming and outgoing energy processes keeps them spinning, keeps them apart, and at appropriate distances. Planets are kept apart at their localizations randomly engendering there from the beginning. As their materials are being depleted they solidify, and move closer to the sun, and this process takes thousands of years. They continue to be kept apart, as various cycles are formed, and still continue. The previous answer also explains part of this question.

Why do planets not veer off their orbits and crash into the sun, or into one another?

Having energy transmuting cycles with the sun, planets rotate around their imaginary axis. All planets/bodies, as existences within/as the fabric of space/energy, orbit the sun at certain distances. Within energy transmuting cycles, energy is being concurrently expelled and replenished, within the same energy transmuting shuttling to-and-fro pathways; therefore they cannot veer of their orbits, or crash into the sun, or into one another.

Planets hold on to their stations, as space is like a spider's web. Once they are localized in a certain area they are stuck, until there is an explosion, or major energy transmuting variations within the web, or externally. However, this analogy is not correct. As planets atmospheres get depleted they solidify, and without energy transmuting cycles, and 'becoming' impotent, they are pulled closer to the sun, and will eventually disintegrate. Mercury is a good example.

ORBITING THEIR PLANETS

What are gas rings around planets? Why do only some planets have gas rings?

Why do moons and gas rings orbit their planets? Why do moon/satellites orbit their planet and not independently the sun (as Pluto), rotating around their axis?

Asteroids, moons/satellites, gas rings, comets and bodies in the Kuiper belt, and the Oort cloud are fragments of an exploded planet, which was localized between Mars and Jupiter. As the gaseous fragments darted out toward the sun, and away from it, they were captured by the inner and outer planets, as gas rings and moons. Any fragment would have to commence an energy transmuting cycle with the sun, in order to be able to exist in close proximity to the sun, and as far away as Pluto. In reality, as these fiery fragments (with appropriate replenishment material) darted to, and away, from the sun, and commencing energy transmuting cycles, they were captured by their planets, as parts of their energy transmuting cycle with the sun. In fact, the capture of these bodies enhanced their energy transmuting cycles, during the maelstrom. It is possible these conditions not only prevented other cycles from disrupting, and/or terminating, but also steadfastly set existing cycles, and the entire solar system.

Gas rings and moons were/are part of their planets' energy transmuting cycle with the sun, and they were, or are continuously degassing or disintegrating. Over the years, some gas rings and moons may have disintegrated.

Presently, gas rings are found only around the Jovian planets. The sun gets enough replenishment material from the inner planets, and as active energy transmuting cycles are not yet actively in force with the Jovian planets, they still have their gas rings, and many moons. Presently, as replenishment energy is decreasing from the inner planets, the sun is actively engaged in extracting Jupiter's moon Io material-its present volcanic activity is evidence.

Mercury, Venus, Earth and Mars may have also had gas rings. Mercury and Venus may have also had a moon or moons. Earth and Mars may have had more moons. All these may have disintegrated over time, as they could not have continued replenishing their depleting atmosphere, and solidified.

The inner planets have no gas rings, and Earth has only a moon, and Mars two. Being closer to the sun with active energy transmuting cycles, they may have lost their gas rings, and also their moons (if they had more than they do presently). The solidification of Mercury, Mars, our moon, and the moons of Mars is solid evidence.

All bodies exist within/as the fabric of energy/interplanetary space. As moons and gas rings have localized in close proximity to their planets, they orbit around their planets set/as in the planets outer atmosphere. Together with their planet they also orbit the sun.

As our moon has completed its degassing, there is no direct energy transmuting cycle between the moon and the sun. Only planets/bodies that are parts of a cycle with the sun rotate around their imaginary axis, and orbit the sun 'individually'. This explains why our moon is not called a planet, as Pluto. The moon however is/was part of the cycle together with the earth, and the sun. Its depleted atmosphere and solidification, explains why the moon does not rotate around itself. It orbits the earth without rotating on its imaginary axis, as it is an integral part of the medium it is in.

Moon and gas rings orbit their planets, being integral parts of their planets energy cycle with the sun. As every existence/body exists within/as the fabric of energy/interplanetary space, it is not the bodies per se that are rotating, or orbiting their planet, or the sun/star, but the medium that they are in. There are no entities.

The moon at times travels faster in its orbit around the earth. The moon's orbit is influenced by changes in the Earth's rotation, its orbiting distance, and speed around the sun, as the Earth in turn is influenced by changes in the orbits, and alterations in the cycles of other planets.

As mentioned elsewhere, Pluto may have initially been a moon to one of the outer planets. It may have been disrupted from its localization to balance the increases or decreases of replenishment material at the sun. Pluto may also have been a comet, but as the decrease in replenishment material was overcome, the sun terminated the attraction, and Pluto went into an orbit en route.

Every sphere can rotate only in a single direction. The sphere is so shaped that rotation is faster at the equator than at its poles, explaining why there is more density at the poles than at the equator. As the atmosphere of the planet near the equator is tenuous, and as this area is faster rotating, gas rings are found orbiting their planets at the equator. This also explains why gas rings are flat, but their flat sides do not face their planet. Gas rings have widths vertical to their planets. Approaching the poles of a spherical body the rotation decreases and density increases. This would disintegrate gas rings with widths horizontal to their planets.

Why do planets have elliptical orbits?

This is answered within revelation 5- why extra solar planets have elliptical orbits.

ROTATIONS

Why do planets orbit around the sun in the same direction they rotate around their imaginary axis?

Bodies/planets having atmospheres to be depleted have energy transmuting cycles with the sun, and rotate around their imaginary axis. The extraction processes generates the rotation around their axis. It is analogous to pulling a thread from a knitting ball. If the corona revolved in the same direction as interplanetary space, energy transmutations between the sun and the planets would not engender, and planets-if they existed- would not rotate around their imaginary axis.

The corona or the outer atmosphere of the sun rotate in a direction opposite to the photosphere. Planets and bodies of our solar system exist within/as the fabric of energy or interplanetary space, which orbits the sun's corona in the opposite direction, explaining why planets/bodies orbit the sun in this same direction.

Why does Venus at some time orbit the sun in the opposite direction than other planets?

The energy transmutation cycle shuttling to-and-fro between the Sun and Earth, interacts/integrates/disintegrates its atmosphere. Radiated heat and pressure/force delves down through the many layered atmosphere to earth's surface, rivers, lakes and oceans, interacting/disintegrating and integrating the water molecule extracting hydrogen, and parts of oxygen into our atmosphere, and outward into space, to the sun. With replenishment material easily available from our oceans, there is more energy transmuting activity between the earth and the sun, than between the sun and Venus.

The constituents of Venus' atmosphere are not appropriate replenishment material required at the sun (also read revelation 6). This explains why it does not possess an active energy cycle with the sun, and rotates very slowly taking 243 earth days for a single rotation. It takes 225 earth days to orbit the sun, as it is closer to the sun than earth. Every 243 days before starting a new rotation Venus rotates in the opposite direction.

Why do Pluto and Neptune's moon Triton and some of Jupiter's moons rotate the opposite way?

There are many possible reasons:

Neptune's moon Triton, and some opposite rotating moons of Jupiter, were bodies in the outskirts of the solar system, extracted as comets at one time, as replenishment material at the sun. Triton being one of the coldest places in the solar system is evidence that it was a body in the far reaches of our solar system. However, after being uprooted by the extraction processes at/of the sun, replenishment energy were sourced from somewhere around the inner planets, or within the solar system and these bodies (being exposed without the extraction force) were captured en route by their planets, transforming into/as their moons. But as the planets were steadfastly set with their own energy transmuting cycles, and as these moons concurrently commenced their own energy transmuting cycles, they rotate in the opposite direction.

The atmospheres of planets, and the adjacent space around these atmospheres rotate in opposite directions. Triton and some of the opposite rotating moons of Jupiter captured later on, localized around their planet's atmospheric layer space and therefore, rotate in opposite directions. This is evidence that they were subsequently added bodies.

It could also be that moons that are opposite rotating than some moons of Jovian planets, have only recently begun active energy transmuting cycles with the sun, as replenishment material was previously available from the inner planets and asteroids.

We can also postulate that some opposite rotating moons of planets are fragments of planets that were stationed closer to the sun than Mercury, which disintegrated only recently.

Why do asteroids in the asteroid belt orbit the sun? Why do/did they not settle at far distances as the comets and icy bodies in the Kuiper belt?

Asteroids are fragments of the exploded planet in our solar system, which was localized at the asteroid belt. As the exploded planet still possessed its fiery interiors, the fragments partially disintegrated, and quickly generating atmospheres around, transformed into asteroids (actually planets). All of these asteroids had individual (or grouped themselves) transmuting cycles with the sun. As these fragments were not thrown out to far distances, they engendered at more or less at the same location of the exploded planet, called the asteroid belt.

As the asteroids were once part of the collective energy cycles of bodies with the sun in the asteroid belt, they will continue to orbit the sun (as was/is the norm, and exist within/as interplanetary space), with or without material to be extracted.

THE MOON

Why does the moon have a circular orbit around the earth?

The moon's orbit around the earth is not perfectly circular. Depending on the extraction/expulsion processes of the sun, within its cycle with the earth, the moon shifts closer, or away from its orbit around the earth. All planets have an elliptical orbit having an energy transmuting cycle with the sun.

Why does the moon not rotate around its imaginary axis, or why does it always keep the same face toward the earth?

We are informed that only one side of the moon is seen 'because' it always rotates-on its imaginary axis-in exactly the same length of time it takes to revolve about the Earth.

Here is a circulating theory.

Quote "the moon does rotate. It rotates exactly once for every orbit it makes about the earth. The fact that the moon is rotating may seem counterintuitive: if it is always facing toward us how can it be rotating at all? To see how this works put two coins on a table, a large one to represent the earth, and a small one to represent the moon. Choose a particular place on the edge of the 'moon' coin as a reference point. Now move the moon around the earth in a circle, but be careful to always keep the spot you picked pointed at the earth (this is analogous to the moon always keeping the same face pointed at the earth). You should notice that as you do this you have to slowly rotate the moon, as it circles the earth. By the time the moon coin goes once around the earth coin, you should have had to rotate the moon exactly once. This exact equality between the moon's rotation period, and orbital period, is sometimes seen as a fantastic coincidence, but in fact, there is a physical process, which slowly changes the rotation period" Unquote.

This explanation is completely preposterous. In the above experiment we are continuously adjusting the moon coin so that the same face faces the earth. The earth rotates around its axis approximately 28 times a month. The moon orbits one complete circle in 27 or 28 days around the earth, and does not rotate around its imaginary axis. If the earth rotates 27 or 28 times around it axis a month, surely during one of these days, somebody, somewhere should see other surfaces of the moon.

As the earth rotates, its adjacent layered atmosphere rotates the other way. This atmosphere has its adjacent layer rotating the opposite way and so on. Only planets/bodies with atmospheres engender/endure an energy transmutation cycle with the sun. Over the years, the moon's interior has completed degassed sustaining the depletion of its atmosphere. Having solidified itself, and now without an atmosphere, the moon does not play any significant part in the earth-sun energy transmuting cycle. Without an atmosphere there is no material to be interacted, disintegrated, integrated and depleted and therefore, the moon does not rotate around its axis-showing its same face always..

Bodies/planets are not 'entities', but structures/processes within/as the fabric of interplanetary space. Atmospheres of bodies rotate in the opposite direction of the body's interiors. Earth's atmosphere also has interplanetary space rotating in the opposite direction. As the moon exists without an energy transmuting cycle (it does not rotate around its imaginary axis), but exists within/as this region of space, it orbits the earth with the same face facing the earth.

See how this works. Take a piece of paper cutting a hole in the middle, and place the earth coin in it (the coin rests on the surface of the table). With the moon coin at the paper's edge (around the hole) move the paper around the earth coin. You will notice that the same point of the moon coin faces the earth. The paper is the medium that orbits the earth, the moon stays at its same location, but it is the medium that orbits the earth. The moon does not rotate around its axis, as it is set in its medium or atmosphere, which is orbiting the earth, and so the same side faces the earth continuously.

THE ROTATION OF THE EARTH

The earth's surface adjacent atmosphere rotates at 1650 km per hour. Half point from the equator, the atmosphere rotation is about 835 km per hour with zero km rotation at the poles.

The atmosphere is the least at the equator as discussed elsewhere, and the total density of atmosphere (and therefore the rotation) increases approaching the poles. As we cannot compare atmospheric density within the fabric of energy this is of no significance. Different layers of earth's atmosphere rotate at different speeds, at different directions, at different areas between the equator and the poles, explaining our tides, currents, winds/breezes and waves of varying characteristics.

The solid earth takes 24 hours to complete one full rotation, and most parts of the earth have nearly equal darkness and light. However, as our earth is tilted, there are regions around the poles which experience more of darkness or light, or longer or shorter days/night during parts of the year.

We can only measure the rotation of an object relative to something else. We can query how fast the earth is moving with respect to other bodies, the sun, the Milky Way galaxy, and other galaxies.

It is postulated, that the surface of the earth spins/rotates around it imaginary axis at 0.5 km/sec, as it orbits the sun at 30 km/sec. All other star fields near our solar system revolve around the centre of our Milky Way galaxy at 250 km/sec. At the equator, the earth's surface moves 40,000 kilometers in 24 hours. That is a speed of about 167 km/hr or 0.5 km/sec (circumference at the equator 40,070 km divided by 24 hours). At the poles this speed is almost zero since the circumference there is zero. The earth orbits at a distance of 149, 60000 km from the sun.

REVELATION EIGHT

A UNIQUE WORLD

The energy transmuting cycles between the sun, and earth's atmosphere and oceans

This revelation elucidates how earth, its atmosphere, oceans, and interiors engendered, and why and how the energy transmuting cycle, continuously interacts and disintegrates, and integrates, our atmosphere and oceans, generating our climate/weather, seasons, and other phenomena.

Important note:

Every existence engenders, exists and endures, as a result of certain degrees of temperatures and pressure/force intensities. Increases or decreases of these forms of energy will alter, or terminate these existences.

Explaining a planet, its atmosphere, surface and oceans

Without understanding reality we create existences (from within our illusory language), and so have problems explaining them. We describe planet earth (oceans, mantle, crust and core) distinct from its atmosphere. The atmosphere, oceans, crust and mantle of the earth are all different types of energy structures containing the molten core (also an energy structure). We cannot define earth as a body/planet without, or with its atmosphere. Energy is everything and everywhere, and earth's molten core, mantle, crust or surface, its oceans and atmosphere, and all bodies/planets are all 'existences' (within our concepts) within/as the fabric of interplanetary space/energy.

Planets could not have engendered, existed, or endured, without their atmospheres. All atmospheres were/are continuously sustained by/with their disintegrating interiors (in the case of earth, mostly its oceans) and therefore, atmospheres are integral parts of the interiors (bodies/planets). However, as the interiors stop disintegrating, the atmospheres deplete, and the solid sphere that was once a planet, awaits disintegration. To understand this revelation we have to continue separating earth (oceans, mantle, crust and core) from its atmosphere.

The engendering of 'earth', concurrently with its original atmosphere

What were the constituents of the original atmospheres of planets?

What were the constituents of the original interiors of planets?

How and why were these original constituents generated?

Normally, fragments from an exploding neutron star would disintegrate within its own star field (without space). Neutron plasma would transmute into/as space, and protons would disintegrate into plasma, and space, amalgamating/drowning into/as free energy, as a black hole. From this swirling black hole a star field may engender.

As an exploding neutron star was in close proximity, the fragments darted through our nascent star field space. These speeding fiery fragments (of neutron plasma and protons) of high heat, and pressure/force and densities, were instantaneously enveloped by our nascent star field space. Contained by space, neutron plasma and protons integrated together, engendering nuclei or deuterium, tritium and helium.

As nuclei continued to be produced, there were increases in heat and pressure/force. The increasing heat incited the dense region within the nascent star field, and concurrently increased the concentrations of/at the dense region, generating higher temperatures and pressure or force, and engendering particles/plasma. Plasma, heat and pressure/force propagating from the nascent sun interacted with the nascent atmospheric nuclei at/of each of the fragments. Particles produced from nuclei interactions, and disintegrations, were extracted back to the nascent sun, as replenishment material/energy, within the same energy propagating shuttling to-and-fro pathways.

Nuclei disintegrating and integrating within space engendered into atmospheres, confining the interiors of the planets. With processes at the atmospheres, nuclei from the interiors were extracted into/as the atmosphere, to be interacted and disintegrated, continuing the energy transmuting cycles, and enduring the atmospheres. As energy transmuted between the dense region, or the nascent sun, and the fragments, the sun became full-fledged, and the fragments transformed into planets, and/as concurrently energy transmuting cycles were now in force.

Initially all the planets were massive spheres of deuterium, tritium and helium. Radiated heat and pressure/force with plasma/protons from the sun, continued to interact, and disintegrate nuclei extracting particles. Deuterium and tritium continued to be disintegrated into protons and neutron plasma. As protons were further disintegrated or extracted, neutron plasma transmuted into/as space. As neutron plasma was/is 'released' there is both heat and pressure/force, and space generated.

At that time, there were a number of energy transmuting steadfastly set cycles, as there were planets. Earth and all the existing planets of our original solar system constituted of deuterium, tritium and helium. The disintegrating interiors continued to sustain the continuously disintegrating atmospheric nuclei.

How and why did earth alter its original atmosphere into its present one?

How and why did the inner planets alter their original atmosphere?

Why and how are earth and other inner planets interiors constituted of heavy elements, molecules and compounds?

We will never know how long our original solar system endured, as radiated heat, pressure/force and protons from the sun continued to interact, and disintegrate the atmospheres of the planets. As the sun could replenish its expelled material/energy, all energy transmuting cycles were steadfastly set. However, planets were at different proximities to the sun, and as radiated heat and pressure/force dwindle with distance, and as energy continuously transmutes, and everything is flux, this original solar system could not endure. There were also other reasons.

There were different forms, and processes of energy transmutations within our solar system with the additional fragments-turned planets from an exploded neutron star. As there were many planets, as there were energy cycles, extraction material to the sun was greater than the amount of material being expelled. Being in close proximity, the inner planets energy transmuting cycles were not stable, and as a result, a planet stationed between Mars and Jupiter exploded. With this event all the inner planets energy transmuting cycles went haywire, subsequently terminating. As radiated heat and pressure/force from the sun was still in force, the complete atmospheres of the inner planets were instantaneously extracted, and depleted, exposing their interiors.

The original atmospheres of all the planets formed, concurrently inciting propagation of radiated heat and pressure/force from the nascent sun. This time, radiated heat and pressure was already existing, or in force. Radiated heat and pressure/force with protons interacted with the planets (exposed interiors), disintegrating nuclei into protons and neutron plasma. Protons continued to be disintegrated and extracted, and neutron plasma amalgamating into/as space, generated more space, increasing the heat and pressure/force intensities.

With the radiated heat and pressure/force from the sun, and the heat and pressure/force generated from disintegrating nuclei, as neutron plasma was being 'released'(forming space), temperatures soared, together with high pressure/force intensities. The additional space facilitated more nuclei to be disintegrated. Concurrently, with increasing space, and with increasing heat, and pressure/force, nuclei integrated. Deuterium combined with neutron plasma forming tritium, and tritium combined with deuterium forming helium. With further integrations, there were further increases in temperatures and pressure/force, and the amount of space, and further integration of nuclei into heavier elements than helium.

There are only two types of nucleons anywhere, a deuterium and a tritium. From these two nuclei, all the elements, molecules, substances, and matter engender.

At certain degrees of heat and pressure/force intensities, deuterium combines with tritium generating helium. Helium is generated, as neutron plasma (which is disintegrating condensed energy, and forever seeking 'release' or disintegration) from a tritium (deuterium-tritium combination) is 'released'. As neutron plasma is released from a tritium, or from a disintegrating deuterium, or helium, deuterium can combine with a tritium, and/or helium or tritium can combine with helium to engender heavier elements.

As neutron plasma is 'released', there is an increase in heat and pressure/force, and as it amalgamates into/as space/energy, concurrently integrations and disintegrations can take place. As neutron plasma was 'released', enhancing the heat and pressure/force intensities, concurrently more space was generated. This space facilitated, and abetted the disintegration of nuclei into protons and neutron plasma, engendering the new atmosphere.

Disintegrations of atmospheric nuclei continued (as the sun extracted its replenishment material), concurrently with the integration of nuclei into heavier elements in the interiors. At each degree increase in temperatures, and pressure or force, intensities, nuclei integrated into heavier and heavier elements. In other words, each nuclei subjected to certain degree of heat and pressure/force intensities integrated together to form their element. As temperatures and pressures increased in incremental steps, one by one, heavier elements than helium were created. The engendering of heavier, and heavier elements, concurrently increased the temperatures and pressure/force, facilitating and abetting the integration of nuclei into yet heavier elements.

Deuterium and tritium combine to form helium, as neutron plasma released generates heat, pressure or force, and space. This is also how we obtain power, or electricity. The integration of deuterium, tritium and helium combinations also produce other elements. Lithium engenders, as two deuterium nuclei conjoins with tritium. Beryllium engenders, as three deuterium nuclei conjoins with tritium. Boron engenders, as four deuterium nuclei conjoins with tritium. With all these integrations, parts of neutron plasma are released.

As boron engenders, it integrates with hydrogen or a proton, generating three helium nuclei, or carbon. But as integration progresses (increasing heat and pressure), boron will integrate with helium instead of hydrogen producing non radioactive nitrogen, and neutron plasma. For boron to integrate with helium requires tremendous temperatures and pressures/force, as concurrently equal degrees of temperatures and pressure intensities are 'released'.

As neutron plasma is condensed disintegrating energy, every element that has more of it is radioactive, and/or unstable. As we go higher in the periodic table, we will notice elements having a greater amount of neutron plasma than proton plasma, or protons, explaining why very heavy elements are radioactive, and/or unstable. If a nuclei/element has too much of neutron plasma it is highly radioactive. An example is caesium which explodes when dropped in water. Neon is not reactive, as it contains equal amounts of neutron plasma and proton plasma, but fluorine is very reactive, as it has extra amounts of neutron plasma.

Importantly, there are multifarious types of elements and matter, as elements have different amounts of proton plasma, and neutron plasma. As neutron plasma has an innate characteristic, seeking to continuously amalgamate/transmute into space, nuclei/elements can integrate forming molecules and compounds, but can also disintegrate with altering temperatures, and or pressures.

As heavier elements were being formed, concurrently temperatures and pressures/force increased, till they reached extreme limits. This is the integration of nuclei as/at the iron element stage, explaining why iron has the most tightly bound nucleus.

The interiors continued to be integrated into iron, until such time a new atmosphere formed. We will learn that the new atmosphere generated water vapor, which condensing into clouds produced precipitation. As precipitation fell down on the fiery interiors, iron exploded (explained following). As a result, iron nuclei engendered with all its nuclei arrayed in certain configuration or patterns, giving them a direction. This explains magnetized iron's innate attractive, repulsive and ‘magnetic’ powers. As magnetized iron-with its one directional nuclei configuration arrays-engendered during this sudden impact, it forever tends to force, or ‘attract’ other similar nuclei in close proximity. Iron also repels other similar nuclei configurations, if their directions confront each other. Forms of energy not only propagate within/as interplanetary and atmospheric space, but also within/as space confining nuclei, and around nuclei in elements, molecules and compounds. Temperatures and pressure/force cannot easily propagate through iron nuclei, which are tightly bound, explaining iron's characteristics.

Heavier Elements

Nuclei integrations producing heavy elements up to iron in the interior continued, until temperatures and pressures reached extreme limits. The incoming radiated heat and pressure with the protons from the sun, continued to disintegrate atmospheric nuclei into protons and neutron plasma. As the disintegrating protons were continuously extracted, and neutron plasma was being 'released', the outer regions formed into a new atmosphere with the generated space, with expansion.

As atmospheric nuclei formed there was an increase in density, and temperatures and pressure/force decreased. As the outer regions reached certain temperatures and pressure/force, integrations continued until oxygen formed. As heat and pressure/force continued to vary within/as the atmosphere, there was continuous integration and disintegrations. The forming oxygen continued to disintegrate into nitrogen, and carbon. However, at certain temperatures and pressure/force, oxygen combined with the departing protons forming water vapor. This changed many things.

The formation of water vapor decreased the temperatures and pressures of/at the atmosphere, increasing densities. This regulated radiated heat and pressure or force from the sun, concurrently decreasing temperatures. With decreasing temperatures water vapor condensed into clouds, and precipitation. As precipitation fell down on the fiery interiors, iron disintegrated explosively. With these explosions, nuclei blasted away bombarding each other, generating heavier elements than iron. Iron engendered with very high temperatures and pressure/force. However, as iron disintegrated explosively, the release of these intensities together with the prevailing high temperatures and pressure/force, integrated nuclei into yet heavier elements than iron. Elements are heavier than iron not consequent to higher temperatures and pressures, but as more nuclei integrated to form heavier elements.

As iron disintegrated with the explosions, the 'release' of neutron plasma, concurrently abetted the bombarding. As precipitation continued, the explosions did not continue for long and therefore, as temperatures and pressures/force decreased subsequently after the explosion, earth's interiors possess only a small percentage of heavier-than-iron elements, some being rare.

Continuous precipitation gradually decreased the temperatures and pressures in the interiors, transforming some of the gaseous nuclei into solids and liquids. As precipitation continued, and as iron was the heaviest element it sank deep into the centre of earth taking with it some sulphur, silicon and carbon. Over time, these elements decayed into protons, and oxygen.

As iron sank down, the outer core formed, and partly solidified, concurrently containing the inner core, explaining why earth has a molten core. As precipitation continued, and as the fiery gaseous elements solidified, the outer core was enveloped with the mantle, and the mantle with a crust, which eventually contained the precipitation as rivers, lakes and the oceans. At that time, the oceans completely covered the crust and the mantle. Over the years, the oceans sustained the continuously depleting atmosphere.

As the oceans gradually depleted, over many years of sustaining the depleting atmosphere, some of the crust was exposed, and the confined molten core (which was occasionally belching out) profusely began spewing out material. Through nooks and openings, trapped gases in the interiors tore apart the lithosphere. All present dormant and active volcanoes, craters and mountain ranges are vestiges of belching out of magma, and release processes of gases from the interiors, as concurrently gaseous elements transformed into liquids and solids.

This transformation trapped lots of different gases/elements in the mantle, crust, and the outer core. This is why and how, as chemical reactions continued over the years, elements, molecules and compounds are embedded in the earth's surface. This is the state of earth today.

The mantle consists mostly of peridotite containing iron, silicon, manganese and oxygen. The continental crust is made up of volcanic lava, and of rock granite containing oxygen, silicon, and aluminum, whereas the ocean crust contains silicon and aluminum.

The periodic table shows that heavier elements than iron have very high amounts of neutron plasma than proton plasma, as compared to elements lighter than iron, evidence of more tritium involvement. As iron disintegrated, less of helium, and more of the deuterium, and tritium were utilized, integrating heavier elements. As these heavier elements than iron engendered with high amounts of neutron plasma, and at varying temperatures and pressures steps, most of them are radioactive and unstable, also explaining why some of these elements are plentiful, while some of them are rare. Over millions of years, much of these heavier elements have decayed and disintegrated.

Oxygen of about nearly 50% is abundant in the earth's crust (trapped by disintegrating iron). Silicon and aluminum are also found in good quantities. All these three elements have a nearly exact mix of neutron plasma and proton plasma, evidence that extra amounts of neutron plasma make an element unstable. Calcium, sodium, manganese and potassium are also found in considerable amounts, as they too have a nearly exact mix of neutron plasma and proton plasma.

Carbon in the hardening permeated formed land, engendered only with helium nuclei, and like oxygen, easily integrates with other elements forming molecules and compounds. The water reservoirs are sinks for the large amounts of carbon, and through chemical and biological processes are locked up in sedimentary rocks as limestone.

Some of the degassing nitrogen would react slowly with the mineral of the crust to form nitrates. Oxygen reacted with iron forming Pyrite and Urannite, explaining why these minerals are easily dissolved out of rocks, under present atmospheric conditions. Deep water deposits contain iron-rich minerals. Iron minerals including iron oxide, iron carbonate, iron silicate, iron sulphide are common in rocks, but do not form today. Red beds are formed, as iron minerals are oxidized, accumulating in the sediment. Over the years, rocks alter, with the weathering by rainfall. Lava flowing from volcanoes, spread over the surface solidifying some parts of land into thin crusts. This crust would have melted and solidified repeatedly, and light compounds would have moved to the surface.

Earth's mantle is composed of iron, magnesium, aluminum, silicon, and oxygen silicate compounds. At over 1000 degrees C, the mantle is solid, but can deform slowly into a plastic manner. The crust is much thinner than any of the other layers, and is composed of the least dense calcium, sodium, and aluminum-silicate minerals. The crust is rocky and brittle, so it can fracture in earthquakes

Our present atmosphere

Engendering our atmosphere

As the oceans formed, the new atmosphere engendered with different constituents. Atmospheric nuclei, and the ocean molecules, are continuously disintegrating, and integrating, again and again.

To sustain this depletion from the atmosphere and the oceans, earth's interiors continuously belches out trapped gases with other elements, molecules and compounds from volcanoes, underwater sea beds, vents and from openings, resulting in plate tectonics and other abnormal phenomena. Nitrogen, water vapor, and little carbon dioxide, with sulphur dioxide, and hydrogen sulphide, some other gases and other elements, continue to spew out of volcanic vents, and other surface areas.

As energy is continuously transmuting, and everything is flux, no existence maintains the same characteristics for any moment of time. Protons, deuterium and tritium everywhere within interplanetary space, within the activating range of radiated heat and pressure/force from the sun (disintegrated within the shuttling pathways), are extracted as replenishment material for the plasma, and other forms of energy expelled. Every existence within our solar system engenders, exists, endures, alters and terminates, resulting from the expulsion, and extraction of material/energy from, and to the sun. As hydrogen or a proton is continuously required as replenishment material for the sun, every proton, element, or molecule constituting our atmosphere, oceans, the exposed surface, and organisms, endure for duration.

No atmosphere is fixed, and our atmosphere is continuously altering its characteristics, as it is interacted, disintegrated, integrated again, and depleted with radiated heat and pressure/force from the sun,, explaining our changing weather/climate, seasons, tides, currents, winds, breezes and other phenomena.

Our continuously generating, and disintegrating, enduring atmosphere

Our multi-layered atmosphere, each rotating at different directions, generating continuous altering temperatures, pressures and densities regulates the intense radiated heat, and pressure/force from the sun.

The interactions and disintegrations at/of our upper atmospheric nuclei from/by the radiated heat, and pressure/force from the sun, generate our visible light.

The continuous interactions, disintegrations and integrations of nuclei/elements, molecules and compounds of/as the atmosphere, engenders our atmospheric temperatures, pressures and densities.

Certain degrees of temperatures and pressure intensities of/as our atmosphere interact, and disintegrate the water molecules in/at our rivers, lakes and oceans. Earth's present continuously altering atmosphere is regenerated, sustained by the disintegration of the water molecule (the energy transmuting cycle) into hydrogen and oxygen. Oxygen and hydrogen produced from the disintegration of the water molecule, rises up into/as the atmosphere.

Hydrogen

As protons are light, and their disintegration into particles is appropriate replenishment material for the sun, hydrogen is extracted at super speeds, most of it disintegrating into plasma at interactions at our upper atmosphere, and also en route to the sun, within the shuttling energy transmuting pathways. As hydrogen is continuously produced, some protons en route-depending on the pressures, temperatures and densities of the atmosphere-combine with nitrogen forming ammonia, with carbon forming methane, and with oxygen forming water vapor.

Oxygen

Oxygen from the disintegration of the water molecule rises into/as the atmosphere. With certain degrees of temperatures and pressure intensities of/at/as the atmosphere, oxygen disintegrates into nitrogen and deuterium. With certain atmospheric degrees of temperatures and pressures, oxygen also breaks down into carbon and helium.

As oxygen nuclei continue to rise up with disintegrations, ozone is continuously produced. Parts of breaking down oxygen conjoin with carbon producing carbon dioxide. Oxygen also combines with the departing hydrogen forming water vapor. Therefore, as oxygen continuously breaks down, and/or some of it combines with other elements/molecules, there is more of nitrogen in our atmosphere than oxygen.

Nitrogen

Nitrogen is continuously produced in the atmosphere with the breaking down of oxygen. Oxygen and hydrogen (departs quickly) are produced in large amounts, with the continuous disintegration of water molecules.

With certain degrees of temperatures and pressures intensities, nitrogen also breaks down into carbon, and other nuclei. Nitrogen combines with the departing hydrogen forming ammonia, but not in large amounts, as hydrogen super speeds away. There is less nitrogen combinations (less disintegrations and integrations) in the atmosphere, as nitrogen stays nearer to the surface. This also explains why nitrogen is abundant in the atmosphere.

As oxygen disintegrates into lighter and lighter elements, it rises higher and higher into the atmosphere. As nitrogen is produced first, it lies low in the atmosphere. Nitrogen lying low is protected from much further disintegration, as radiated heat and pressure/force from the sun is regulated by the multi layered atmosphere. Lying low in the atmosphere, it does not undergo much reactions or condensation, and is therefore, not chemically active. Without appropriate temperatures and pressures, nitrogen does not dissolve in the oceans.

Carbon and carbon dioxide

Under certain degrees of temperatures and pressure intensities, carbon formed from disintegrating oxygen, conjoins with oxygen nuclei forming carbon dioxide, and with the departing hydrogen forms methane. Carbon also further breaks down into helium and deuterium, as carbon with/of 3 helium nuclei easily disintegrates with radiated heat and pressure/force from the sun. For this reason, as it combines with oxygen to form carbon dioxide, it is not found in the atmosphere in great amounts. Furthermore, it is produced from disintegrating oxygen, which first breaks down into nitrogen, explaining the lesser amounts of carbon dioxide. Carbon dioxide being more stable and heavier, is found only in small amounts in earth's atmosphere. As creatures inhale oxygen, and exhale carbon dioxide, it adds to in/as the atmosphere.

Carbon, nitrogen and oxygen have equal amounts of proton plasma (protons) and neutron plasma, and therefore, with departing hydrogen they easily combine with each other, and other elements to form molecules and compounds. Life is possible only with carbon, hydrogen, nitrogen, and oxygen combinations. If carbon engendered, and endured, as a gas, it would have been in the atmosphere, and cells would not engender in the oceans-there would have been no life.

Over the years, carbon in the surface is pushed down deeper into the earth by tectonic forces. As it heats up, it melts, and combines with surface embedded/trapped oxygen forming carbon dioxide, which rises up back to the surface, and then into/as the atmosphere. It enters the atmosphere violently through volcanic eruptions, or more gradually in seeps, vents and carbon dioxide-rich hot springs.

Other lighter elements like lithium, beryllium and boron (constituted of deuterium and tritium nuclei) do not constitute the atmosphere as gases, as they easily disintegrate into deuterium and protons, which are easily and quickly extracted. Tritium also easily disintegrates, explaining why these elements are not found in/as the atmosphere, as other gases.

Other gases

As hydrogen is swept away at super speeds it does not easily combine with nitrogen, oxygen and carbon nuclei, explaining why there are small amounts of methane, ammonia and water vapor.

Ozone

At certain degrees of temperatures and pressures, as oxygen rises up from the disintegrating water molecule, it disintegrates into nitrogen and carbon. With altering temperatures and pressures at different atmospheric layers, and at each disintegrating step, oxygen rises up further (getting lighter). The remaining disintegrating oxygen nuclei form ozone, explaining why ozone is found higher in the atmosphere, and nitrogen occupies the lower atmosphere. As with all atmospheric elements, molecules and compounds, ozone too is continuously disintegrated and generated.

Our atmosphere, or air, or nuclei continue to disintegrate, and integrate with the prevailing temperatures and pressures/force. It is consequent to the oceans that our atmosphere is continuously engendered, and endures, sustained with rising oxygen, and hydrogen. Oxygen disintegrates into other lighter elements, and protons depart at super speeds to the sun.

Important note:

Revelation 12 and 13 details the engendering and endurance of living cells, and life. The continuous outflow of hydrogen, and the continuous breaking down of oxygen into nitrogen, carbon, other nuclei and hydrogen engendered living cells. Creatures in the sea, on land, and in the air, evolved respiratory mechanisms breaking down atmospheric oxygen into carbon, and other nuclei, and protons.

Regenerating the continuously depleting atmosphere and the oceans

Our present atmosphere is constituted of nitrogen, ammonia, methane, oxygen, ozone, water vapor and carbon dioxide. Nitrogen is about 79 percent, and oxygen about 20 percent of the atmosphere. There are also large amounts of hydrogen, as the water molecule in/of the oceans, are continuously disintegrated, and hydrogen, and oxygen, rise up into the atmosphere. Hydrogen cannot be measured in our atmosphere, as it is continuously departing at super speeds. Most of the protons are disintegrated speeding through the atmosphere, and immediately leaving our upper atmosphere.

Composition of the atmosphere varies little with altitude except ozone, water vapor and carbon dioxide. Water vapor is almost totally absent above 8 km. Unlike water vapor, most ozone is produced at high altitude and therefore, its concentration increases with altitude. The relative composition varies from place to place on the surface of the earth. Due to the presence of water vapor, tiny liquid droplets such as fog, or tiny solid particles such as ice crystals, smoke, sea salt crystals, dust, and volcanic emissions are suspended in the air.

Hydrogen and Oxygen

Oxygen is in abundant supply in/as the atmosphere, as water molecules in the rivers, lakes, and oceans are broken down for the extraction of the hydrogen/proton. In other words, the present atmosphere endures as the oceans endure, as the water molecule is continuously disintegrated. Hydrogen is extracted, departing at super speeds, leaving oxygen in its free state. Oxygen rises up, and disintegrates, forming most of the constituents of our atmosphere.

At certain atmospheric temperatures, and pressures, atmospheric oxygen again combines with the departing hydrogen forming water vapor. With certain temperatures and pressures water vapor condenses into clouds, and there is precipitation, falling down into/as rivers, lakes and the oceans. The atmosphere and the oceans are intimately linked, and endure, as the cycle continues.

Although precipitation replenishes the lakes and oceans, hydrogen is continuously extracted, and therefore, both the atmosphere, and the oceans, gradually deplete. To sustain this depletion, earth's interiors continuously belches out gases, and other material through seeps, vents, openings and volcanoes, explaining why there are so many active volcanoes. At times, with abnormal energy transmutations, material from the interior of the earth violently erupts from dormant volcanoes, or increases the amount from presently active ones, balancing the increase in depletion.

As earth altered its original atmosphere, and was in the right proximity to the sun, we experience our present type of atmosphere. No existence is fixed. Earth’s atmosphere is continuously interacted, disintegrated, and integrated. Every compound disintegrates into molecules, and every molecule disintegrates into nuclei, and every element also breaks down into nuclei, and nuclei disintegrate into protons, and neutron plasma. Neutron plasma amalgamates into/as space, and protons further disintegrate into particles, extracted to the sun as replenishment material.

The continuously altering temperatures, pressures and densities, disintegrates oxygen into other elements forming molecules and compounds, and concurrently also disintegrates the water molecule in rivers, lakes and oceans regenerating the atmospheric disintegrations. The continuously altering atmospheric temperature, pressure and densities, also facilitates and abets other disintegration, integration, and other processes, which manifest our weather, climate, seasons, tides, currents, sea breezes/wind, waves and other phenomena.

Why are inner planets' atmospheres and interiors of different constituents than earth?

As the planet exploded, and the inner planet's interiors were exposed to radiated heat and pressure/force from the sun, all the inner planets took the same transforming pathways as earth.

With different proximities to the sun, experiencing varying degrees of temperatures, and force/pressure intensities, and different interactions, disintegrations and integrations, Mercury, Venus, and Mars' interiors, produced a different mix of elements, molecules and compounds.

The new atmospheres of Mercury, Venus and Mars, may have produced precipitation, falling down on their gaseous interiors, as was the case with Earth. However, experiencing different forms of energy, their elements combined, and/or disintegrated differently, generating different constituents in their atmosphere, and their interiors. As a result, their oceans may have depleted much quicker than earth. It is also possible, that one, or all of these planets did not possess a molten core to sustain their continuously depleting atmosphere and oceans, over the years, explaining Mercury's and Mars' present solidification, and their nearly completely depleted atmospheres. However, as our atmosphere and oceans deplete, our molten core will gradually deplete, and Earth will follow the same fate as Mercury, Mars and our moon.

Earth's new atmosphere brought down continuous precipitation, and with the formation of the mantle and crust, which contained the precipitation into/as oceans, earth's interiors was completely covered with water. This subdued the violent maelstrom of the interior. Over the years, the disintegrating water molecules of the oceans sustained the depleting atmosphere. Moreover, occasional precipitation, and cloud cover, maintained the gradual evaporation of the oceans. At other inner planets there was a different scenario, depending on the intensities of radiated heat and pressure/force, the types of atmosphere, and other conditions explained above.

It is incorrect to speculate that it is 'because' Venus has high temperatures, carbon dioxide remains in its atmosphere. It is also incorrect to speculate that Mars with cooler temperatures has carbon dioxide in its remaining thin atmosphere. There is no causation. These are their present states.

All the inner planets may have engendered heavy elements as earth. All the inner planets may have altered their atmosphere like earth to produce water vapor, clouds, and precipitation engendering their oceans. However, their proximity to the sun with different degrees of temperatures, and pressure intensities, produced a different mix of elements. As precipitation generated their surface, there was a different mix of elements embedded both in their surface, and the new atmosphere that was forming.

At Venus, radiated heat and pressure/force may have quickly disintegrated its water molecule in/at its oceans. However, the oxygen rising into/as the air would disintegrate into nitrogen (as it is with earth), but with higher heat and pressure/force, nitrogen would disintegrate into carbon. Moreover, at closer proximity to the sun, with higher temperatures and pressure/force than at Earth, hydrogen departing at faster speeds would not easily combine with oxygen producing water vapor, to regenerate the depleting oceans. Oxygen continued to disintegrate into carbon, and carbon continued to combine with oxygen producing carbon dioxide.

As water vapor did not form, or if it formed did not condense into clouds (with high temperatures), there was no precipitation, and the oceans-if formed-soon depleted. Without an ocean, and appropriate elements in the atmosphere to be interacted, and disintegrated, Venus does not have an active energy transmuting cycle with the sun, explaining its very slow rotation. With carbon dioxide in the atmosphere, the fiery core continues to heat Venus, and concurrently as Venus also receives radiated heat and pressure/force from the sun-as it exists in close proximity-its present temperatures is at around 500 degrees centigrade. As its atmosphere is not disintegrated easily, it is 100 times denser than earth's atmosphere.

More material was extracted from Mercury's atmosphere and oceans (if it generated any), as it is in close proximity to the radiated heat and pressure/force from the sun, explaining its present solidification, and presently depleted atmosphere.

Mars is further to the sun than earth, yet it has completely solidified, and has very little atmosphere. As explained earlier, Mars experienced different degrees of radiated heat and pressure/force intensities, and accordingly, elements, molecules and compounds generated, were of a completely different mix than the other planets. Mars' oceans may have also depleted over time. Unlike the gradual extraction from earth's oceans, and interiors, which are regulated by its multi layered atmosphere, and it oceans, extraction from Mar's interiors were extracted with greater force over a shorter duration, not in the distant past, explaining its deep chasms, and massive volcanoes.

How did water vapor form to constitute earth's atmosphere?

How did earth's oceans form?

As the planet exploded, the original atmosphere of deuterium, tritium and helium was instantaneously depleted, exposing earth's interiors to solar radiated heat and pressure/force.

Nuclei disintegrated into protons, and the 'release' of neutron plasma generated space, and heat/force, and concurrently more nuclei could rise from the interiors to be disintegrated. As disintegration continued, temperatures and pressure/force intensities increased, and deuterium and protons integrated forming more deuterium, tritium and helium. At every incremental step, increases of temperatures and pressure/force, deuterium, tritium and helium, integrated to form heavier elements up to iron. At the integration of iron, temperatures and pressures/force were at the extremes.

As integration of heavier elements continued in the interiors, deuterium, tritium and helium continued to be engendered. With solar radiation and pressure/force, concurrently these nuclei also disintegrated externally, forming a new atmosphere.

At certain temperatures and pressures, atmospheric nuclei integrated to form oxygen. Interacted oxygen disintegrates, but some of it combines with the departing hydrogen forming water vapor. With the drop in temperatures and pressures, water vapor engenders, and condenses into clouds, bringing about precipitation. Rain began falling on the fiery interior, generating yet more water vapor, and more cloud formation.

Continuous precipitation on the fiery gaseous interiors transformed some of the iron to disintegrate explosively, forming heavier elements. Continuing precipitation solidified, and liquefied the elements into the inner and outer core, the mantle and the crust, which contained the precipitation as/into rivers, lakes and oceans. The molten core was now contained by the outer core, the mantle, crust, and the oceans. Precipitation eventually decreased, as concurrently degassing of/at the interiors decreased, and as the mantle and crust was completely covered with water.

The depleting atmosphere could now be replenished by the disintegrating water molecules of the oceans. The atmosphere and the oceans were now intimately linked (presently both exist/endure under these conditions), subduing the volcanic eruptions, and controlling the rapid depletion of the core's magma. Presently, earth's interior does not have to belch out much of its magma, explaining why Earth still holds onto its magma, and is presently larger in size than Mars, which is further away from the sun.

With the continuing energy propagating/extracting cycle, the hydrological cycle formed, and the atmosphere and the oceans now sustain each other. At present, the water molecule is continuously disintegrated into hydrogen and oxygen. Hydrogen leaves the ocean, and the atmosphere. Earth continuously looses hydrogen from the atmosphere, oceans and everywhere. Oxygen disintegrates into other constituents of/as the atmosphere (regenerating it), and again combines with the departing hydrogen to form water vapor, and precipitation, replenishes the continuing depleting oceans.

Hydrogen and oxygen disintegrate in the lakes and the oceans, to again recombine in the atmosphere. This changing of localization of hydrogen and oxygen as/in the ocean, and in the atmosphere, besides other events/conditions, is what produces our climates, weather (dryness, coldness, wet, dry, humid, clear or dull skies etc), tides, currents, wind, breeze, temperatures, pressures and humidity.

There may have been water at other inner planets. However, the sudden, and massive disintegration, and the production of the appropriate elements from earth's interior, to adjust/rectify the quick depleting atmosphere over a tumultuous period, altered the atmosphere that engendered our oceans.

Other inner planets and moons atmospheres may have engendered water vapor, clouds and precipitation, but without any, or all of earth's atmospheric and interior conditions, the water contained may have boiled off early altering their atmospheres. As the other inner planet's oceans evaporated faster, there was greater and faster extraction of material from Mercury, Mars, the asteroids, and some inner moons of planets, explaining why Mercury and Mars have shrunk faster than earth in size, and have solidified.

On earth, sea temperatures never reach more than 30.5 degrees C, and therefore, water vapor that is formed in the atmosphere does not disintegrate easily, but with the right temperatures and pressures and densities condenses to form clouds, and precipitation. At other planets this would not have been the case, explaining why earth still possesses its oceans. In reality, we ask what limits sea temperatures. There is no causation, as everything is cyclical, and engenders/exists with another existence of different characteristics. Sea temperatures manifest and alter, as atmospheric temperatures manifest and alter. Some areas of the sea are warmer than the others generating hurricanes, or abets storms, as the energy transmuting cycle is more active at sea bed areas with easier extraction material.

Why do the Jovian Planets still hold on to their original atmospheres?

The explosion of the planet disrupted the original energy transmuting cycles of the inner planets, and not the Jovian planets. Radiated heat and pressure/force from the sun, propelled the gaseous fragments from the exploded planet, toward the outskirts of the solar system. Jupiter, Saturn, Uranus and Neptune absorbed these gases, becoming larger, and some fragments were captured transforming into moons and gas rings. With these processes, they did not loose their atmosphere, and if they did, it was soon reconstituted again. This was not the case with the inner planets, as after the explosion, their exposed interiors partly disintegrated, generating new atmospheres.

As there is still plenty of replenishment material from the inner planets, the Jovian planets do not have active energy cycles with the sun, explaining their size, their faster rotation, and their original type of atmosphere.

Our solar system would have continued to exist with all the planets having energy transmuting cycles with the sun. The sun would continue to expel plasma, and protons would interact with the nuclei of atmospheres, and over the years planets in close proximity would gradually deplete, and subsequently terminate one by one. Heavier elements than helium, molecules, or compounds, or matter, would never have formed, if a planet in our earlier solar system did not disintegrate. This would explain that matter can engender, and exist only at places with violent upheavals, and these types of events, experienced by our inner planets. In reality, if that planet did not explode, matter would not have engendered, and we would not have existed. **Heavier elements than helium, life, and humanity, each are accidents.** There had to be matter and water first, facilitating life to evolve, and life has to undergo hundreds of evolutionary steps, facilitating humanity to engender.

How does a planet hang on to its atmosphere?

It is assumed, that a planet's ability to retain an atmosphere is determined by its gravitational field, and a gravitational field in turn depends on a planet's mass. In the case of our planet, earth's mass is great enough to keep most gases (except for very light gases like hydrogen and helium) from escaping. This is incorrect as there is no such thing as gravity (revelation 10).

As energy transmutes everything is flux, and all existences are continuously changing their characteristics. Atmospheres of bodies/planet exist/endure, being continuously interacted, disintegrated and integrated, by radiated heat and pressure/force within the energy cycle between them and the sun. These processes extract lighter particles (from disintegration) as replenishment material, for the expelled forms of energy at the sun, continuing its functioning/existence. With interactions atmospheres (and in the case of earth, our oceans also) are gradually depleted.

Planets are not 'separated' from their atmospheres. As we learnt, a fiery fragment would not engender into a body/planet, if its atmosphere did not concurrently engender, sustaining the existence/endurance of each other, as parts of the same cycle.

Planets and bodies with atmospheres, continue their energy transmuting cycle with the sun. As these planets/bodies belch out their interiors to sustain their depleting atmospheres, they solidify. Over the years, most of the inner planets being in close proximity with the sun have solidified. Mercury, Mars and our moon have lost their atmosphere, and are nearly completely solid.

The characteristics of water

The water molecule is made up of oxygen, and two hydrogen nuclei (protons). Liquids and gases are both fluids. As water is a viscous fluid, how exactly it flows is a function of its viscosity. Water will always take time to move any distance due to its own viscosity, or the interaction of its viscosity with its surrounding. No matter how fast you pour out a bucket of water, it will always take some amount of time to empty the bucket, always.

An element is a solid, as its nuclei configurations are tightly bound. This means that there is less space surrounding the nuclei configurations than liquids and gases. Therefore, space propagating as pressures/force or temperatures (revelation 9) cannot easily loosen this bond.

An element is a liquid, as its nuclei configurations are less tightly bound than in/as a solid. This means that there is some space surrounding a liquid's nuclei configuration. Therefore, space propagating, as pressure/force, or temperatures, motions the nuclei, explaining why it is a liquid.

An element is a gas, as it nuclei configurations are even less tightly bound (more surrounding space than a liquid) than in/as a liquid, so temperatures and pressures/force easily expands it, explaining why it is a gas.

As water is a liquid, heat (propagating within/as space) moves it around a lot faster. Heat and/or force are forms of transmuting energy, and heat constitutes pressure/force. In a liquid the nuclei configurations are more spread out (more space around) and heat/force expands these configurations generating the faster motion. When water reaches 100 degrees C with heating, it turns into a gas called water vapor, and this happens, as the nuclei configurations expand. When they expand, nuclei are more susceptible to the energy transmuting cycle extraction processes, and getting lighter, rise up, explaining why water as a liquid, transforms with heat into a gas we call water vapor, and why it also rises up.

With continued boiling or heat, all the water will turn into water vapor, and as nuclei disintegrate, it will turn into plasma, and amalgamate into space/energy. As light materials are extracted they rise, eventually 'escaping' into/as the air. In reality, they amalgamate into/as space/energy, as they are extracted to the sun.

Evaporation happens, as hydrogen, and other light disintegrating nuclei/particles, are being extracted from interactions, and disintegrations (within the energy transmuting cycles) from water in the rivers, lakes, oceans, surface materials, and also from organisms. Depending on the extraction and expulsion processes of the energy transmuting cycle, our many layered atmospheres continuously alter its pressures, densities and temperatures. Appropriate temperatures and/or pressures incite evaporation.

Evaporation and Transpiration

Evaporation is the process where nuclei/molecules in a liquid or solid state (the substance sublimes) being interacted, and disintegrated, become susceptible within/to the energy cycle and therefore, easily extracted, and or transmuted into a gaseous state (expanding nuclei). Evaporation increases with higher temperatures and pressures/force. This is evidence, as the altering atmosphere incites processes, as varying degrees/intensities of temperatures and pressures, resulting in all the phenomena observed and experienced. In other words, higher temperatures and pressure/force expand nuclei, and as the nuclei configurations of elements, molecules and compounds expand, they transmute into a gaseous state, and rise up, or evaporate. Increases of 10 degrees of temperature will approximately double the rate of evaporation. The humidity of surrounding air will also influence evaporation. Drier (warmer) air will increase evaporation than humid moist air. The presence of wind (pressures) will also increase evaporation. Clearer skies (easier penetration of atmospheric heat) will also hasten evaporation.

Evaporation and transpiration are two such processes, where disintegrating water transforms (transmutes) into a gaseous state. This cannot happen when the humidity of the atmosphere is less than the evaporating surface-100% relative humidity. Therefore, with clear skies, and with more intense temperatures, more evaporation takes place, as there is more extraction.

Wind also increases the potential for evaporation and transpiration. On still days, water evaporating to the air remains close to its source, increasing the local humidity. As the moisture content of the air increases, evaporation will diminish. A steady flow of air, or wind, will remove the newly formed vapor. High temperatures, and other factors, are necessary for evaporation. Evaporation is faster at higher temperatures, and in liquids with lower surface tension. We can also force evaporation with heat.

Air does not hold water vapor, but the dry gases in the air, coexist with water vapor being one of the gases. It is temperature (mainly), and pressure (secondarily), which affect the rate of water vaporization. If the air already has a high concentration of the substance evaporating, the given substances will evaporate more slowly. Also, if air is already saturated with other substances, there will be a lower capacity for the substance evaporating. If the substance is hotter, evaporation is faster.

At 100 degrees C (the boiling point of water) all water will rapidly be turned to vapor, as the energy (heat propagating through space) supplied to the water, breaks apart all the molecular bonds in water. Temperatures and pressures influence the conversion of water into vapor

Wet any part of your body at any time, and that part feels cool. This happens, as within the continuous energy cycle, extraction processes disintegrating lighter particles are being pulled away-this is evaporation. Blow on the wet part, and there is more cooling, explaining pressures and temperatures playing a key role in evaporation.

Dogs hang their tongue out during hot days, or when they are tired. Evaporation breaks down the water molecules on their tongue extracting hydrogen, with some parts of oxygen inhaled. This serves two purposes; cooling the body, as well as facilitating breathing in oxygen concurrently.

Transpiration is the process of water loss from plants. The extraction processes of the energy cycle extracts, or transport nutrients and water from the soil into the roots, and carries them to the various parts of the plant. As water extracted from the roots reach the leaves, the water molecule is disintegrated, and hydrogen extracted, explaining why most leaves have evolved flat surfaces. These processes also keep tissues from becoming overheated, and at the same time there is growth. With these processes, trees grow tall, otherwise they would be scattered over the ground. With evaporation, and also other conditions (revelation 12/13) creatures keep growing, or there is life, as cells continuously replicate.

Water Vapor

If you leave water in a container anywhere, for some days, at the right temperatures, and/or pressures/force, all the water will eventually disappear. Over time, the energy transmutation cycle gradually disintegrates (separates nuclei) the water molecule, extracting all the hydrogen nuclei, as oxygen rises up into the air.

As water is boiling (a faster process, as boiling heat is more intense than radiated heat at earth's surface), the water molecules keep on disintegrating faster, until all the water rises up leaving nothing.

Everywhere on land, or the oceans, water is continuously subjected to the atmosphere-ocean/land surface cycle. Our oceans, lakes and rivers, are continuously being agitated by atmospheric temperatures, and pressure/winds, facilitating processes within the energy cycle, to disintegrate the water molecule, extracting hydrogen.

As water vapor transmutes into its/as gaseous state, pressure/heat is manifested, and this is called humidity. At a given temperature, an increase of water vapor (or concurrently generated pressure) in/as the air, corresponds to an increase in the humidity of the air. Water vapor is supplied to the atmosphere by evaporation of water from oceans, lakes, rivers, wet lands, from vegetation (transpiration) and other organisms. Rising water vapor obstructs the incoming energy transmutations, and therefore, incoming energy in the form of temperatures and pressure will be more intense in a drier atmosphere, explaining an increase in evaporation.

The air we breathe has water vapor in it, besides other gases. Water vapor is harmless, as eventually at the right temperatures and pressures, it will conjoin with oxygen in the atmosphere, turning back into water again. The presence of water vapor in the atmosphere is one of the things that make earth livable for organisms.

Water vapor in the atmosphere is continuously being subjected to the atmosphere-ocean/land surface cycle. It is the biggest contributor to the natural greenhouse effect, and varies most in the atmosphere. Cold air can hold little water, and so the atmosphere over the Polar Regions contains very little water vapor. In contrast, air over the tropics is very humid, and the atmosphere can contain up to 4% water vapor. This explains why an increase of normal water vapor in the atmosphere is important in climate change.

Temperatures, pressures and densities, result from the many different processes in the interactions/integrations/disintegrations of the multi-layered, opposite rotating atmosphere. Water vapor and some other gases, play a role in temperature changes. Moisture controls, and influences the heat and pressure of the atmosphere.

If earth did not have an active energy transmuting cycle with the sun, there would have been more hydrogen in the atmosphere, and a different mix of gases. With a different mix of gases, our atmospheric temperatures, pressures and densities would also be different, and life would not emerge. However, with a different mix of gases, water vapor would not engender in the atmosphere, and we would not have had our oceans. Without water, cells and life would not emerge. Without water vapor, the gaseous heavy elements in the interior would not disintegrate forming matter. Matter, life, and us, engendered with water vapor in our new atmosphere.

The hydrological cycle

Earth has a unique multi layered atmosphere each with different characteristics, sustained by the oceans, and its molten core. The energy transmuting cycle, between the sun and earth's atmosphere/oceans continues, as forms of energy propagating from the sun continuously interact, disintegrate and again integrate our atmosphere, oceans, and everything in the biosphere depleting it gradually. Changing atmospheres (altering temperatures, pressures and densities, generating various phenomena) and disintegrating water molecules, of/in the oceans, generate the hydrological cycle. This hydrological cycle continuously regenerates the continuously disintegrating/depleting atmosphere, and the oceans. The atmosphere endures, as protons and nuclei extracted from the oceans (disintegrating water molecules) reinforces/reconstructs the depleting, disintegrating atmospheric nuclei.

At certain temperatures and pressure/force intensities (depending on the extraction and expulsion of energy within the cycle), the departing hydrogen and oxygen recombine to form water vapor in the atmosphere. Again, depending on temperatures and pressure or force, water vapor forms clouds, generating precipitation, replenishing part of the evaporating water of/from the oceans. The hydrological cycle therefore, endures the depleting atmosphere and the oceans. The gradual depletion of the atmosphere and the oceans, are sustained by the belching out of magma from the molten core, which is also gradually depleting.

Water as a solid, liquid and gas plays a great part in the hydrological cycle. The hydrological cycle takes place in the hydrosphere-the region containing all the water in the atmosphere- and on the surface of the earth. The cycle is the movement of water in its solid, liquid or gaseous state through this hydrosphere. This process is divided into condensation, infiltration, runoff, evaporation and precipitation. Water changes states with temperature and pressure, and this explains why extraction/evaporation takes place. Water changes into water vapor (a gas), explaining why it rises, and may condense to form clouds, which precipitates replenishing the water that is extracted/evaporated.

The extraction/expulsion processes within the atmosphere-ocean/land surface cycle, alter atmospheric temperatures, pressures and densities. Without atmospheric temperature, pressure, and density alterations there would be no evaporation, or extraction. Molecules would not disintegrate, and water vapor would not form. There would then be no clouds, and no precipitation. Planet earth would not endure today.

The cycle

Depending on the extraction/expulsion requirements of the energy cycle, temperatures rise over water in lakes and oceans. As surface water heats up, molecules are disintegrated, and nuclei are extracted (changing into gas or water vapor), a process we call evaporation. Warm air current rising from the earth's surface, lifts this water vapor up, where it circulates within/into/as the atmosphere. Evaporation of warm surface water increases the amount of moisture in the colder drier air, flowing immediately above the water surface.

As the temperatures are different higher up, the gas or water vapor (as moist air is lifted it cools) condenses, clinging onto other lighter particles in the air. When an air package rises and cools, the water vapor condenses back to liquid water around particulates like dust, called condensation nuclei. Initially, these condensation droplets are much smaller than raindrops, and are not heavy enough to fall as precipitation. These tiny water droplets create clouds. As the droplets continue to circulate within the clouds they collide, and form larger droplets. As the air gets more and more moist, the droplets forming clouds grow larger, and eventually get so big that the swirling atmospheric winds can no longer hold them up, and the droplets fall from sky, as precipitation in the form of rain, snow, sleet or hail, depending on atmospheric conditions, such as temperature.

Though the amount of precipitation varied widely over the surface of the earth, evaporation and precipitation are globally balanced. If evaporation increases, precipitation also increases. If temperatures increase, there would be a world wide increase in evaporation from the oceans, leading to higher overall precipitation.

Evaporation results in condensation and precipitation. This supports, and maintains the biosphere, and/or the ecological existences, explaining why evolution of life took place.

As the oceans are the largest reservoir of liquid water, that is where most evaporation occurs, and since the oceans cover around 70% of the earth's surface, most precipitation fall right back into the ocean, continuing the cycle. Some precipitation falls on land where it is taken up by soil and plants, and some runs off into streams and lakes, some percolating into the ground water reservoir, some falls on glaciers, and accumulates as glacial ice.

The hydrological cycle replenishes the constituents of the atmosphere and the oceans, but they are gradually depleting within the energy transmuting cycle. The molten core belches out material sustaining them, but this will not continue forever.

Clouds and water vapor

Clouds are accumulations of water droplets, or solid ice crystals, which float in the earth's troposphere (the lowest part of earth's atmosphere) moving with the wind.

Clouds form, as water vapor (through evaporation) condenses (turns into liquid water or ice) with other particles in the air. This occurs, as cool air can hold less water vapor than warm air, and excess water condenses into either liquid or ice. Warm air laden with moisture rises into the atmosphere, and when it gets to a certain height cools. At cooler temperatures, it can no longer hold all its moisture in the form of water vapor, and so the extra moisture changes in small drops of water, or bits of ice, and this forms clouds.

No two clouds are alike, and there are different shapes. We have different types of clouds, as cloud formation takes place at different heights/layers of atmosphere, and clouds will be composed of different nuclei/elements/molecules depending on pressures and temperatures.

Low thick clouds confront temperatures from the upper atmosphere, cooling the surface of the earth. High thin clouds work just the opposite, allowing high temperatures, but at the same time some of the material extracted is prevented from escaping outward, thereby warming the surface of the earth.

Whether a given cloud will heat, or cool the surface, depends on several factors, height, size and the constituents of the cloud. Clouds are a source of rain, snow, thunder, lightning, hail and other weather phenomena.

Cooler waters and warmer land surface, and vice versa

Evaporation is why the land is cold when the oceans are warm, and the oceans are cold when the land is warm. There is greater evaporation without clouds, as the energy cycle easily disintegrates the water molecule extracting hydrogen. The water is cooled with this disintegration/evaporation, and the land gets warmer.

The atmosphere-land-ocean cycle is continuous. Warm air over the oceans disintegrates the molecules, and the subsequent evaporation cools the previously warm area, and so there is cooling, and heating subsequently.

There is less evaporation with clouds, as the clouds condense to eventually precipitate. With clouds temperatures are therefore, cooler, and less evaporation takes place. After precipitation there are clear skies, and the cycle goes over again.

How are bubbles formed? Why do bubbles rise?

Water condenses as it cools, and/but starts expanding when it reaches 4 degrees centigrade. Different liquids have molecular structures with varying nuclei configurations, explaining their condensation and expansion attributes.

Bubbles engender, when/as water or liquids are agitated. Molecules of water, or a liquid, are constituted of integrated nuclei configurations. As the molecules disintegrate releasing the nuclei bonds, or the nuclei configurations, hydrogen is set free. Extracted and rising hydrogen (protons confined by space) through a liquid, or water, form bubbles. Bubbles also engender with pressures/force, and different forms of temperatures/radiation on liquids.

As water is being heated, bubbles start appearing at the bottom (that is where the heat is, and molecules are disintegrating), and bubbles increase with more heat, rising to the top, all trying to reach the surface at once. The bubbles push out to the surface, and jump out of the surface-water is boiling.

Heat, pressures, and cold temperatures, in some cases, integrate and disintegrate nuclei configurations. In this case, bubbles contain, or are per se the separated hydrogen. As they are now susceptible (in their new state) to the energy cycle, they rise up, and at the surface they disintegrate, as space/bubbles amalgamate within/into atmospheric space-or bubbles break.

Larger bubbles rise faster than smaller bubbles, as they contain more amounts of lighter material being extracted. If you inflate two balloons, one more than the other, and attach a valve between the two, you would expect the gas/air to flow from the smaller to the larger or vice versa, but eventually both will be the same size. Things within the biosphere are constituted of different materials, but are structured to conform to the extraction and expulsion processes of the energy transmuting cycle.

Earth's many layered atmosphere of different characteristics

The surface of the earth is surrounded by a continuous engendering, or enduring or terminating, or integrating/disintegrating blanket of air or atmosphere. This atmosphere reaches over 560 kilometers from the surface of the earth. This envelope of gas surrounding earth changes from the ground up. Most of the atmosphere (density) is within 16 km from the surface of the earth. The atmosphere is less dense from the ground up, as going further up, the atmosphere spreads out to a greater area than at the surface.

The troposphere starting at the surface of the earth extends 0 to 18 kilometers at the equator, and 8Km at the poles, and is most dense. The temperature in this region generally decreases, as altitude increases. The pressure at the top of the troposphere is only ten percent than that at sea level.

Between the troposphere and the stratosphere is the tropopause, which is a transition layer. The stratosphere starting from over the troposphere extends further than 50 kilometers, and is dry and less dense. The temperature here increases with altitude rising gradually from -3 degrees to a maximum of 32 degrees F or 0 degrees C. The earth's ozone layer is located here, extending between 20 and 60 km. Only the highest clouds like cirrus form here. Most of the water vapor lies here. The stratopause separates the stratosphere from the next layer.

The mesosphere starts over the stratosphere extending to 85 kilometers, and here the temperature again falls as low as –93 degrees Celsius as you increase in altitude. The air here is in a very excited state. The mesopause separates the mesosphere from the thermosphere.

The thermosphere starts above the mesosphere, and extends 560 kilometers high. The thermosphere is a thermal classification of the atmosphere, as there are large temperature variations. The atmosphere is tenuous with very few molecules. This area is empty. However, the few nuclei and molecules here are easily ionized by interactions of the radiated heat, and protons propagated from the sun. This is the area where the space shuttle, the Hubble telescope, and many other earth observing satellites reside. Here long distance radio communication is made possible with less density. Here also, the temperatures go up-as high as 1727 degrees Celsius as you increase in altitude, as the atmosphere starts decreasing in consistency and the orbital rotation is faster.

All these drops and increases in temperatures, are within a span of 600 kilometers from the surface of the earth. The tropopause and the troposphere are known as lower atmosphere. The regions of the stratosphere and the mesosphere are called the middle atmosphere, and the thermosphere region is called the higher atmosphere.

Atmospheric regions are also named in other ways.

The thermosphere includes the exosphere and part of the ionosphere. The ionosphere begins at about 75 Kilometers above the surface of the earth, and continues to about 650 kilometers. This is the region where protons, radiated heat and pressure/force from the sun interact and disintegrate nuclei. The ionosphere is the region where departing hydrogen or protons clash with incoming protons, and these interactions generate our visible light. Beyond the thermosphere the exosphere starts (the outermost layer) and continues merging with interplanetary space where particles extracted, and protons expelled are the prime components.

Why are there increases and decreases in temperatures, pressures and densities at different atmospheric layers?

Every sphere including the earth rotates in one direction. As the earth is a sphere and spins, it bulges at the equator. As it spins, the atmosphere at the equator is less dense since it spins faster. Approaching the poles the atmosphere is thicker and/or denser. As a result, approaching the poles the atmosphere rotates at a much slower speed than the atmosphere at the equator. This is not all.

Earth has a many layered atmosphere. Each atmosphere rotates in an opposite direction to each adjacent layer of atmosphere. For example, the troposphere (the adjacent layer of atmosphere next to the surface of earth) rotates in an opposite direction to the surface of the earth (or earth). The troposphere may not actually be rotating, but as the surface of the earth is solid, and rotates, it would appear that the troposphere rotates in the opposite direction. But there is more.

The uppermost layered atmosphere rotates faster than the lowest atmosphere, and the solid earth. So a spherical body like the earth will have:-

-the uppermost atmosphere over the equator rotating much faster than the lower atmospheres, and the surface of the earth

-the uppermost atmosphere over the equator rotating much faster than the uppermost atmosphere over the latitudes and the poles

-the next adjacent lower atmosphere over the equator rotates slower than the upper atmosphere over the equator, and faster than the same layered atmosphere over the latitudes and the poles.

Each layered atmosphere will rotate differently at the equator, latitudes and the poles. Each layered atmosphere will rotate opposite to its adjacent atmosphere, and so on.

These different layered atmospheres rotating at different direction, different speeds at the equator, latitudes and poles, and different speeds at different altitudes, is what make our many layered atmosphere continuously altering to different temperatures, pressure/force and densities. Importantly, the incoming and outgoing energy transmutations alter the temperatures, pressures and densities at different regions of/as the atmosphere, explaining why we experience changing breeze/winds, different types of tides, currents, climate/weather, and many other types of natural phenomena.

Temperatures and pressures manifest, as elements, molecules, or compounds are disintegrated into nuclei, and nuclei into protons and plasma. Temperatures and pressures also manifest, as plasma and protons integrate into nuclei, and nuclei into/as elements, or molecules, or compounds. The earth's molten core throws out magma through volcanic activity. This also alters atmospheric density, pressures and temperatures.

Summarizing temperatures, pressures and densities

Temperature and the atmospheric layers

The temperature of the earth's atmosphere varies with altitude; the relationship between temperature and altitude varies between the different atmospheric layers, and at different places from the equator to the poles.

Troposphere-0-17 km-temperatures decreasing with height

Stratosphere-17-50-km temperatures increasing with height

Mesosphere-50-85-km temperatures decreasing with height

Thermosphere-85 to 640 km and above-temperatures increasing with height

The average temperature of the atmosphere at the surface of earth is 14 degrees C. The temperatures on earth range from -88 degrees C to 58 degrees C-the coldest recorded at Vostok, Antarctica 1983 and the hottest at Libya, Africa 1922.

The troposphere is a dense area with large amounts of nuclei disintegrating, integrating and expanding, generating continuously altering temperatures and pressures, explaining why all the weather occurs here. Here clouds form, precipitation falls, wind blows, and humidity varies from place to place, as the atmospheric temperatures and pressures interact with the surface/oceans below. Carbon dioxide and water vapor forming here do not rise up maintaining the troposphere's temperatures. Water vapor keeps our surface temperatures from -40 to 55 degrees C. Without water vapor we would experience -100 degrees C, as with Mars. Temperatures decrease with altitude at a rate of about 6.5 degrees

C per kilometer, and are about -100 degrees C at the uppermost boundary of the troposphere.

At the stratosphere the continuous engendering and disintegration of ozone generate temperatures, which increase in altitude. However, pressures must also increase, but this is not the case here explained following.

The mesosphere is less dense with less interactions, or disintegrations, and therefore, temperatures decrease with altitude.

The thermosphere is an area where incoming radiated heat and pressure/force and protons, interact with the outgoing extracted protons. These interactions not only generate our visible light, but also increase temperatures. Temperature increases are also enhanced with the incoming radiated heat and pressure/force from the sun interacting the mesosphere nuclei.

It is evidence here that all existences exist with another existence of different characteristics. An adjacent atmospheric layer would have different temperatures and pressures. If this was not the case, there would be no adjacent different type layer.

As the earth (surface) rotates, the gaseous adjacent troposphere rotates the other way. The stratosphere rotates in the opposite direction than the troposphere, and so on. This explains, why our continuously altering atmospheric temperatures, pressures, and densities, manifests the multifarious phenomena observed and experienced.

As mentioned earlier, spin and space are the keys to every existence engendering, enduring and terminating. As earth rotates, and also orbits the sun, our altering atmospheric temperatures, pressures and densities manifest our seasons, climate/weather, tides, currents, breeze/winds and other phenomena.

Atmospheric pressure

With a multi layered atmosphere with different characteristics as explained earlier, atmospheric pressure varies with location, altitude and time. Atmospheric pressure drops by 50 % at an altitude of about 5 km (equivalently about 50% of the total atmospheric mass is within the lowest 5 km). The average atmospheric pressure at sea level is about 101.3 kilo Pascals (about 14.7 pounds per sq inch). As altitude increases the air pressure decreases. At an altitude of 10,000 feet, the air pressure is 10 pounds per square inch (and there is less oxygen to breathe).

As incoming energy 'pushes' down toward the surface of the earth, density and temperatures increase, and there is more pressure. However, there is proportionate outgoing energy with the extraction of material/energy, as other forms of energy transmute, and therefore, pressure decreases with altitude. This also explains the density decrease at higher altitudes. Lower pressure translates into high speed winds.

Density

The density of air at sea level is about 1.2 kilograms per cubic meter. This density decreases at higher altitudes at approximately the same rate that pressure decreases (but not quite as fast). 90% of air is located in the troposphere and stratosphere. 70% of the atmosphere exists within 11 km of the planetary surface.

The earth's atmosphere changes in pressure and density with altitude. Its density is greater near the earth's surface, and thins out at higher altitudes, as space spreads out. The atmosphere extends above the surface of the earth to a distance of many thousands of kilometers, but 95 percent of its total mass is found within 20 kilometers of the surface.

With changing temperatures, pressures and densities, the energy propagating/transmuting cycle continues altering the atmosphere characteristics, generating movement like winds and breezes, processing water vapor, clouds and precipitation, and the many other cycles, continuing or enduring all existences.

Why does air pressure and density decrease with altitude?

The outermost atmospheric layers are tenuous at the equator and latitudes, as they rotate faster. In other words, the outer most layers spin faster, as they are tenuous. The upper atmosphere spreads out into interplanetary space, explaining why density decreases with altitude.

The pressure/force together with radiated heat from the sun interact the upper atmosphere. Within the same shuttling to-and-fro, energy transmuting pathways, there is an equal outgoing pressure/force explaining the decrease of air pressure with altitude.

We would assume that temperatures, pressures and densities should increase toward the surface of the earth, as temperatures and pressure/force propagate from the sun. This is not so, as energy is everything everywhere, and only energy transmutes. Nothing is a one-way phenomenon. There is equal energy expended and replenished, or received and extracted, and an equal incoming and outgoing pressures/force. Many forms of energy propagate from the sun, and concurrently other forms of energy propagate to the sun, as energy transmutes shuttling to-and-fro within the same transmuting pathways.

Why and how does rotation generate high densities and low temperatures at the poles, and high temperatures and low densities at the equator?

A rotating body eventually becomes a sphere. Rotation is only in one direction. With a spherical body, there is less dense atmosphere at the equator, as spreading out into space, it is tenuous. As a result, a spherical body is denser at the poles, than the atmosphere over the equator.

A sphere can either be a gaseous, or a solid body. Solid bodies have no atmosphere, as within the energy transmutation cycle it has depleted. Gaseous bodies, or partially solid bodies like earth, possess an atmosphere. As the sphere spins faster at its equator the atmosphere is tenuous, and the rotation here is faster than as we approach the poles (where it is zero). The nuclei/molecules in a tenuous atmosphere have more room to motion, and with rotation/momentum, increase the temperatures and pressures/force. Therefore, temperatures and pressures/force are higher with faster rotation, and a tenuous atmosphere at the equator, and lower with more density, and slower rotation at the poles.

Climate and Weather

Weather is the short-term cycle, and a season is the long-term cycle, and one on a much longer time scale we recognize the cycle of climate. Weather happens every day, and is made up of many parts. One part is temperature-how hot or cold the air is. Another is precipitation-water that falls from the sky as a liquid or solid, or a mixture of the two as rain, snow, sleet and hail. Another is wind-gentle breeze or a strong tornado.

Weather is the state of the atmosphere. The environment can be wet or dry, hot or cold, clear or cloudy, rain, lightning or thunder, calm or storm. Climate is the average weather conditions at a place over a period of years. The climate of a region rarely changes significantly in a single lifetime. Changes in climate have a more profound effect.

There are three major climate zones, tropical, temperate and polar, and they each differ in temperatures and precipitation. Tropical climates are usually warm and wet most of the year whereas polar climates are usually cold and dry.

What generates climate and weather?

Our weather, or climates, or seasons are not experienced as a direct result of the heat energy and pressures from the sun. Our multi layered continuously altering atmosphere, regulates/controls the radiated heat and pressures from the sun, continuously engendering different temperatures, pressures and densities. This results in weather/climate and seasons. There is no causation.

If radiated heat from the sun propagated directly to the troposphere without our upper atmosphere, we would not experience temperature, density and pressure alterations and therefore, no weather or climate changes. The high degrees of radiated heat and pressure intensities, as experienced at our upper atmosphere would terminate all forms of life instantly.

Our multi layered atmospheres rotating at different directions, different speeds at the equator, latitude and poles, and different speeds at different altitudes, manifest different temperatures, pressure/force and densities.

As the energy transmuting cycles continuously interacts, disintegrates, integrates, and depletes these layers of atmosphere, temperatures, pressure/force and densities alter. Moreover, our rivers, lakes, oceans and surface are continuously being interacted, disintegrated, depleted and integrating again. Earth has over 500 active volcanoes, and their eruptions also effect climate and weather changes.

Normal radiated heat and pressures interactions generate our usual climate/weather, seasons, currents, tides and wind/breeze at the same time. Abnormal interactions, or an increase in extraction processes, generate changes in normal climates, weather, seasons, and currents, and so on. Furthermore, there are catastrophes such as increasing volcanic activity, resurrection of dormant volcanoes, plate tectonics, earthquakes, tsunamis, hurricanes/tornadoes etc.

The atmosphere is continuously integrating, disintegrating, and depleting. As different areas of atmosphere around the solid earth alter with temperatures, densities and pressures, different regions experience varying phenomena at the same time. Clouds form, there is precipitation, humidity, temperature changes, dryness, wetness, wind, breezes, tides, currents and so on.

Areas transmuting/transforming to other states, forms, structures or processes, will alter yet other areas into other states, forms, structures or processes of energy.

The alteration and disruption in the normal climates/weather, temperatures and other phenomena we sometimes experience can also result from energy transmuting cycle disruptions within our solar system.

If atmospheres enveloping the planet's interiors would not continuously alter, our sun/star would not engender, and if it did, it would be short lived.

Earth's atmosphere has formed a unique energy transmuting cycle with the oceans, and the earth's surface, and both cannot exist without the existence of each other. It is the consequence of cycles within oceans, within the interior of the earth, within the biosphere, within the atmosphere, between the atmosphere and the oceans/land, and between the sun and our atmosphere, that we have seasons, climates/weather, temperatures changes, changes in pressures, producing phenomena such as precipitation, tides, currents, waves, breeze and so on.

Why is weather near the equator much warmer than the weather near the poles?

The first incorrect explanation

At the equator the sun is straight overhead, but at the poles it is lower in the sky. We are informed, that the sun rays travel further through the atmosphere, and lose more of their energy and also, that the sun rays are spread over a larger area, so the energy per square meter is lower. It is stated, that 'because' of the positions of the earth and the sun, the sun shines directly on the areas near the equator which receive more solar radiation than the poles that receive less.

The second incorrect explanation

It is stated, that the sun's rays hit the equator at a direct angle between the north and south latitudes, and radiation reaching the atmosphere here, is at its most intense. In other cases, the rays arrive at an angle to the surface, and are less intense. The closer a place is to the poles, the smaller the angle, and therefore, the less intense the radiation.

The correct explanation

Why should the sun rays shine directly only at equator areas? The sun is 150 million kilometers away from the earth, and is much larger than the earth, and heat energy does not converge at the equator and therefore, this explanation is not correct.

Energy in all forms of radiation/temperatures/light and pressures propagate (spread out) from the sun, and does not 'travel' in rays or beams.

Each atmospheric layer rotates differently than its adjacent layer, with the troposphere rotating in the opposite direction than the partially solid earth. As the earth is a sphere, and obviously bulges out more at the equator (faster rotating) the atmosphere is less dense than the atmosphere approaching the poles, in relation to the surface of the earth. Heat is dispersed more easily through a tenuous atmosphere, making regions around the equator warmer. Moreover, as the atmosphere rotates much faster at the equator, there is more heat generated. At the poles the atmosphere is dense, and with lesser interactions there is less heat generated. With a denser atmosphere heat is not easily transferred to the surface, and therefore the surface and its adjacent atmospheric layer stay cool.

Why are there more thunderstorms in summer?

Trade winds north of the equator blow from the northeast. South of the equator they blow from the southeast. This results from multi layered atmosphere rotation in different directions, which is faster than the solid earth, rotating slower.

The trade winds of the two hemispheres meet near the equator resulting in the air to rise. As the rising air cools, clouds and rain develop, and cloudy and rainy weather near the equator create tropical conditions. Moisture and rapidly rising warm air generates thunderstorms, explaining why thunderstorms would occur more often in spring and summer, particularly in humid areas such as the south-eastern United States. The high humidity in combination with warm summer temperatures, generate large amounts of warm moist air rising into the atmosphere where it easily forms a thunderstorm.

There are approximately 1800 thunderstorms occurring at any given time or about 16 million thunderstorms each year. Most of the thunderstorms last about 30 minutes, and are typically about 24 km in diameter.

Wind and Breezes

Wind or breeze is air, or nuclei in momentum. Winds develop as a result of pressure, densities or temperatures differences between two locations within earth's troposphere.

Sea breezes for example, develop due to differential heating of land and sea at the coast during warm sunny days. Winds also blow out from high pressure regions into low pressure regions. Winds do not blow in a straight line, but follow spiraling paths, as large movements of air are deflected by the Coriolis force (explained following), as a result of the rotation of earth's different atmospheres. At a global scale, the temperature and pressure differences across the latitudes of the earth, generates the global wind belts.

As different layers of the atmosphere rotate in different directions, there are winds and currents in the oceans moving temperatures and/or pressures, water vapor and winds in different directions. Winds and currents move from the lower latitudes to the poles, and vice versa.

The layers of atmosphere rotate at different speeds, and at different directions than the surface earth which rotates slower. The atmosphere over the equator and latitudes is less dense, and faster rotating than the atmosphere approaching the poles, which is denser and slower rotating. This results in interactions engendering surface winds, which blow west to east in the temperate regions of the northern and southern hemisphere, creating the 'prevailing westerlies'. In equatorial regions, the pattern is reversed with east to west easterlies. Stagnant air masses such as doldrums form at the equator between these major air currents.

Existences within the same cycle have different characteristics to support/maintain each other, explaining why winds and currents have different directions, and are also different both in the northern and southern hemisphere. Westerlies blow from the southwest on the Northern hemisphere, and from the northwest in the southern hemisphere. Westerlies propel storms from west to east across middle latitudes. Both westerlies and trade winds blow away from the 30 degrees latitude belt. Over large areas centered at 30 degrees latitude, surface winds are light.

Air slowly descends to replace the air that blows away. Any moisture the air contains evaporates in the intense heat. The tropical deserts such as the Sahara of Africa, and the Sonora of Mexico, exist under these conditions.

Unlike the Pacific and Atlantic oceans, the Indian oceans is completely enclosed on the northern side-a configuration that gives rise to drastic seasonal change in the winds and currents.

Land and Sea Breezes

A steady wind-the sea breeze-is a result of the uneven heating during the daytime between the land, and the adjacent water.

During the day, there is a greater interaction/disintegration and integration of the many layered atmospheres, exposed to direct radiated heat and pressures. Altering temperatures and pressures from these interactions disintegrate the water molecules of the oceans, generating an increase in temperatures over the surface of the oceans. Air rises over the heated land surface, and concurrently air over the surface of the oceans move to supplant this displacement of air. This generates the wind/breeze over the oceans toward or onto the land (a sea breeze). The wind or breeze generated, concurrently facilitates, and abets extraction or evaporation.

As the temperature difference between the land and water increases throughout the afternoon, the winds/breeze increases in strength, reaching a maximum in the middle to later afternoon. At nights without atmospheric interactions (the atmosphere is not exposed to direct heat and pressures) there are less alteration in temperatures and pressures, and the land cools down. As there is no warm air rising over the land surface, there is no wind/breeze from the oceans. However, as the water molecule disintegration and extraction is continuous, there is interactions/extraction only over the oceans at night, and the air over land moves over the oceans (a land breeze).

Altering atmospheric nuclei movements over land and the oceans continuously alters temperatures and or pressures. As the wind/breeze move over the oceans toward the land, and from the land toward the oceans, the altering pressures increase or decrease. The altering low and high pressure completes the circulation of air/wind/breeze. In other words, the heating, or cooling, or the changes in temperatures and/or pressures, result in the circulation of air/wind/breeze. Low pressures translate to high speed winds.

Whenever large land and water bodies are adjacent to one another sea breezes develop, generating thunderstorms, for example Florida's abundant summertime rainfall is a result of sea breezes.

Daytime warming and night time cooling

Why is the warmest part of the day between 3 and 6 pm?

During the day parts of the atmosphere is directly exposed to radiated heat and pressure/force, and heat is generated. Heat that reaches the surface surpasses that which is emitted from the ground, with whatever extraction that is taking place. This energy imbalance last from the morning, and throughout the afternoon, and the ground warms throughout the day.

As the adjacent layered atmosphere gets warmer, air near/over the ground increases its temperatures. Between 3 and 6 pm, the ground is warmer than earlier in the day. Eventually, the density of the warm air decreases, and convection develops. Under calm winds conditions, this convection does not mix with the warm air very much. However, if there is a moderate, or strong wind, the air will be mixed upward throughout a significant portion of the atmosphere.

Night time cooling

At nights, our overhead atmosphere is not directly exposed to solar heat, and with less interactions, less heat is generated. During a clear calm evening (one without clouds or wind); the ground begins to cool throughout the night. As the ground becomes cooler than the air, the air above transfers some heat toward the ground, but the ground immediately overcomes this heat. This process continues throughout the night, as the surface continues to cool. However, this process proceeds slowly, and by morning the coldest air is usually found near the ground.

Why are evenings and early mornings cooler than during the day?

At sunsets and sunrises, our upper atmosphere is less exposed to the radiated heat and pressure/force from the sun, explaining cooler evenings, nights and early mornings.

Tides

Are tides 'caused' by the moon, or the sun, by their gravitational pull?

Tides are periodic rises and falls of large bodies of water. As with other existences/phenomena, tides are an integral part of the energy propagation/transmuting cycle, between the sun and the atmosphere, and also the cycle between the atmosphere, and the land/oceans, and are not 'caused' by the moon. There is no cause or effect, or causation.

There is no such thing as gravity (revelation 10) and therefore, no gravitational 'pull' of the sun, or the moon to 'create' the tides on earth. The processes of/at our continuously interacting and disintegrating, and integrating atmosphere, generates our different forms of weather/climate, and other phenomena. The disintegrating, and extracting/integrating, processes of the ocean molecules alter atmospheric temperatures, pressures, and densities, generating winds, currents and waves, which move large areas of water known as tides.

The solid earth rotates slower than its adjacent atmosphere. There are different atmospheric interactions, as parts of the many layered atmosphere directly face the sun. During the day, oceans facing the sun have more of material extracted than at night. If the earth did not rotate, there would be no tides, but other types of existences/phenomena to endure energy transmuting cycles. The rotation of the earth (oceans/land), the speed of rotation of the many layered atmosphere, their many different types of interactions, and the myriad processes taking place within the atmosphere and the oceans, enable existences to exist/endure as parts of their cycles. Therefore, tides and many other phenomena occur, facilitating existences to exist, and endure.

High tides and low tides correspondingly, balance the expulsion/extraction of material within the two energy transmuting cycles. Large bodies of water are disintegrated to extract hydrogen and oxygen, and these processes generate tides. However, as bodies of water rise, other adjacent areas subside, explaining high and low tides. High and low tides occur on a 26 hour period, evidence that during night, and day, there are different intensities of interactions/extractions, as half of the atmosphere is either directly facing exposed to, or away from the sun's propagating radiated heat and pressure/force.

We are incorrectly informed, that tides are 'caused' by water being pulled in the direction of the moon. If the relationship were this direct, only one high tide (at a given place) each day would take place in relation to the earth's rotation. Water would be pulled toward the moon, when those areas were at a position facing the moon. Yet, two cycles of high and low tides occur each day, roughly thirteen hours apart.

High tides occur both where the water is closest to the moon, and on the opposite side of the earth, where it is farthest from the moon. If the moon directly influences tides, how can we explain the tide on the opposite side of the earth not facing the moon? There is no such thing as gravity, and even if we assume there is a gravitational pull from the moon, logically we know that earth's gravitational pull is far stronger than the moon for it to influence the tide on the opposite side. Obviously, at points in between where the water has rushed away from the shores, there are low tides. This happens, as everything is part of a cycle (existences must be of different characteristics) as adjacent to a high tide, there will be a low tide, and so on. During a twenty-six hour period, each point on the surface of earth's oceans moves through a series of tides: high, low, high and low again.

Why are there two tides a day and not just one?

If tides were generated by the gravitational effects of the moon, and or the sun, there would be only one tide per day, as explained earlier. The explanation that the moon exerts a gravitational force on earth is completely preposterous, as this would be causation. We are informed that the strength of the gravitational force decreases, with increasing distance, and 'because' the surface of the ocean is closer to the moon than the sea floor, the surface water is attracted more strongly to the moon. On the other side of the earth, with the sea floor being closer to the moon, the force is more strongly toward the surface water, and therefore there is another tide. This is incorrect, as there is no such thing as gravity or gravitational force. If this is the case, there should be two tides of equal 'heights' in a day. In reality, many parts of the earth do not experience two tides of equal 'heights' in a day. Explaining this we are informed, that the moon's orbit is at an angle to the earth's equator, so one tidal bulge may be in the northern hemisphere while the other is in the southern hemisphere. It is also postulated, that except Antarctica, the shape of the earth's continents prevent the tidal bulges from simply following the moon. These are all incorrect explanations.

Each ocean basin has its own individual pattern for tidal flow. In the South Atlantic oceans, the tides travel from south to north taking about 12 hours to go from the tip of Africa to the equator. In the North Atlantic, the tides travel in a counter-clockwise direction going around once in about 12 hours. In some parts of Mexico, there is only one high tide, and one low tide a day. The Pacific oceans have two high tides, and two low tides each day, but are of unequal height.

All the above phenomena occur, as there is more land mass in the northern hemisphere than in the southern hemisphere, which has more oceans. There is more extraction from the southern hemisphere oceans explaining the timings, directions, levels and other different characteristics of tides. If the moon 'causes' the tides, how do we explain these occurrences? Extreme snowfall can disrupt tidal currents, so how do we explain the effects of the moon in these circumstances?

Every existence exists with another existence of different characteristics. Therefore, there cannot be one tide but two, one high and one low at the same time in adjacent areas.

Why is it that during full moon and new moon phases we have extremely high and extremely low tides-spring and neap tides, if the moon does not 'cause' tides?

The moon as all bodies/planets engendered, as they enveloped themselves with an atmosphere. Obviously, over the years, it could not support its atmosphere-now nearly depleted-and has solidified. At first, with its atmosphere, it played a significant part between the earth, moon and the sun energy transmuting cycle. Now it is an impotent existence within this cycle, explaining why it does not rotate on its imaginary axis (no atmosphere for extraction and rotation) showing the same face toward earth. It therefore, plays an insignificant part in some of earth's phenomena.

At full moon, and new moon phases, the moon is either in front of the earth, or behind it, in a straight line to the sun. At these positions, the normal energy propagating/transmuting cycle processes interacting with our atmosphere, and oceans alter, as the moon gets directly in front, or behind the earth (becoming an additional body directly within the energy cycle pathway, and closer to the earth), generating abnormal higher and lower tides.

There are no entities. The moon is a solid body within/as interplanetary space/energy, and as at close proximity to earth plays a minor role within the propagation/extraction cycle, in the same way an object traversing light casts a shadow.

As the normal extraction process increase, the normal expulsion processes decrease, and as the normal expulsion processes increase, and the normal extraction processes decrease, there are extremely high and low tides at full and new moon phases.

What are tidal currents?

The atmosphere turns much faster than the solid/surface earth, and the solid/surface 'earth' turns much faster than water can flow around/with it. By the time the water really gets moving in response to the tidal forces, these forces have changed in direction and strength. With these processes, as forces act in one direction subsequently forces act in an opposite direction.

The surface of the earth (land and oceans) rotates slower than the adjacent atmosphere, and also in the opposite directions. As a result, as the atmosphere at different areas is of different densities and pressures, and rotate at different speeds at the equator, and approaching the poles, tides and/or tidal currents have different directions, currents and strengths.

Moreover, the expulsion and extraction processes altering temperatures, pressures and densities, continuously alters wind/breeze, currents and other related phenomena. Furthermore, these forces are also different depending on the depth of water.

Ocean Waves

Wind, breeze, or air movement, tides, and currents (water movement), are all the result of the Sun-Earth energy transmuting cycle, and the atmosphere-ocean-surface cycle.

Energy as radiated heat and pressure/force with protons from the sun interact our atmosphere. These interactions are propagated in/as different forms of energy, delving down to the water molecules in/of the oceans. Concurrently, disintegrating nuclei are extracted from the oceans, and these processes generate waves, and other atmospheric phenomena.

As the wind blows over water it tries to draft the surface of the water with it. The surface cannot move as fast as air, and so it rises. When it rises, there is pressure again-within the atmosphere-ocean cycle-pushing the falling water momentum below the surface. The extraction processes in a reaction pulls this swell back up again. The tug of water between extraction, and the incoming pressure constitutes wave motion.

Waves of water do not move horizontally, they only move up and down. A wave does not represent a flow of water, as energy is in momentum within the medium, which is water. Everything everywhere is energy, and waves are only energy transmutation as momentum. In other words, water is not moving from here to there, but force/wind as a form of energy is in motion.

Revelation 9 details, that it is not light, temperatures, pressure/force that motions, but as everything is energy, these forms of energy manifest as space waves, or transmutes into/as these forms of energy. Therefore, water waves, as it transmutes into/as wind or force.

Winds or air movement generates waves on the surface of the water/oceans and lakes. As everything is energy, friction between air and water molecules generates waves. Waves on the ocean can also be caused by tides, interactions among waves, submarine earthquakes or volcanic activity, and atmospheric disturbances (alterations in pressures/densities).

Wave size depends on wind speed, wind direction, and the distance of water over which the wind blows. The longer the distance the wind travels over water, or the harder it blows, the higher the waves. Stronger winds (like storm surges) cause larger waves. Earthquakes or plate shifting under the oceans generate huge waves or tsunamis. Tsunamis (sometimes called tidal waves) are different from surface waves, and usually result by underwater earthquakes, volcanic eruptions or landslides.

Lakes do not produce strong waves, as they are normally enclosed as/in smaller areas. Oceans are spread out great distances, and large areas are affected by the constantly changing direction of air or atmospheres.

The Coriolis force/effect

As the earth (mantle/crust and interiors) rotates in one direction, the adjacent atmosphere rotates in the opposite direction. Observed from above, the atmosphere over the surface rotates clockwise, and the mantle/crust/oceans (solid earth) rotate anti-clockwise. Obviously, facilitating rotation, both the surface, and the adjacent layer of the 'solid' earth cannot rotate in the same direction.

As the atmosphere is continuously integrated and disintegrated, there are continuous temperature, density and pressure changes throughout each atmospheric layer at different areas. Pressure differences tend to push winds in straight paths, yet winds follow curved paths across the earth. This is called the Coriolis force/effect.

As air begins flowing from high to low pressure, the earth rotating under it (and the opposite directional rotation of the troposphere), makes the wind follow a curved path. In the northern hemisphere, the wind turns to the right of its direction of motion, and in the southern hemisphere it turns to the left. The Coriolis force is zero at the equator. The northern hemisphere, and the southern hemisphere have different characteristics, but are parts of the same cycle-the rotating earth.

The earth is a partially solid body, and its atmosphere spreading outward gets tenuous. It is the atmosphere, and the oceans, which are more affected by the energy radiation/transmutation cycle than the solid earth.

Winds are just bodies of particles, and are subject to the Coriolis effects basically driven by the changing atmospheric conditions.

Changing atmospheric conditions result from interactions within the energy radiation cycles, and this has the effect of producing three major convention zones in each hemisphere.

As a result of the Coriolis effects, water in a bathtub flows out in a counter clock direction, when being emptied in the northern hemisphere, and clockwise in the southern hemisphere. If you travel in a spacecraft observing the earth from below the 'south' pole, and as the 'solid' earth rotates from west to east, you will notice the earth rotating clockwise, and if you observe the earth from above the 'north' pole, you will notice that the earth spins counter clock-wise. North, south, east, west are human constructs for human minds.

The water flows out of the bathtub differently in the hemispheres (counter clockwise in the northern hemisphere, and clockwise in the south), as water (in the bathtub) is part of the solid earth, and not part of the atmosphere, flowing out turning in the direction the earth turns. When we look at the bathtub we are seeing it from above (our vision is within the atmosphere).

The immediate atmosphere over the surface of the earth rotates the other way, as everything is part of a cycle, and have different characteristics, explaining why flowing out water also rotates. This also explains why water being filled in a bucket, or water leaving a wash basin, swirls. When the water is in the bath tub it does not rotate, as the surface of the earth is not felt rotating, as we are standing on it.

Creatures have evolved mechanisms (to avoid attention to certain phenomena, as this would interfere with other more important requirements of foraging and survival) engendered with/from the environments (land/ocean and the atmosphere), and do not feel the earth their feet are on, rotating one way, and the air (their bodies within/as) rotating the other way.

If the Coriolis force turns winds to the right in the Northern Hemisphere, why do winds go counter clockwise around large systems, such as hurricanes, north of the equator?

It must clearly be understood that there is no causation. All natural phenomena (as humans understand them) result from the normal expulsion, and/or extraction energy requirements of the energy transmuting cycle and therefore, all disasters, or catastrophes (with our understanding), result from abnormal energy transmutations.

Hurricanes begin as tropical storms over the warm moist waters of the Atlantic and Pacific oceans. As moisture evaporates, it rises until enormous amounts of heated moist air are twisted high in the atmosphere. The winds begin to circle counter clockwise (as that layer of atmosphere spins this way) north of the equator, or clockwise south of the equator.

The centre which is peaceful is called the eye. Around the eye of the hurricane winds speed between 100 and 300 miles per hour. As long as the hurricane remains over warm waters, it continues to pull moisture from the water, and grows in size and force. This explains why high temperatures and pressures, facilitate easier extraction. If it crosses into land, or cooler waters, it looses its source of power (warm waters and high pressure), and the wind gradually slows until it is no longer a hurricane (force/pressure).

Large systems such as hurricanes are not normal phenomena. In fact, these unusual phenomena result from temperature, density and pressure variations within atmospheric layers, generated by temperatures, pressures and density variations somewhere else-abnormal processes, or increases in extraction material. Unusual phenomena could occur either in the atmosphere around the globe, within the sun-earth's atmosphere transmutation cycle, or outside of it, or within the atmosphere-ocean cycle, the latter most likely.

With the northern and southern hemisphere forming a cycle, the equator is a neutral zone, and the most unlikely place for density, temperature and pressure variations. Over the equator, the atmosphere is tenuous, and thus faster rotating, and vice versa. This type of atmosphere does not alter temperatures, densities and pressures easily. This explains why tropical cyclones such as hurricanes, and typhoons, will not form on the equator, even though the other factors there such as warm ocean water would make them likely.

As explained earlier, if you look at the earth from above the North Pole, and as the earth rotates west to east you will observe the earth rotating counter clockwise. In the bathtub example, we learnt that water leaving the bathtub will rotate the way the earth rotates (depending on which hemisphere you are in). Water in the bathtub is part of the solid earth (not the atmosphere) that is turning counter clockwise, as we look from above the North Pole We assume that the Coriolis force turns winds to the right, as we are looking at the atmosphere from below (part of the surface of the earth), in the northern hemisphere. If we are looking at earth from the North Pole, the surface of the earth would be spinning counter clockwise, and the atmosphere adjacent to it would be spinning clockwise. If we are looking at the earth from above the equator, we would observe the atmosphere to be spinning clockwise and therefore, the Coriolis force should be turning winds to the right, or the opposite of the atmosphere rotating direction.

Why do hurricanes begin at the tropical areas of the Atlantic and Pacific Oceans?

Tropical cyclones/storms exceeding 74 mph in the Atlantic are called hurricanes and typhoons in the Pacific. Eastern and Central Pacific storms are called hurricanes, and storms west to the international dateline are called typhoons.

A hurricane/typhoon has winds blowing in a large spiral around a relatively calm centre called the eye, about 30-50 kilometers wide. The entire wind movement around this eye may extend to about 500 kilometers.

As we learnt, the earth is tilted, as when the energy transmuting cycles of our solar system initially formed (or perhaps during the course of their existence) a greater amount of energy was/is extracted from our southern hemisphere, with less land mass, and more oceans. During the summer months, as the southern oceans are not directly exposed to the sun, the energy cycles alters its extraction processes to areas around the tropical areas of the Atlantic and Pacific oceans. This explains why the earth continues orbiting the sun tilted.

As earth is tilted during its orbit around the sun in summer and early fall, more of the northern hemisphere is exposed to the sun. These changes in the atmospheric extraction areas, generate high humidity, and light winds (the atmosphere over the tropics, and the equator is tenuous), facilitating extraction from these areas. As extraction continues, the gradual disruption of the normal current, wind directions, speed, temperatures and pressures engenders thunderstorms. Thunderstorms engender, as the surface of water gets warm with extraction, so higher temperatures concurrently increase the intensities, or strength of a hurricane. As extraction continues the hurricane moves across the oceans, and its strength increases, as more and more evaporation, or extraction takes place.

As the earth gets closer to the sun, global warming will result in stronger, and more frequent hurricanes. The strength, size, and wind speeds of hurricanes will increase in the future, unless replenishment material for the sun is sourced elsewhere. If this happens, earth will alter its tilt, and there will be major changes to our seasons, and climates/weather.

Why do rising sea surface temperatures account for hurricanes only in North America, and how do we explain no such effects in other parts of the world with those temperatures?

Why are hurricanes becoming more intense, but not frequent?

Hurricanes occur in the tropical regions of North America, as the Gulf of Mexico, and surrounding areas, contain easily disintegrating material for extraction, as described with the Bermuda triangle mystery.

In the past, hurricanes beginning in mid-Atlantic approached North America, but broke up taking a turn way up north. Now with the exposed sea beds they continue toward, and around the Gulf of Mexico. As earth's interior continues to belch out magma, the mantle and crust sink further, tearing apart the sea beds, generating more exposure of the sea materials. This extra and faster extraction results in more intense hurricanes, especially around the tropical regions of North America.

Hurricanes are not more frequent than in the past, as there are energy transmuting cycles between many bodies, and the sun, within the solar system, where replenishment material can be sourced. Presently one of Jupiter's and Saturn's moons is engaged in violent volcanic activity.

Starting 1995, the surface water in the mid-Atlantic oceans has become steadily warmer. Hurricanes breed and expand over warm water, and during the decade they have been increasingly numerous and intense. A warming world produces more of these immense lethal storms.

Hurricanes are expected to batter the southern United States, and the Caribbean, for years, as a result of changes within earth's interiors, altering the normal expulsion/extraction cycle processes, and changing atmospheric temperatures and pressures, or resulting in warmer weather.

We have been informed that hurricanes/typhoons come in cycles lasting up to about 35 years. Hurricanes and Typhoon patterns are linked and opposite. When hurricane activity is up, typhoons are down. Fewer hurricanes mean more typhoons. This all depends on the availability of extraction material around earth, and also from sources from within the solar system.

Climate change does not guarantee more tropical cyclones. The storms do require warm water. They generally do not develop unless the surface is at least 26.5 degrees C, and the hotter, the better. Even more crucial, they need warm air at about 12 kilometers above the surface. Climate change would meet those demands. But a storm grows powerful only if the air is relatively calm. Strong winds in the atmosphere blow them apart. During the El Nino era, there were not many hurricanes due to its strong winds. Warming may lead to high altitude winds, and would mean fewer hurricanes.

CLIMATE CHANGE AND GLOBAL WARMING

Elucidating earth's atmospheres

As atmospheres are depleted, planets/bodies endure, as parts of an energy transmuting cycle with the sun. If this was not the case, bodies would not have existed as planets, but as impotent bodies like some asteroids, and our moon, existing within/as the fabric of interplanetary space/energy. As our solar system has much energy transmuting cycles, as there are planets, our solar system, the sun, the planets, and bodies, continue to exist/endure.

The continuous disintegration of the ocean molecules, and the belching from earth's interiors, continuously regenerate the many layered, and differential rotating (of different characteristics) depleting atmosphere. These regenerating processes alter temperatures, pressures and densities, in/at/of different regions of the multi-layered atmosphere. In other words, the continuously engendering atmosphere exists/endures, as continuously altering temperatures, pressures and densities.

Our atmospheric temperatures, pressures and densities engender, as a result of the interactions at/of our atmosphere from solar radiated heat, and not from the sun directly, explaining why nearer the poles it is cooler than areas in the tropics, and at the equator. At the equator, and the tropics, as the earth is a sphere, the atmosphere rotates faster than areas of atmosphere, as we approach the poles. The atmosphere over the equator is tenuous than approaching the poles. The 'earth' with its molten core rotates much slowly than its atmosphere. Faster rotating atmosphere means less density, and temperatures (from interactions) can propagate more easily to the surface, explaining warmer temperatures at the equator, and the tropics. At the poles with less rotation there is more density. Here temperatures and pressures are correspondingly less, and can not easily reach the surface, explaining cooler temperatures.

Nuclei of/as our atmosphere are continuously disintegrating and integrating, and therefore, cannot 'hold' temperatures or pressures. Our atmosphere, or atmospheric nuclei, exists/endures, as continuously altering temperatures, pressures and densities. Temperatures and pressure/force from these altering nuclei (processes) are propagated within/as atmospheric space (revelation 9), explaining different regions manifesting their characteristics. Temperatures, pressures and all forms of energy, propagating within/as space, as it waves, dwindle with distance, and density, also explains different regions with altering temperatures, pressures and densities.

No such thing as greenhouse gases

We are informed that the dirtier the earth's atmosphere the less sunlight also gets through, so less heat is radiated back from the earth's surface, and the less the greenhouse gases can do their warming work. We have also been informed that the sun delivers energy in the form of sunlight to earth, with two thirds heating the surface of the earth, and one third bouncing back in 'space'. The key is how long it takes, for the energy to work its way back up through the layers of atmosphere. Too fast, and it will be cold for life, and too slow, and it will be hot for anything to survive. The

14 degrees average temperature is experienced, as the exit times are just right, and these depend on the concentrations of greenhouse gases in the atmosphere. Atmospheric temperature rise and fall accordingly, producing changes in the winds, precipitations, cloud -cover humidity and the rest. These are all incorrect explanations. There is no causation, and everything continuously alters.

We are informed, that in a greenhouse, closed glass results in heat to become trapped inside. It functions in a similar manner on the scale of an entire planet. Greenhouse gas/effect occurs when a planet's atmosphere allows heat from the sun to enter, but refuses to let it leave. It is postulated, that any gas hindering the flow of heat back into space is a greenhouse gas. As explained earlier, the atmosphere is comprised of many gases, which are continuously integrating and disintegrating, generating different temperatures, densities and pressures, so greenhouse gas effects have no meaning. Greenhouse gas and its influences are human concepts.

We cannot postulate that any particular gas is a greenhouse gas, or any particular gas like ozone, prevents radiation from delving through the atmosphere regulating heat, as every gas is comprised of nuclei. Why and what makes ozone prevent radiated heat from the sun reaching earth's surface? What about the prevailing other gases?

The density of any area of the atmosphere, or any particular gas, does play a small role dictating, or regulating temperatures and pressures, as these forms of energy propagate only through atmospheric space (revelation 9). With clearer skies, and also depending on temperatures and pressures, these increased intensities-as atmospheric space waves (revelation 9)-easily delves through the atmosphere, disintegrating more of the water molecules on the surface and oceans, explaining why there is more extraction.

Radiated heat and pressures from the sun, are regulated by the different directional multi layered rotating atmosphere, resulting in different temperatures, pressures and densities at each layer, and also at different places at the equator, and at different latitudes.

The continuous interaction, disintegration and extraction of the water molecule, and the material continuously being belched out from earth's interiors, also play a key role in the continuously altering atmospheric temperatures, pressures and densities. The rise and fall of atmospheric temperatures, pressures and densities is what manifests our weather/climate, seasons, breeze/winds, precipitation, humidity, tides, currents, and other normal and abnormal phenomena.

It is wrong to assume, that our atmosphere 'allows' or 'traps' the sun's heat like the greenhouse glass. It is incorrect to state that 'heat' is 'trapped', or 'bounces' back into space. Heat/temperatures, and pressure/force, or nuclei, are all forms of energy. It is wrong to speculate that some postulated green house gases are good at absorbing sunlight, and converting that energy into heat. It is wrong to think that these gases sometimes do not release this heat into space at nights.

Is the ozone depleting?

We are informed

-Climate changes speeds up ozone depletion because the stratosphere cools as the earth's surface warms. The cooler temperatures encourage formation of ice crystals in stratosphere clouds that provide a surface where chlorine can attack the ozone.

-Colder arctic winters also produce the stratosphere clouds, which require the spring sun to spark the ozone destruction.

-Sulphuric acid clouds from volcanoes provide a surface for the ozone-destroying chemical reaction, which could lead to 100 percent depletion of Arctic zone under certain conditions.

-A combination of cold winters and volcanic plumes, thick enough to create an Arctic ozone hole, could happen as many as five times over the next 30 years.

As explained earlier, air, or nuclei, or the atmospheric gases are engendered, alter or terminate, with rise and fall, or alterations in temperatures, pressures and densities, and vice versa. The atmosphere as temperatures, pressures and densities is continuously altering and therefore, the above explanations are incorrect.

As discussed above, why and what makes only ozone prevent radiated heat, and pressures, from delving through the atmosphere? Ozone makes up less than 1 percent of the atmosphere, and even without these 'greenhouse' gases, temperatures can increase/decrease, as everything is energy. If the ozone is being depleted by an abnormal transmutation of an energy state, process, structure or form, another form, structure, state or process of energy will supplant it. Ozone is made of up nuclei, and as ozone is a form of energy, temperatures and pressures are also forms of energy. Therefore, altering temperatures, pressures and densities can do the work of ozone-if it is being depleted. In other words, slight increases or decreases in temperatures, pressures and densities, will regulate the intensities of radiated heat and pressure/force from the sun-the same work done by an amount of ozone in the atmosphere. As the ozone hole gets larger for two months in a year, and/or as ozone is less produced, altering (increases or decreases) temperatures, pressures and densities compensate for these abnormal processes serving the same purpose, or maintain the norm.

Many gases comprising the atmosphere continuously alter their characteristics. How can we correctly measure an increase, or decrease, of these postulated greenhouse gases from a constantly changing disintegrating and integrating atmosphere? Temperatures, pressures and densities are the atmosphere per se, constituted of continuously integrating and disintegrating nuclei, and cannot be calculated or measurable.

All the sun needs is light particles to replenish its expended energy, and so it will continue to propagate forms of energy to interact, disintegrate and integrate atmospheres. If atmospheres do not have appropriate material other cycles will alter to provide it, and if not, cycles will terminate.

The ozone depletion over the years, and the larger ozone hole like all other abnormal phenomena, and disruptions, and changes we observe, and experience, is the consequence of abnormal energy transmutations, resulting from abnormal extraction/expulsion processes within our star field, or other adjacent star fields, continuously taking place as everything is flux. We have not stopped to think, that the larger ozone hole, or its depletion over the years is perhaps an alteration in the structure of the atmosphere, adjusting its composition to maintain normal temperature, pressure and densities. It is the atmosphere per se, which manifest/exhibit temperatures, pressures and densities.

No part or constituent of the atmosphere, or a particular gas, prevents/controls sunlight, or heat energy, or pressure/force, from the sun, as there are continuous interactions, disintegrations and integrations of/at the multi layered atmosphere.

It can be postulated that as Mercury and Mars have now solidified with very thin atmospheres, the sun is now extracting more protons from our atmosphere (and oceans). The increasing intensities of hurricanes, and other catastrophes, are evidence. As a result, more oxygen rises into the atmosphere, and concurrently with its disintegration, more carbon dioxide is being produced.

Altering normal degrees of temperatures and pressure intensities, in turn, has altered the processes that generate ozone. Ozone is not being depleted, but is being produced less, or carbon dioxide and/or oxygen have increased in the atmosphere. Or in other words, the alterations of either temperatures, or pressure/force, or density of/in/at the atmosphere, prevent oxygen from further disintegration into ozone. However, this additional, or extra extraction, does not take place throughout the year, explaining why the ozone hole gets larger only for some months. The hole gets larger, possibly with an increase of hydrogen being swept away.

The ozone hole is around the South Pole, with a greater ocean surface than the northern hemisphere and therefore, more extraction takes place. It can also be postulated, that extraction may be continuous throughout the year, but during some parts of the year, other areas of the earth face the sun during its orbit at different proximities, experiencing different degrees of radiated heat, and pressure intensities. Therefore, other abnormal phenomena like hurricanes occur, altering extraction processes explaining why the ozone hole gets larger only for some months. It can also be postulated that during the southern hemisphere's winter, the ozone hole does not face the sun, and the decrease of extraction of material from the southern seas, generates the larger ozone hole.

Are humans responsible for the depletion of ozone and changing the climate?

It is postulated, that there is an increase of carbon dioxide, methane, and nitrous oxide in the atmosphere. Carbon dioxide is mostly in balance since life appeared on earth, but it is assumed, that humans' other activities have increased the carbon dioxide, and triggered climate changes.

Burning fuels and other material will obviously cause smog, and other forms of pollution within the biosphere, depleting oil, natural gas and coal. With the gradual depletion of the earth's mantle, and upper core, humans are depleting the earth's reserves. It is not significant, but altering cycles will alter the atmosphere, and with the atmospheric adjustments, there are minor density, temperature, and other changes.

Whether we burn fossil fuels, or operate huge power stations, the production/increase/decrease of these gases is too little, and insignificant to alter/deplete the ozone. If the ozone is depleted, temperatures, or pressures and/or densities would alter, enduring existences at the norm.

The destruction of forests which naturally absorbs carbon dioxide further acts to increase it in the atmosphere. Carbon dioxide is also produced by the burning of wood, natural gas, coal, furnace oil and gasoline. However, carbon dioxide can be engendered/disintegrated by other atmospheric processes. With more extraction from the oceans, more oxygen rises up into/as the atmosphere, and oxygen disintegrating into nitrogen and carbon, generates more carbon dioxide.

The ozone layer depletion, or the ozone hole, is not the result of CFC, and other related material over-usage. If humans are blamed for the ozone depletion, why is there an ozone hole only over the Antarctic region where there are no industrialized countries or populace?

The dramatic thinning of sea ice in recent years-widely cited as evidence of global warming- has occurred before in the 1700's. The sea ice extent was the same then, as it is now. Ice shrinking is a part of a natural cycle, rather than the result of human so called green house emissions. The activity of humans and the burning of fossil fuels, and the utilization of CFCs, and other material are insignificant, and too little to be even considered. Removing oil or fuels is a human method, or an unnatural way of disintegrating the interiors. Global warming is not human induced.

Why is there climate change, and global warming?

Why are we experiencing more weather disasters?

During the past as Mercury and Mars had active energy transmuting cycles with the sun, replenishment material from the earth was not actively extracted. Presently, with these bodies having nearly depleted their interiors (and their atmospheres), the sun seeks more of its replenishment material from earth, and presently earth has a very active energy transmuting cycle with the sun. As a result, more of our oceans in the southern hemisphere are interacted, and disintegrated, and extracted, than ever before. This would explain why solar activity correlates with changes taking place with earth's climate and weather.

More of hydrogen and oxygen is extracted from the southern seas (more water than the northern hemisphere's land mass). As hydrogen and oxygen rise up, more carbon is produced, as oxygen disintegrates, and as carbon readily joins with oxygen more carbon dioxide is in the atmosphere. The atmosphere warms up, as a result of more carbon dioxide in the atmosphere, explaining climate change, and/or global warming.

The year 2005 experienced a climate of record extremes. It was the hottest year with the most Arctic melting. This year experienced the worst Atlantic hurricane season with the most named tropical storms (26), most hurricanes (14), most top-category hurricanes (5), and most expensive hurricane damage. The Caribbean waters were the warmest with weeks-long high temperatures causing extensive bleaching of coral reefs. It was the driest year for many decades in the Amazon where a continuing draught may surpass anything in the past century. Globally it is getting warmer at nights. These record extreme were striking, as they happened in widely separated parts of the world involving different physical forces.

Global warming

Global warming is the worldwide rise in surface and atmospheric temperatures. The upper layers as well as the deep oceans are warming up. There is rapid retreat of glaciers in Greenland, Alaska, and the Himalayas, and thinning and disappearance of sea ice in the Arctic oceans during the summer months. The permafrost in Canada, Alaska and Siberia are melting, and there is a rise in sea levels, and extreme weather changes.

How will ecosystems respond to global warming? Should we be concerned?

As energy transmutes, and everything is flux, all existences continuously alter. With so many energy transmuting cycles, as there are planets, there are abnormal events. Sometime the energy emitted varies between too strong and too subtle. It has been happening after every 11 years-the sunspot cycle. Numerous sunspots show that solar output is highest at these times, and when they are less, the value is lowest. When there are no sunspots (as all existences are parts of a cycle) there are extraordinary cold spells on earth. When there is an increase in solar activity, the ice regions melt, and it is hot enough to grow grains.

The last major change was at the end of the last ice age, where small changes in the earth's orbit around the sun, reduced the amount of radiated heat earth's atmosphere received, and earth cooled enough for vast ice sheets to move south from the Arctic.

We are presently experiencing more ice, snow, cold, heat, fire, drought, flood, rain, wind, and dust. There is talk now that humans are to blame (discussed earlier), as carbon dioxide warms the atmosphere, generating all the abnormalities.

Global warming and climate change are not the same. Climate amounts to the weather in a region averaged over a long period of time. Climate changes have profound effects on global average temperatures. Climate changes alter wind patterns, and we may experience more, less or no hurricanes/typhoons. Climate changes can also alter precipitation, and there could be droughts, or more snowstorms. Climate change may also alter ocean currents, humidity, more clouds, and trigger melting of ice, sometimes on a smaller scale, or around the globe.

2005 was the hottest year on record, edging out 1998. We assume that El Nino boosted global temperatures, and 2005 had no such help, but what we fail to note is that El Nino was the result of global warming and therefore, 1998 was less warmer than 2005. El Nino is part of the atmosphere, or global warming, and not the 'cause' of it.

Global warming could melt polar ice rising sea levels, and weather all around the world would go haywire. There will be new kind of diseases, and threatening tropical diseases. Tropical and equatorial regions will change into deserts, and people there will face drought, and famines.

As temperatures change on land and the oceans, there would be more water vapor in the air, and the ground may dry out. Even if rainfall increases, the lands would dry out, and there would be less crops growing with a risk of famine.

Significant tropical warming will not only bring drought, but also increasing severe storms and hurricanes, as water vapor from warm oceans condenses into rain. New types of hurricanes would engender becoming more intense, and wander further into temperate latitudes affecting more of the populace.

Sea levels will also rise, partly because water expands as it warms, and partly because more of the polar ice will melt. The oceans will flood with freshwater from melting ice, altering coastlines. With global warming, sea levels will rise at first, and later on ice streams in Greenland and Antarctica will move quickly, as temperatures rise, but will be partially offset by increased snowfall. This may last for some time. Even if sea levels rise by more than 25 meters this is of no significance, as increases in evaporation depends on temperatures, and ocean surface areas. Ocean surface areas will not increase, but temperatures can. If temperatures increase, there will be more evaporation, and more oxygen in the atmosphere, increasing the amount of ozone, carbon dioxide and water vapor, raising temperatures even further. Eventually, the ocean levels will fall.

But we also have to look at the bright side for some more years in the future. With more snow there will be more types of winter sports. Changes bring about new lifestyles. Atmospheric changes may also bring about increased life spans, and better health. Atmospheric temperature, pressures and density changes may also bring about myriad benefits to organisms, transforming the world into a more colorful, healthier and pleasurable place.

Abnormal climate or global warming, and all other catastrophes result from the abnormal functioning of the our energy cycle, or density variations in other energy transmuting cycles, or the instability of other solar system cycles or adjacent star fields, and not human induced processes, or energy utilization in any form or commercialization. The earth warms up, and cools down without any help from humans. Climate is changing, as the sun is growing 'stronger' or 'weaker' (depends how you want to comprehend it), as earth and the planets get nearer to the sun, and is therefore, unstable.

There are concurrent changes both at the sun, and within all member bodies of the sun's different cycles. These changes being gradual in our lifetimes, and in the past generations, there is no record how environments have changed, as humankind began existing only a few hundred generations ago to read, or write, and keep archives. However, we observe the evidence in the different species, which have evolved from few cells in the oceans to organisms like us.

Life has evolved and continues to exist, as a result of certain temperatures, pressures and densities, or types of atmospheres which were prevalent when life 'originated', and continues to exist. Ozone depletion may alter conditions terminating life on earth. However, more significant are abnormal energy transmuting variation within our cycle, or other cycles, that can wipe out life, the earth, or the entire solar system. Ecosystems may change the types of cycles and existences, but living things will terminate with any extreme alterations in atmospheric densities, pressures or temperatures.

We live in these smooth times when humankind began thinking, and archives have been kept for the past few hundred years. The constituents of the atmosphere will change, and living things may alter, evolving mechanisms to conform to these changes, but eventually the altering constituents of the atmosphere may not be suitable for our kind of life.

It is only as humans have observed, and experienced smooth times, continuing existing with the cycles that have endured, that we postulate that the depletion of the ozone, and global warming are of significance, as we are concerned with these disruptions to our lifestyles and survival. We live in smooth times assuming there is constancy. Energy continuously transmutes, and there is flux. Anything anywhere can happen at the next second, or any time in the future. If any one of the sun's transmuting cycle terminates, this assumed constancy, smoothness, and/or tranquility, will end, and our solar system will metamorphose, or transform into free energy in/as a black hole, and/or another type of star field.

EARTH'S OCEANS

Why are the oceans salty? How do the oceans remain salty? Why does the salinity of sea water not deplete with evaporation?

The main chemical constituent of seawater is common salt-a very simple chemical compound. As the elements of the interiors have been degassing, and/or disintegrating over millions of years, ocean water contains every natural occurring element known, together with numerous compounds and minerals.

The salt in the oceans have accumulated over 4 billion years having rivers flow into it, carrying dissolved salts from soil erosion. However, the oceans contain some elements in abundance that are not found in river water; chlorine, bromine and sulphur, for example. Rivers are not the only source of dissolved salts. Under the sediments on the ocean floor are different types of material.

Hydrothermal vents are places on the ocean floor where sea water has seeped into the rocks of the oceanic crust, becoming hotter and dissolving some of the minerals from the crust, flowing/mixing into/with the ocean. With hot water comes a large complement of dissolved minerals. As volcanoes erupt under water, it dissolves some of the mineral constituents.

Water evaporates from the oceans to fall again as rain, but evaporation is a process where only hydrogen and oxygen is extracted, and the salts remain in the ocean. As rainwater passes through soil, and percolates through rocks, it dissolves some of the minerals, a process called weathering. We drink this water, but we cannot taste the salts, as its concentration is too low. Eventually, this water with its small load of dissolved mineral of salts reaches a stream, flowing into lakes and the ocean.

The salt content of seawater does not change, as new minerals are forming on the sea floor at the same rate, as salt is added. The annual addition of dissolved salts by rivers is only a tiny fraction of the total salt in the ocean, which has accumulated over millions of years. Thus, the salt content of the sea is at a steady rate.

As a result of the huge volume of the oceans, hundreds of millions of years of river input were required for the salt content to build up to its present level. Some parts of the ocean are much saltier than average. The red Sea contains the saltiest seawater. There is also variation in salinity in the oceans. It all depends on the extraction processes, and climatic conditions. The Northern Indian Ocean is much less salty during the monsoon season, as rainwater moves into the oceans.

Why are some lakes salty?

Lakes salty, or not, have less intense agitated waves. In fact, waves are generated as evaporation takes place, and vice versa, as everything are parts of a cycle. In other words, it is easier for the water molecule to be disintegrated and hydrogen extracted, when there is agitation (temperature and pressure changes generating the agitation).

There is little or no evaporation from lakes, as they are not spread out as oceans, generating wind flow (explained earlier). Rivers and streams bring water to the lakes, and other rivers carry water out of the lakes. Thus, lakes are really only wide depressions in a river channel, which have filled with water. Water flows in one end, and out the other.

The Great Salt lakes and the Dead Sea are salt lakes (ten times more salty than sea water), which have no outlets. These lakes are salty, as rivers over time carried salts to the lakes. After years and years of river inflow the salt content increased, building up to the present level.

If there is evaporation from the oceans, why is rain water not salty?

The continuous evaporation of water (extracting hydrogen and oxygen) from the sea, rivers and lakes do not carry off the salts from ocean water. Fresh water in rain, rivers and even ice is not salty, but carry some traces of it. Fresh water keeps coming in, and so the sea remains at the same level of saltiness.

How do our lakes and oceans manage to keep the same levels of supply?

The energy transmuting cycle disintegrates the water molecule in the oceans, extracting hydrogen and oxygen. As evaporation takes place, oxygen in the atmosphere combines with the departing hydrogen, forming water vapor (at certain temperatures and pressures). Water vapor condenses to form clouds, and precipitation. Precipitation replenishes the water that has been extracted or evaporated.

The oceans do not keep the same levels, as hydrogen continuously departs, and the atmosphere is continuously sustained by the oceans, which gradually deplete. Humanity was not around to archive how much of our oceans have depleted over time. To sustain this gradual depletion, the interiors of the earth belch out material from volcanoes, seeps, vents and other openings. Eventually, our inner and outer core will completely deplete, and earth will solidify like our moon, Mercury and Mars, and eventually disintegrate.

AS THE EARTH ORBITS THE SUN

EARTH'S SEASONS

Weather changes in cycles, and cycles of weather changes are called seasons-winter (the coldest), summer (the hottest), spring, autumn, or fall. As our earth orbits the sun one complete rotation in a year, and at the same distance, we have our different seasons at the same time every year.

It is postulated that the earth has its many seasons, as a result of its elliptical orbit around the sun, and also as a result of its 23.5 degrees tilt. We are informed that the sun emits rays that hit earth's surface at different angles, and these rays transmit the highest level of energy when they strike the earth at a right angle. Temperatures in these areas tend to be the hottest place on earth. Warmest temperatures at a particular location on the earth occur when that location is tilted toward the sun during the summer. In addition, the number of daylight hours is greater, increasing the time available for solar heating. Other locations where the sun's rays hit at a lesser angle tend to be cooler.

We are also informed that as the earth rotates on its tilted axis around the sun, different parts of the earth receive higher and lower levels of radiant energy, and this creates the seasons. Winter occurs when that part of the earth's surface is tilted away from the sun, the angle of the midday sun is lower, and there are fewer daylight hours. Consequently, summer and winter occur at opposite times of the year in the Northern and Southern hemisphere. All the above explanations are incorrect.

The expulsion of material/energy from the sun to the earth, and the extraction of different processes, states, or forms of energy from the earth's atmosphere, land and oceans toward the sun, as replenishment, is proportionately more or less equal.

Planets rotate around their imaginary axis orbiting around the sun, as a result of the expulsion/extraction energy transmuting cycle. We have learnt, that existences are observed/experienced, as they have randomly structured, and were/are able to sustain another existence (which also randomly concurrently structure), and vice versa, as parts of the same cycle. So existences have certain characteristics, which made/make it possible for them to engender and endure.

The tilt and the elliptical orbit of the earth are existences (processes) within/between earth's atmosphere, and the sun's energy radiation/propagation cycle and therefore, they are why earth has its seasons, and not 'because' of it. There is no such thing as causation.

Atmospheres of planets are not solid bodies disintegrating with time, but continuously engender, or integrate, and disintegrate. Seasons, climates, or weather, are per se the alterations, or the alterations in/as the temperatures, densities and pressures, as/of the atmosphere. Earth interiors is not affected (does not have seasons, or climates/weather, or temperatures/pressure density changes).

As our troposphere and or biosphere are the atmosphere we/organisms experience/observe these phenomena. In other words, organisms as the atmosphere, experience these atmospheric changes. When we say the weather is dry, or humid, or hot, or cold, we, as the density (structures) of the atmosphere are experiencing these atmosphere altering states.

Heat energy and the distant sun

Does heat energy from the sun directly give us our seasons? There is no such thing as sunlight, as explained in the next revelation. However, temperatures, other forms of radiation, and pressures from the sun, propagating through interplanetary space, interact with our multi layered atmospheres, continuously disintegrating, and integrating it.

If direct heat energy from the sun engendered our seasons, the earth would not have a tilt in the first place, and we would not have seasons. If direct heat energy from the sun was responsible for our seasons, earth's tilt would always adjust further away from the sun, as it orbits around the sun. This would facilitate the sun to continuously extract material from the oceans (the southern hemisphere has more oceans).

Every six months earth's tilt either faces the sun, or is away from it. A tilt of 23.5 degrees should bring about a slight change-if heat energy was directly responsible-but not a completely different season change. The earth is not static (does not show the same face to the sun continuously, as the moon does to earth), but rotates once every day, exposing parts of the Northern hemisphere (in winter when it is closest to the sun, and the northern hemisphere is tilted away from it) to the sun, every day.

If the surface of the earth were affected directly by sunlight, and/or heat energy was direct from the sun, the slight earth's tilt would not engender winter in the north, with summer in the south. Moreover, throughout the year moderate temperatures around the equator would not be the norm.

In the Northern hemisphere, January is cooler than July, even though the sun is closer to the earth. This should not be the case, as the sun is 150 million kilometers away from the earth and therefore, a slight tilt should not bring about a season change. The closeness of the earth (in its orbit) to the sun should not bring about winter in the Northern hemisphere, as it would balance the amount of heat lost, as the northern hemisphere tilts away at this time. As the earth is closer to the sun during the northern hemisphere's winter, the areas near the tropics of Cancer should receive sufficient amount of heat since the earth is rotating every day. This heat should balance the amount of heat not received, as a result of the tilt. The temperatures should not change so much, as to bring about a new season.

The earth is not a square to consider unequal heating of different areas, and the angles at which the sun ray's hit the earth (there is no such thing as a single ray of heat or light, and/as energy propagates in all directions from the sun-no such thing as angles). The earth rotates and therefore, heat energy directly bringing about our seasons cannot be considered, as excluding the areas of the North pole (when the earth is closest to the sun and is tilted away) most of the Northern hemisphere areas are exposed each day.

The intensity of the sun rays do not represent the warmest or coldest days. This postulation would be correct if the earth was not a sphere, and was not confined by an atmosphere. It is important to note, that it is our atmosphere that is interacted, and disintegrated, altering and engendering again. This manifests our seasons, climate, weather, and other phenomena. Temperatures depend not only on the amount of heat the atmosphere receives from the radiated heat of the sun, and the area of atmosphere that is heated (interacted), but also on the amount it looses, due to the absorption of this heat by the land and the oceans. It is not until the ground and oceans absorb enough heat to reach a certain level (with the temperatures generated by/of the continuously altering atmosphere), that we feel the coldest days of winter, or the hottest days of summer.

Our seasons, climate/weather changes, temperatures, pressures, densities, winds/breezes of the atmosphere, and tides, currents, and waves of the oceans result from the following:

1- Atmospheres exposed to direct radiated heat, and pressures, during day and night.

2- The depletion of nuclei from the atmosphere, and the oceans.

3- Interactions generating temperatures, pressures, and density changes, at different areas of the multi layered atmosphere.

4- Oceans exposed to atmospheric temperatures, pressure and density changes, and the extraction/evaporation of lighter material from the disintegrating water molecules, and also the temperature and pressure alterations from these interactions.

5- As Earth is a rotating sphere, and as its multi layered atmosphere rotates at different speeds, and at different directions, and as the atmosphere is tenuous at the equator increasing in density approaching the poles, there are different temperatures, pressures and densities engendering at different locations, within the atmosphere, and at the surface of the earth.

These processes 'regulate' the intense radiated heat and pressure/force from the sun from reaching the surface of the earth.

Why is earth's orbit elliptical and not circular, or why does it come closer or further away from the sun after every six months? Why is the northern hemisphere tilted away when the earth is closest to the sun? Why does it tilt toward the sun when it is furthest away from it?

Why does earth have a tilt?

We define earth as consisting of the surface/land/mantle, crust and the molten core, different from its atmosphere/oceans. However, as we now know, it is earth's atmosphere which is interacted, integrated, and disintegrated continuously that manifests seasons, climates, weather, temperatures, pressures and densities. Our defined earth (interiors) does not experience these phenomena.

Earth did not have its tilt at the beginning. Earth's oceans completely covered the crust, and the mantle-there was no land. However, the crust was not an even surface, and as the oceans gradually depleted, the land/ surfaces at the northern, and southern hemisphere, were exposed-land. With more oceans, there was/is greater extraction from the southern hemisphere. As a result, earth tilts, facilitating and abetting the extraction processes. As the tilt is maintained through its orbit around the sun, earth's atmospheric temperatures, and pressures, alters to the many seasons/climate and weather. If earth had no tilt, the atmosphere would manifest one, or two seasons.

As the earth orbits the sun during the northern' hemispheres winter, the southern oceans face the sun, resulting in more extraction from the oceans, explaining earth's tilt away from the sun. With more extraction from the southern oceans, earth comes closer to the sun, explaining earth's elliptical orbit. As there is more extraction from the southern oceans, the southern hemisphere has less dense atmospheres at this time (there is more extraction with clearer skies). There is also more atmosphere (nuclei disintegration) extraction with more atmospheric interactions. With more heat generated, the southern hemisphere experiences summer. As there is less interactions/extraction from the Northern hemisphere it experiences winter, although earth is closer to the sun.

After six months, as the northern parts (the tilt) of the earth faces the sun with its more dense atmosphere (approaching the poles), there is more interactions/extraction from/at the atmosphere than at the oceans (less oceans in the northern hemisphere), resulting in increases in temperatures, and summer in the northern hemisphere. With more extraction of the atmospheric nuclei, and less extraction from the oceans, earth does not come closer in its orbital distance during the northern hemisphere's summer.

At different distances, and with certain areas of land and sea, exposed to radiated heat and pressures from the sun, the atmosphere accordingly alters temperatures, pressures and densities resulting in seasons, climate/weather changes.

We have studied earth's tilt only during our lifetimes. In reality, tilts of planets change over periods of time, and/as the atmosphere changes accordingly, engendering different seasons, climates and weather, and other phenomena over the years.

Earth's tilt is presently 23.5 degrees, and at these conditions, we experience our type of seasons, climate and/ or weather. Earth 23.5 degree tilt is not as a result of causation. Existences endure, facilitating energy transmuting processes and/or cycles. Existences are observed and/or experienced, as they could form cycles with other existences, sustaining one another.

Why there are temperature contrasts in the mid latitudes between winter and summer?

Maritime climates in the mid latitudes are moderated by the influence of extraction of material from the oceans, but within the continental interiors winter temperatures can fall to -40 degrees, whilst summer temperatures often exceed 30 degrees C. In the polar region the summer temperatures remain close to, or even below freezing. For six months there is continuous light, and for the next six months in darkness temperatures can often drop to -70 degrees C.

As the earth is a sphere the atmosphere at the tropics, and equator, spreads out. The atmosphere here is tenuous and therefore, faster rotating, or vice versa. Rotating tenuous atmospheres facilitates and abets the interaction/disintegration, and extraction of material. As we approach the poles the atmospheres get denser, rotate slowly, and without much interaction there is less heat generation, and the areas are cooler. Within the tropics the seasonal patterns of temperature changes are not evident. At the equator, and the tropics, as the earth is a sphere, and bulges out at these latitudes, the atmosphere is less dense. With this tenuous atmosphere the heat generated with interactions is easily transferred to the surface.

The year is divided up into wet and dry seasons. Wet seasons occur at certain months of the year at different places, as the atmosphere alters, generating significant vertical uplift, or convection of air that is accompanied by the almost daily formation of large thunderstorms.

This zone of convection is called the Inter Tropical Convergence Zone ITCZ, which moves with the seasons north and south of the equator, between the tropics of cancer (Northern hemisphere) and Capricorn (Southern hemisphere). Near the tropic of Cancer the ITCZ approaches only during June and July, and climates at these latitudes generally experience only one wet season, and a prolonged dry season throughout the remainder of the year. Near the tropic of Capricorn the short wet seasons occurs during December and January. In some parts of the world, for example India, the special pattern of atmospheric pressure, and wind, which accompanies the wet season, is known as monsoon.

What causes ice ages?

We are informed that there is something about the way the planet tilts, wobbles and careen around the sun presumably brings on the ice ages every 100,000 years or so. This is incorrect.

Humanity was not around to think, or to keep archives, and so this information is refutable. Abnormal changes in natural conditions result from a decrease, or transitional increase of extraction or expulsion material to/from the sun. There are much energy transmuting cycles, as there are planets. Any alteration in these cycles will alter normal conditions.

Why do some areas have total daylight or total darkness at some time during the year?

At noon on the summer solstice the sun is directly overhead at 23.5 degrees north latitude (tropic of Cancer). During this time, the tilt of the earth's axis result in locations north of 66.5 degrees north latitude (the arctic circle) to be facing the sun 24 hours giving them 24 hours of daylight. At noon on the winter solstice the sun is overhead at 23.5 degrees (tropic of Capricorn), and the Arctic Circle is in total darkness 24 hours of the day.

OUR LIVING WORLD

Why is our earth unique?

Over the years, as a result of the continuous disintegrating or degassing of the molten core (as the core shrinks, and the mantle and crust sink) plate tectonics have shifted huge land masses or continents. As these processes continue to take place, there are earthquakes, resurrection of dormant volcanoes, tidal waves, tsunamis, and other catastrophes.

Earth's geology is unique, and cannot be applied to any of our planets, as each of them has their own geology. Earth is unique among our solar system's rocky planets. There are no mountain chains on other planets. The mountains chains on earth are vestiges of violent volcanic activity in the later stages, as the upper layers of the outer molten core solidified with land/surfaces, which covered the molten core, and contained the oceans.

It is the only planet still volcanically and tectonically active. As our atmosphere and oceans are depleting, and the molten core belches to sustain them, it shrinks in size and therefore, the earth's surface and its interior are constantly in motion. The depleting molten core results in the mantle, crust and outer core to sink. At some depth the mantle is partially molten, and can flow albeit slowly in response to steady forces applied for long periods of time. Solid rock in the mantle when subject to heat and pressure in the earth's interior, over millions of years, alter the compositions creating plate movement, earthquakes and other catastrophes. We live under the threat of earthquakes waiting to happen.

Continents keep on drifting across the earth's surface, and the ocean floor spreads. Earthquakes take place, as there is a sudden release of energy in the earth's crust or mantle, as the tectonic plates jostle against each other becoming distorted, as the surface sinks. The zones where the rocks are weakest suffer sudden violent shocks, destructing nearby surfaces. Continental plates continue to move, and they will always move, as energy transmutes, and everything is flux.

Volcanism occurs where magma from the earth's interior is able to force its way through a weak zone in the lithosphere. Weaknesses are most likely to occur in the vicinity of plate boundaries. Continental volcanoes occur in zones, and regions, where continents are colliding. Volcanoes will keep erupting material until the core solidifies. Some volcanoes explode killing thousands of people such as Japan's Unzen volcano in 1972; some may have a gentle flow of lava, as Hawaii's Kilauea volcano-a tourist attraction.

Venus seems to also possess a molten core, but it has an atmosphere more of carbon dioxide-a material not actively disintegrated/extracted. This type of atmosphere would not require its molten core to spew out material to replenish its atmosphere, as often as earth's molten core was/is required to do so. Without a continuous belching molten core Venus, Mercury, Mars and our moon, have no tectonics plates to collide, and no great earthquake fault zones, and not many volcanoes to form chains of mountains. Mars, Mercury and our moon have depleted their atmospheres over the years, and have long solidified to inert spheres.

Cyclical processes involving water, oxygen and carbon dioxide are essential to plant and animal life. These cycles sustains/supports a biosphere where life like ours could evolve, and endure.

Within the hydrological cycle, ocean and lake waters evaporate creating new rain clouds. Rain washes away rocks, and soils, gradually creating sand, and silt. Rivers flow carrying those sediments from hills and mountains to the valleys below. When the climate turns cold, water is taken up in huge ice sheets, which spread out from the poles, and sea levels fall around the world. With warmer weather the ice sheets retreat, and the water flows back into the sea.

The water cycle

The earth holds a finite amount of water, and never runs out (but will eventually), as water moves through cycles constantly being used, and constantly being replenished.

Earth holds about 500 billion gallons of surface water in its oceans, lakes, rives, icecaps, groundwater and atmosphere, and a large amount is locked up in minerals, its crust and mantle.

All interactions between the oceans, and other parts of the water cycle, take place at the oceans surface, with rivers and rain adding to the top surface, while surface evaporation returns water to the air.

The water cycle can be depicted, as water evaporates from the oceans forming clouds that rain on the land (and the oceans), and finally collecting in streams and rivers that return to the sea. Of course, this water cycle is simple, but the complete cycle is much more complex, involving many interlocking cyclical processes, occurring over time, from a few hours to millions of years.

The oceans

Only a few hundred yards deep, the water in the oceans absorbs light, and the deeper you go into the water the colder, and saltier it gets. The pressure also increases.

With appropriate prevailing temperatures, the upper part teems with life with micro organisms to giant fish, and sea mammals. The deep oceans are dark, and dense, subject to tremendous pressure, and chilled to only a few degrees above freezing. Shallow water provides seafood, while the deep is a kind of desert, inhospitable and spare of life. Much of the deep water comes from melting ice, which circulates spreading out slowly.

The carbon cycle

Natural processes generated by the energy cycle releases carbon dioxide into the atmosphere. Combustion, decay, and respiration also add carbon dioxide to the atmosphere. Carbon dioxide is removed from the atmosphere by plant photosynthesis and by carbon dioxide being dissolved directly into the sea-the huge oceans store carbon dioxide. Rain that is naturally acidic reacts with carbonates in rocks, as carbon dioxide in the atmosphere dissolves rain forming carbonic acid. Chalk and limestone release carbon dioxide when they react with acid rain. Volcanoes also release carbon dioxide, as chalk and lime stones.

The nitrogen cycle

Lightning recombines nitrogen with oxygen forming nitrous oxides. Fixing, or making nitrogen into compounds occurs naturally by the action of certain bacteria in soil, essential for plant growth.

The rock cycles

As precipitation solidified some of the elements, all rocks at first were igneous, or fire formed. The crust of our planet began as molten rock. Rocks that have been changed since they first formed are called metamorphic rocks. Sedimentary rocks are thick deposits, layer upon layer of igneous rocks, which are buried, baked, and turned to stone.

The rock cycle began when glowing outer layers began to solidify. Today, volcanoes are the processes producing new igneous rocks. Magma or molten form of rock is sent through volcanoes in air, or underwater. Ocean tides, flowing rivers, and streams, weather these rocks. All types of rocks weather to form more sediments, as they melt, or metamorphose, and start the rock cycle anew. The cycles of weathering and erosion of rocks, sedimentation, and creation of new rocks, has gone on since the beginning of the earth, and still continues.

With the many cycles formed, continuously generating temperatures, pressures and other conditions, the biosphere engendered, and endures, where life like we know it could proliferate. We experience 'moderate' temperatures and pressures or force, as our form of life could only engender from these conditions. In reality 'moderate' and 'suitable conditions' are our concepts. Life like ours could engender, and endure only with those conditions, but there are other organisms that cannot live with our moderate temperatures and conditions.

We live in these smooth times, without any major catastrophe involving abnormal atmospheric changes. We assume that the present conditions have always been, and will continue for much of the future. We postulate our earth and the solar system is 'unique', imagining within our own thoughts/concepts. As energy continuously transmutes, everything is flux, and existences endure, eventually terminating the next second, day, week, month, a year or millions of years.

Why did Pangaea break up into many continents?

With continuing heavy precipitation, the oceans completely covered the mantle and the crust. Concurrently, with the interiors completely covered by water, the all around maelstrom of volcanic activity subsided, and the molten core, and the outer core, were confined. Earlier there was no land that could be seen.

Our new atmosphere could endure with its constituents, as the majority of its depletion was sustained by the oceans (even today). Over millions of years, the oceans continued to sustain the depleting atmosphere. However, over time, the oceans gradually depleted exposing parts of the crust, and land, and the oceans, made up earth's surface. This massive surface area, or land, is called Pangaea.

Even as the oceans covered the mantle, and the crust, there was continuous spewing out activity from the core, through the mantle and the crust seeps, and other openings. Logically, Pangaea or the huge land mass area should not have separated, or broken up into continents, as the mantle and the crust sink. In fact, it should have remained more compacted.

As parts of the crust were exposed, the surface exposed area of the oceans decreased, concurrently decreasing the amount of extraction (parts of which replenish the depleting atmosphere). Therefore, the interiors once again had to resurrect the belching out processes.

Magma blasted out through the mantle, and the crust, through sea beds, and the engendering of volcanoes. The breaking up of Pangaea was the result of these processes, all over the surface of the earth. The shape of the present continents are evidence, that each of them are massive surface areas of a huge land mass called Pangaea. These processes also engendered the myriad sea bed structures, islands, plateaus, mountains, mountain ranges, and volcanoes, dormant and active. It is incorrect to state that erosion built mountains.

As the continents drifted apart, flora and fauna also separated. After Pangaea broke up, migration of creatures was not possible, as there was no means of navigation. Over time, as flora and fauna experienced different climate/weather, and other living conditions/environments, they are observed today with different characteristics at different continents of the earth. The emu, ostrich and rhea are the same bird living in different continents, but with some distinct characteristics. Even within humans we observe different color, hair, face, eyes, body size, shape, customs, culture and languages (revelation 33).

What is an earthquake and why do they happen?

A fault is a fracture, or zone of fractures, between two blocks of rock. If the blocks move rapidly there is an earthquake, and occurring slowly, it is a form of creep. There are normal, thrust and strike-slip faults which may range from a few millimeters to thousands of kilometers. As the inner core gradually shrinks, the outer core gradually sinks, so most faults produce repeated displacements over geologic time.

Plate tectonics is the continual slow movement of the plates. As the cores continue to belch out disintegrating materials sustaining the depleting atmosphere, it shrinks in size. As a result, the outer core, mantle and crust gradually sink, resulting in plate shifting, and subsequent earth quakes all around the globe.

An earthquake results from a sudden slip on a fault. The sinking of the mantle and outer core, pushes the sides of the fault together. Stress builds up and the rocks slips suddenly releasing waves of energy, which travel though the earth's crust generating the shaking we feel during an earthquake. An earthquake occurs when plates grind and scrape together.

Although it may seem we are having more earthquakes, some of magnitudes 7.0 or greater they have remained fairly constant during this century, and have actually decreased in recent years. A temporal increase in earthquake activity does not mean a large earthquake is about to happen. Similarly, quiescence or the lack of seismic activity does not mean a large earthquake is going to happen.

Are there earthquake precursors that can lead to useful predictions?

It is incorrect to pinpoint fault zones, or earthquake regions considering plate shifting, experienced in the past. As explained elsewhere, earth's interiors continue to belch out magma, and the inner core is shrinking. As a result, the mantle, crust and outer core sink, as the earth solidifies into being far smaller than from how it began. As plate tectonics happen just about anywhere, it would be impossible to predict when and where the next earthquake will take place. If an earthquake happens under sea, there may be large waves or tsunamis on the surface of the oceans.

Can we predict earthquakes by tracking sunspots?

A sunspot hypothesis suggests that certain changes in the sun-earth environment affect the magnetic field of the earth, which can trigger earthquakes in areas prone to it.

Scientist's state, that energy mass from the outer periphery of sunspots is directed toward the earth. This disturbs the magnetic field of the earth, and causes changes in the atmosphere, ionosphere and geosphere. It is like when a magnet is placed some distance away from an iron with a sheet of paper in between. The magnet is able to exercise an effect even with the paper in between. Such changes also result in fluctuating water levels. These changes are not the cause of the quake, but they can trigger quakes in areas where there are active faults. These changes take place about 24-36 hours before the earthquake, so they can definitely help predict it. This hypothesis is only partially correct.

First of all there is no magnetic field surrounding earth. Energy is not directed toward the earth only at sunspot occurrences. As discussed earlier, presently, as replenishment material from the inner planets decrease, radiated heat and pressure/force extracts more energy from earth. As a result, the energy transmuting cycle between the sun and the earth alters frequently. Coronal activity or sunspots are correlated with abnormal changes in earth's atmosphere. Sunspots arise from abnormal energy transmutations.

As our earth presently has a very active energy transmuting cycle, many disasters occur. With the continuous belching of earth's magma the molten core is shrinking and eroding, and accordingly, the mantle, crust, the ocean floor, and the outer core gradually sink. This has generated the seismological active areas of the earth. To replenish the atmosphere, and the oceans, during the decrease in replenishment material for the sun (generating sunspots and solar flares), our interiors belch profusely generating conditions triggering earthquakes. Earthquakes will take place depending on the conditions of those seismological active areas, but will not always happen at all sunspots times.

Why are some areas deserts, grasslands or another a rain forest? Why are there different forests and deserts, and why are there different types of life in each area?

It is the result of climate-a characteristic condition of the atmosphere near the earth's surface at a certain place on earth. It is the long term weather of that area (at least 30 years). This includes the region's general pattern of weather conditions, seasons, and weather extremes like hurricanes, droughts, or rainy periods. An area's climate is determined by air temperature and precipitation. Climates and/or weather in a region dictates/engenders/endures the environments.

World regions are controlled by climate. The climate of a region, will determine what plants will grow there, and what animals will inhabit it. All three components, climate, plants, and animals, are interwoven creating the fabric of a biosphere.

Some plants and animals survive only in deserts, as they evolved mechanisms enabling them to continue living under those conditions. If those desert animals could 'think', they would query, as to how other creatures including us humans can live in environments not as habitable as deserts. It is how things engendered or structured, and how they exist/survive and endure.

How does earth's interior work?

If the earth's core is so hot why does it not melt the mantle and the crust?

The earth's core is very hot perhaps 10,000 degrees C. There is a great distance between the core, and the surface.

The earth is composed of a rocky crust, a mantle of hot plastic rock, and outer core and an inner core. At that depth, pressures and densities are so great that the outer core is solid, whereas the inner core is still liquid, despite temperatures to be in the range of 10,000 degrees C, or about as hot as the surface of the sun. With the continuous degassing of the core throughout the 4.5 billion history of the earth, the earth's composition has altered, and temperatures have fallen by only a few hundred degrees C.

The outer core confining the innermost core is of the same material, but of a different state of matter. The outer core is confined by the mantle and the crust. As pressure/force continues to delve down towards the surface of the earth, there is continuous pressure on earth's interiors. As you go deeper into the ocean, there is more density, and even greater increasing pressures, and this prevents the heat from dissipating, and also confines the inner core.

Within an energy transmuting cycle if everything everywhere is energy, and energy transmutes into states, processes, structures or forms, and cannot be created, destroyed or "depleted', how and why does our atmosphere and oceans deplete? We can understand earth's interiors being depleted, but why do star field bodies and earth shrink over time?

Some forms of energy (protons) are propelled to the far reaches of the solar system. As temperatures/radiated heat, and pressure/force from the sun dwindle with distance, these forms of energy cannot interact with those materials, to form energy transmuting cycles, and therefore, interstellar cold dark matter/energy keeps 'accumulating'.

In the case of earth, protons are disintegrated and extracted from nuclei within our atmosphere, and from the molecules of water. It is the gradual depletion of earth's hydrogen, or in other words, the disintegration of earth's elements into hydrogen/protons (degassing of earth's interiors) that everything alters, and earth will shrink, solidify, and eventually terminate.

Eventually, over millions of years all the bodies/planets will solidify, and disintegrate, and the entire solar system excluding the sun will be filled with nebulae. As there will be no planets/bodies with atmospheres to continue an energy transmuting cycle, the sun will die. It would probably expand into a red giant with the nebula in our solar system, and eventually collapse into/as a black hole.

Concluding:

Within the sun-earth energy transmuting shuttling to-and-fro cycle, forms/states of energy not only interact with our atmosphere, but delve through it to our rivers, lakes surface and especially our oceans. Our oceans are a different type of atmosphere. Humans have conceptualized the existence 'ocean' being different than an atmosphere. In reality, we have an atmosphere enveloping the solid earth, and also the ocean (as another type of atmosphere) enveloping parts of the mantle and crust of the earth's interiors. Both 'atmospheres' have nuclei/elements and molecules, which contain hydrogen or protons. 'Individual' protons are easily disintegrated into particles extracted at supper speeds, which are continuously required for/as replenishment material at the sun.

All matter, or objects, are constituted of the proton and therefore, all matter, living or non-living are disintegrated, some instantaneously, some after a certain duration, and some gradually decaying. This is why; there is eventual termination of living things, and decay of non-living things.

Constituted of particles, which make up the protons, which make up nuclei/elements, and molecules, which make up our body, we are not aware, and do not feel incoming and outgoing pressure/force, as we are constituted, and conform to the atmosphere, and as the atmosphere, we exist/endure to/as continuously temperature, pressure/force and density changes. We are not aware of the disintegrating water molecule (hydrogen and oxygen), the disintegrating atmospheric atomic nuclei, and departing hydrogen. Constituted of/as/by the atmosphere, we are not aware of it rotating at hundreds of kilometers per hour.

C - Elucidating Our World

REVELATION NINE

NO SUCH THING AS SUNLIGHT

Elucidating energy transmutations-light, temperatures, pressure/force, colors and sound

The first nine pages are recapitulated enabling readers better understand this revelation. These may be omitted.

NO SUCH THING AS EXISTENCES

Inadvertently, accumulating sounds which unfolded into a language, humans think or conceptualize, imagine and query within their language minds, which in reality is an illusion. As we can imagine, we query existences. Querying existences we create them. As these existences are from within an illusory language, we cannot explain our own creations. In other words, we cannot explain reality (existences) from within our own generated language.

Everything everywhere is energy. Existences are forms, states, processes or structures of energy, as energy continuously transmutes. In reality, there are existences, as humans describe energy transmutations.

As energy continuously transmutes, everything is flux. As everything is energy, and as all existences exist, or endure, and eventually terminate, within/as the fabric of energy (there are no entities), every 'existence' continuously alters its characteristics, with/at every moment of time. Every existence 'fixed'/static, or propagating, or in motion, alters its characteristics continuously, as everything everywhere is energy, which is continuously transmuting into forms, states, processes or structures or existences.

As we have segregated the myriad forms, states, process or structures of transmuting energy into/as existences, we then have to understand how, and why they exist, and why they continuously alter. So after we create existences we try to describe, and explain each of them. Obviously, we end up with unanswered questions, more questions, unsolved problems, more problems, and more theories. This revelation describes some existences, as it transmutes, enabling us to 'understand' our creations.

As energy is everything, everywhere, no existence is an 'entity'. Every existence engenders, exists and endures, concurrently with another existence of different characteristics sustaining each other. A particle, proton, nuclei, molecules, compounds, objects, astronomical bodies, planets, stars and galaxies, all exist within their space (which is an integral part) sustaining each other.

TRANSMUTING ENERGY AS/WITHIN A STAR FIELD

Free energy condenses into particles forming a dense region, concurrently generating space-a nascent star field. As concentration continues, and the periphery is shed off, a nebula of protons engender, which subsequently form an energy transmuting cycle with the dense region-a star field. Within this energy transmuting cycle, all forms of energy, as temperatures, pressure/force and plasma/protons are expelled from the star. Within a shuttling to-and fro pathway, these propagating forms of energy interact with other forms of energy, extracting yet other forms of energy to the star, continuing its functioning. Forms of energy transmute/propagate, as concurrently other forms of energy transmute/propagate.

The dense region would not transform into a star without an energy transmuting cycle. In other words, if other existences within the nascent star field do not concurrently engender, facilitating and abetting energy transmutations, no existence including the star field would engender.

Light/heat/all forms of radiation, or temperatures, and pressure/force, and sounds (existences as humans understand them) manifest, as energy transmutes into forms, states, processes or structures. For example, as space transmutes, waving to certain frequencies and wavelengths, there is light. Without space (waving) there would be no light, and energy would not propagate, and other forms of energy would not manifest.

No origin, source or target

Energy in the form of light, all forms of 'radiation', temperatures and pressure/force 'propagates' (transmutes), but not from a conceptualized postulated 'source'. In reality, there cannot be a 'source' or origin, as everything is comprised of energy in states, processes structures or forms. Forms, states, processes or structures of energy, are continuously transmuting into yet other forms, states, structures or processes of energy.

The word 'propagation' is a human concept, as in reality, as there is no source or origin, and no ‘entities’, energy does not 'propagate'. Energy continuously transmutes into yet other states, processes or forms, and this is postulated as energy 'propagation'. Humans have given names/words for different physical phenomena (existences) trying to understand their creation, and or theories.

We have been informed, that for billions of years the sun has been expelling energy in the forms of temperatures/radiation and pressure/force. Where did all this energy come from? Where does all this energy go to? It is postulated, that the sun contains 'hydrogen' enabling it to burn for certain duration. Again, how and from where does this hydrogen come from? Where does it go?

The sun or a star is not the 'source' of light, or forms of 'radiation', temperatures, pressures or force. The sun does not 'generate' light', or temperatures, or pressures. The sun per se is light, or temperatures, or pressure. There is no causation, but only transmutations of energy. The sun is an existence, as a transmuting form of energy process/structure, and not a distinct body. Energy transmutations, or processes at a dense region (the star/sun) generate these forms of energy (as humans create and need to understand existences).

A house lamp is not the 'origin' of light. Electricity (a form of energy) makes the lamp glow, and this light (a form of energy), or electricity (another form of energy), 'originates' from another energy state, process, structure or form. There are no sources, origins, or beginnings, or terminations, or targets, but only cycling/transmuting energy.

Within a star field, or our solar system, energy from/as the star/sun propagates/transmutes, as/within star field or interplanetary space, as forms/states/processes/or structures of energy. Concurrently, within shuttling to-and-fro pathways, energy as other forms/states/processes and structure of energy are extracted (propagate/transmute) back to the star/sun, as replenishment for the expelled energy. Nothing is a one way phenomenon.

Every existence, or being, whether it be a particle, nuclei, molecule, objects, cells, people or organisms, atmospheres, nebula, planets or stars are processes, states, structures or forms of continuously transmuting energy.

Forms of energy transmuting/propagating

Everything everywhere is energy, and existences are structures of energy, within/as the fabric of existential energy expanse. Therefore, energy cannot be created or destroyed, lost, absorbed or gained, but only transmutes. Some forms of energy 'propagate', as energy continuously transmutes from one form, structure, state or process to yet another form/structure/state or process. In reality, as everything is energy it does not 'propagate' from one place to another, but only alters in states, processes, structures or forms. As we create existences, there are states, processes, forms and structures of energy; and/as we try to describe/explain them.

Forms/structures/processes and states of energy such as temperatures/all forms of radiation, pressures/force and plasma from the star/sun, 'propagate' within/as space-also a form of energy. This propagation is not a one way phenomenon. Energy shuttles to-and-fro from the star/sun to bodies within star fields or interplanetary space, continuously propagating/transmuting from one state, structure, form or process to yet another or any state, structure, form or process.

Energy transmutes or 'propagates' as space waves

Without space nothing would be able to motion. Light and/or all forms of radiation/temperatures, and pressures/force, would not engender in the first place-there would have been no existences. As space is an integral part of any particle, plasma, proton, object, planets, star or galaxy, forms of energy can transmute/propagate within/as space.

Heat or temperatures, pressure or force, and other forms, states, processes and structures of energy, propagate, or radiate, or transmute only through, or within/as space, as it waves. However, energy transmuting/propagating is not a one way phenomenon, nothing is. Forms of energy propagate within a shuttling to-and-fro pathway, explaining the waving of space. The space wave or waving of space per se is momentum. Momentum is also a form of energy. Therefore, waving space, as/in momentum, also changes its characteristics to other forms/states/processes, or structures of energy, as well as concurrently alters other forms/states/structures or processes of energy, propagating/transmuting through, as, or within it. In other words, it is the momentum, or propagation generating (transmuting) space to wave, or vice versa.

If you stand at the edge of the seashore you will feel water waves approaching you, and at the same time moving away from you. You can observe/feel the water coming toward you, yet concurrently you seem to be pulled away, as the water recedes. This incoming and outgoing processes generates the waves, or vice versa. Energy (in the form of momentum or motion) is transmuting as waves. In reality, water is not moving from here to there, but as water per se as/is also a structure of energy it stays where it is, but the momentum (transmuting energy) is observed, and/or felt, as the water waves. What you feel is energy in momentum, as everything stays at the same place (as everything everywhere is energy, and there are no 'entities'). The waves manifest energy in momentum. The waves are the breeze/force/momentum. We know now everything is energy and therefore, water as a structure of energy, as well as the waves-are structures/processes of energy, and momentum-a state of energy.

Energy in/as a state, structure, process or form, continuously transmuting into yet other states, structure, process or form engenders momentum, as space waves. This makes it appear that energy is 'propagating', or in momentum.

A particular part of water is not moving from there to here, but it is energy manifesting momentum. As you drop a pebble in water there are ripples or waves propagating outwards from the point from where the pebble hit the water. This is propagation in a two dimensional context. As everything everywhere is energy, and there are no entities throughout/within the fabric of space, space waves propagating from a point 'source' outwards in all directions, in a three dimensional context.

We can observe a ripple on the surface, but cannot do so within water. If you are within water, and create a disturbance there are no waves, as the disturber, and the medium of water are the same thing-structures of energy.

We assume water 'moves', but it is not the molecules of water moving, but the fabric of water/energy. A paper boat does not 'travel' along/with the waves or ripples. Even though molecules join together to form water, each molecule (of nuclei configuration) is confined by space. Water waves strike a floating piece of wood, but do not carry it along. The water does not really strike it, but the phenomenon is generated, as the wave's motion. Wave's motion per se is the force/pressure/momentum, or transmuting energy.

All forms, states and processes of energy, momentum or propagate within or through/as space. Space waves at different intensities and or frequencies manifesting light, all forms of 'radiation'/temperatures, sound, and other forms, states, structures or processes of transmuting energy. The wavelength, frequencies, and or intensities of each 'propagating' wave, through/as space, will manifest the different degrees of heat (temperatures), different intensities of force or pressure, the different forms of radiation (light, gamma rays, ultraviolet rays, x-rays etc) or in other words, the different states, structures, states or processes of energy.

Continuously altering characteristics, or dwindling intensities

The outer atmosphere of the sun (the corona) extending for millions of kilometers is continuously disintegrated and integrated, from the outgoing energy from the sun, and incoming energy from other bodies within the solar system, as energy transmutes.

All forms of radiation/temperatures and pressure/force cannot maintain their intensities, propagating throughout a star field, or interplanetary space. In other words, existences do not maintain their characteristics, as energy is continuously transmuting, and everything is flux. Energy propagates/transmutes between bodies within/as a star field/solar system; as energy transmutes from one form, state, process or structure into yet another form, state, process or structures.

Energy in forms, states, process or structures is 'expelled' from the sun, and 'extracted' (as other energy forms, states, process or structures), as replenishment energy from other bodies toward the sun, within the same shuttling to-and-fro pathways.

As all forms/states/processes and structures of energy 'propagate', or transmute within, or as through/by space, they alter with each space wave. Existences (as human concepts) change their characteristics, or dwindle with momentum, and/or distance, and the diverging of space waves. In other words, propagating with distance disperses the intensity/degree of any form of energy (transmuting). Moreover, as momentum is also a form of energy, this 'propagation' per se decreases the intensities of existences through distance from the 'source'. Existences will alter to different degrees of temperature, different forms of radiation, different intensities of pressure/force, and sounds, depending and proportionate to the intensities produced by processes at the 'source', and the distance propagated/transmuted.

Light as a form of energy dwindles (transmutes) with distance, 'propagating' through/as (diverging three dimensionally) space-also a form of energy, as it transmutes. This 'light' is not apparent as light, propagating after a certain distance (depending on intensities at ‘source’). Energy is continuously transmuting from one state, structure, form or process into yet another state, structure, form or process, and therefore, light as a form of transmuting energy, alters its characteristics, propagating through distance.

Space waves decrease their wavelengths, altering/decreasing the intensity of light with distance, even if the temperatures, power/force or intensity at the 'source' continues, and is constant. Light transmutes to different forms of radiation with distance.

As light propagates through/as space waving through density, it activates particle and nuclei configurations. Light propagating through density, or air, or atmospheres (interactions) is reinforced with temperatures and pressure/force generated with the increase in particle vibrations. Reinforced propagating light through density will therefore, 'travel' to greater distances.

As we have conceptualized, and segregated the different forms/states/processes of transmuting energy, we have named each existence (transmutations of energy) differently, enabling us to understand these phenomena.

Energy is everything everywhere, and in reality, cannot 'dwindle'. Dwindling is in reality, energy continuously transmuting from a form, state, structure or process, into yet another form, state, structure or process. For example light, and heat (their characteristics alter-as humans conceptualize existences) dwindle with distance, but energy stays the same (only the intensities or characteristics of these existences shuttling to-and-fro within the same energy transmuting pathways alter).

Light and heat are human constructs, explained/described to distinguish the forms of energy. In reality, both heat and pressure or force, constitutes light, or in other words, light is apparent as certain degrees of heat and pressure/force intensities propagate.

Earth's atmosphere does not receive light from the sun, but other forms of radiation, or decreased, or dwindled temperatures and pressure/force intensities. Obviously, we can not exist with those extreme temperatures and pressures/force produced at the sun.

Proof that as energy 'propagates'/transmutes, all forms, states, processes, structures of energy dwindle with distance, or existences continuously change their characteristics

-A light bulb stays lit, being continuously supported by another structure of energy-electricity. It will stay lit, as long as electricity (a form of energy in momentum, or force/pressure) continues to force/heat the filament. The area that it lights up will depend on the intensity of energy/electricity/force at the 'source'.

-A street lamp will light up only a definite area.

-Light from the sun dwindles with distance, changing its characteristics to radiated heat, and is not light anymore, explaining why interplanetary space is in darkness. There is light only around the sun.

-Radiated heat from the sun interacts, and disintegrates our upper atmospheric nuclei, engendering our visible light.

-Planets and other bodies can be observed, as heat radiated from the sun interacts with their atmospheres producing light. With different types of atmospheres, planets are observed having different hues. Light from the sun does not reach any planet.

-Extreme temperatures and pressures/force at the sun dwindle in intensities reaching earth. We cannot exist with those energy extremities.

-Light/heat and pressures/force from the star/sun dwindle with distance and therefore, the bodies at the outskirts of star fields/solar system are dark and cold/icy.

NO SUCH THING AS SUNLIGHT

As we have 'individualized' the different forms/states/processes and structure of transmuting energy, existences, or phenomena need to be explained/described.

Light produced at/by the sun processes propagates from the sun, but dwindles in intensity, or changes/transmutes its characteristics before reaching Mercury, and is no more sunlight (light). Therefore, we do not receive light from processes at the sun, as this light transmutes to radiated heat, and lesser intensities of pressure/force. We could not survive with those energy extremities produced at the sun. Earth's atmosphere receives many forms of radiation, such as ultraviolet, infrared and others.

We describe light as white light, and visible light (within/with which we see). Light and each form of radiation (infrared, ultraviolet, gamma rays etc), are each constituted of certain degrees of heat, and force/pressure intensities.

Radiated heat, with pressure/force, interacts with earth's atmospheric nuclei, and interactions with outgoing energy, produce our visible light. Light produced at our upper atmosphere is then propagated within/as atmospheric space to the surface of the earth (explained following). Light at the sun is produced in the same way. As there are interactions at the upper atmosphere of the sun, or the corona, with outgoing and incoming energy, there is light. We see this light at the sun. Light from the sun does not reach any of the planets of our solar system. We see light at the stars, and light does not propagate to us from the stars.

PROOF THAT THERE IS NO LIGHT (SUNLIGHT) FROM THE SUN

Orange/Red sun during sunrises and sunsets

We are informed that as we are always at the other end of the horizon, both at sunrise and sunset, and observe the sun through a greater volume of atmosphere than when the sun is overhead, we see the sun as a reddish orange glow. Depending on the density of our atmosphere, these sunrise and sunsets colors change into different hues. This is incorrect, as if this was the case, we should also be seeing a different colored moon at the horizon. Yet the moon's 'color shine' is the same at any time, anywhere.

The dominant gases constituting our atmosphere are nitrogen, oxygen and hydrogen. Hydrogen is produced in large amounts (disintegrating of the water molecule); nearly twice as much as oxygen, but it is continuously quickly swept away within the energy expulsion/extraction cycle. As radiated heat from the sun (not light from the sun) interacts with our atmospheric gases, nitrogen is manifested as a reddish/orange color, hydrogen blue, and oxygen green. Our atmosphere should have been reddish in color during the day (as nitrogen is the dominant gas). However, nitrogen remains low in our atmosphere (explained following), and oxygen as green color is not significant, as it constitutes only 20% of the atmosphere.

As hydrogen is swept away, our day skies look blue in color. This blue enhances the light at the sun, through the radiated heat pathway, and the sun appears silvery white, or sometimes it is golden yellow (depending on atmospheric density, and other factors). Therefore, hydrogen is insignificant in altering our dominant lower atmospheric reddish colors (the blue in the sky is far off in space-our upper atmosphere). Observed from above, the sea look greenish-blue, or cyan, as oxygen (green) and hydrogen (blue) constitutes the water molecule.

We have our visible light for many reasons (also explained following). The light at the sun (not from the sun) dominates over our atmosphere. There is no nitrogen at our upper atmosphere, and therefore, no colored red/orange light. The interactions from/by radiated heat, pressure/force, and protons from the sun, activate upper atmospheric nuclei, which are ionized, producing light. This light is propagated within/as atmospheric space, explaining why we experience light even at the surface of the earth (read following). As the light at the sun (not from the sun) completely dominates over our atmosphere (during the day), and these three combination colors of red, green and blue alter to a single 'white' color, or white light, space waving propagating white light is enhanced.

At sunrises and sunsets we observe the sun from the other side of the horizon. Although radiated heat from the sun interact parts of the upper atmosphere at the horizons, the light produced at those areas does not reach us, as there is more of ‘unlit’ atmospheric space, and atmospheric nuclei between us, and the upper atmosphere at the horizon.. During sunrise and sunsets, hydrogen at the horizon is being extracted from the atmosphere facing the sun, not overhead where we are stationed. As we observe the sun at the horizon through more of our nitrogen dominant colored atmosphere (with more of our upper atmosphere not producing light), we see the sun through mostly reddish/orange colors. As hydrogen departs through the same incoming radiated heat pathway, we see the altered silvery white color of the sun during the day. If we see the sun through a dark glass we see the sun red, as the dark glass cancels all the atmospheric colors allowing only the dominant nitrogen (we are within the nitrogen dominated atmosphere), preventing the light from the sun to alter to white light.

The colors of our atmosphere observing the horizon sun is different, depending on the continuously changing density of the atmosphere, and the localization of the setting, or rising sun. In reality, the color at/of the sun, and also at the moon, does not change at any time, anywhere.

During darkness, or in other words, there is darkness, as there is no radiated heat to 'light' up our atmospheric gases, and also no over head light (light at the sun through the same radiated heat pathway) to 'light' up these colors (as we see through our atmospheric nuclei colors).

As detailed earlier, the atmospheres of planets/bodies light up interacted with radiated heat, not light. The moon's atmosphere has depleted over the years and therefore, without an atmosphere, there are no nuclei to be activated to produce light, even though radiated heat from the sun (not sunlight) reaches the moon. The moon's surface contains shiny material, which 'lights' up from propagated light produced at our upper atmosphere.

There is no light, or heat produced, or propagated from the moon, and therefore, our atmospheric nuclei do not alter manifesting their respective colors, explaining why we see the moon without the colors of our atmosphere. Light at its surface is not dominant over our atmosphere. With these conditions, and as the moon is in close proximity to earth, we see the moon at its same luminosity overhead, or at the horizon. The luminosity at/of the moon will increase, or decrease, depending on the density of our atmosphere, and its distance from earth.

We observe light at the sun, and at the stars, and at the moon. Light and all forms of energy dwindle with distance and therefore, we do not receive sunlight, or light from any star, or from the moon.

Light at the stars

If light propagated to us from the stars we should be observing them at different hues, and colors, depending on their distance, luminosity, processes and size.

We observe light at the stars through our atmosphere, interplanetary space, our interstellar medium, their interstellar medium, and their star field space. Some of these star fields have different nuclei, molecules and compounds comprising their interstellar medium, explaining why some of them are observed at different hues.

If light 'traveled' from the sun it would have lit up interplanetary space, and we would never be able to see any stars at night. We do not see stars during the day, as the light at the sun dominates our upper atmosphere, preventing us from seeing the stars. We see stars at night by their light around them, and not as a result of their light traveling to us. Galaxies are observed red, as light from star fields eventually dwindles, transmuting to other radiation forms with distance.

Observing planets without sunlight

The light from/at the sun does not 'reach' any of our interplanetary bodies/planets. As radiated heat interacts with their atmospheres producing some type of light, these bodies/planets are observed. If there was sunlight reaching Mercury, Venus or our moon, we would observe these bodies as dark orbs, as the moon is observed during our solar eclipse seeing their dark sides. As these planets orbit the sun, we should be seeing phases of their lit up parts (as we see our moon) during parts of their orbits.

The whole of interplanetary space should be lit up

When one side of the earth is not facing the sun that part of the atmosphere is not interacted by direct radiation, and there is no visible light, but darkness. If sunlight reached earth, we should be seeing light on either side of our horizon at night. Light produced at the sun, exists only at certain distance around the sun, and therefore, interplanetary space is in total darkness.

If sunlight propagated throughout interplanetary space, the planets and other astronomical bodies would be conspicuous, and stand out. We observe the incandescent at planets by interactions, and disintegrations of their atmosphere by/from radiated heat from the sun, and not by sunlight. We observe bodies/planets in different hues/colors, as radiated heat interact their differently constituted atmospheres.

Our entire galaxy should be lit up

There are billions of stars in our galaxy, and our sky should have been flooded with continuous light if sunlight, and light from the stars, propagated without dwindling in intensities, and changing characteristics.

We see light at the sun, and light at the stars, including their lit up interacted nebula, or atmospheres of bodies within their star field. This may explain twinkling stars.

The light at the moon

If sunlight reached the moon, it should have been observed at different luminosities orbiting earth. The moon's surface is laced with certain type of material, which 'lights' with earth's propagating upper atmospheric light, as we are observing it, and as it is in close proximity to earth.

The moon cannot be lit (interacted) up by sunlight (or radiated heat), as it has a very thin atmosphere. If sunlight reached the moon, we would not observe its surface dark areas, which are seen both during the day and night, and also at full moon. If sunlight propagated to the moon, we should continuously see a dark orb (as in an eclipse) when the moon is seen during the day. During some days, as its reflective surface is continuously 'lit' up from the light at our upper atmosphere, we see parts of its face at certain angles. The moon should not be seen sometimes during the day, as light from the sun, if it propagated to the moon would have made it invisible.

Icy/cold regions

The Kuiper belt region objects/bodies at the outskirts of our solar system are freezing cold and icy, evidence that light, and all forms of energy dwindle with distance. Therefore, sunlight does not propagate throughout interplanetary space, and even radiated heat and pressures from the sun dwindle with distance.

Surface temperatures on the moon

The surface temperatures of the moon change from 225 degrees F (107 C), during the day, to -243 degrees F (-153C) at night. Without an atmosphere, if sunlight reached the moon, the surface would experience much hotter temperatures, and we should have been seeing a brighter moon. The same face of the moon is observed from earth throughout its orbit, so what is day and night at the moon?

The extremities in temperatures result from radiated heat from earth's upper atmosphere, and not from the sun. As the moon does not rotate around its imaginary axis, the same side facing the earth (the day side) is continuously being heat radiated from the light produced at our upper atmosphere, which is continuous, as at every hour some areas of earth's upper atmosphere is heat radiated from the sun, explaining the moon's 'day' temperatures. The other side does not face the earth, explaining the cool 'night' temperatures of the moon.

The full moon

At full moon we see a larger surface, as it is lit up by light from our upper atmospheric light from both sides of the horizon. The moon is much smaller than the earth, and if it was lit up by sunlight, we would not be seeing the moon at the full moon position for many days in a month. We would be seeing solar eclipses every month. The full moon on a clear night will be much brighter, both as a result of no dust, smog or clouds, on either side of our horizon (which partly obscuring light at the horizons), and also our observation of the full moon through it directly overhead.

Eclipses

All forms of energy dwindle with distance, and light as a form of energy dwindles with distance, changing its characteristics. In other words, propagating to a certain distance, light is not light anymore, but alters to radiated heat, or other forms of radiation, and therefore, light from the sun does not reach the moon.

If sunlight reached the moon, it would have been observed as an eclipse-as a dark orb-every day, everywhere from earth during daylight, not only at eclipse time, and at some places on earth. Only some places on earth can view it due to its angular orbit at certain days of the month.

Eclipses are strong evidence that light (sunlight) produced at the sun, and from the sun does not reach earth. Eclipses happen only when the sun, moon and earth are all positioned in a straight line. This situation does not occur often, as the plane of the earth’s orbit around the sun is at a different angle from the plane of the moon's orbit around the earth. Therefore, at most orbits, the moon is usually located just above, or below the imaginary straight line connecting the sun and the earth. Only about every six months do the planes of the earth, moon and sun all intersect creating the conditions needed for an eclipse.

As the earth's orbit changes in distance from the sun, the moon also alters its orbit around the earth. It not only comes closer or goes further away, but also orbits the earth at a different angle, just above or below the imaginary straight line, explaining eclipses. A solar eclipse happens when the moon passes between the sun and the earth, seen only during daylight. The lunar eclipse happens when the earth passes between the sun and the moon, seen only at nights.

We inherited and evolved mechanisms to see reality, as we desire it. The sun is 400 times larger than the moon; the moon is four hundred times closer, or the sun's diameter is 400 times that of the moon, and is 400 times further away, yet both the sun, and the moon, appear to be the same size in the earth's sky. During a solar eclipse, with this amazing coincidence of our visual reality, the moon completely covers the sun. The traversing moon, for some time, stops the radiated heat and pressure/force from the sun, and the upper atmospheric nuclei are not actively interacted. At this time, the light at the sun (not from the sun) is shielded by the traversing moon. As we see the sun through our colored lowered atmosphere, and also through the upper atmospheric nuclei which are not interacted (parts of it not ionized), the moon appears as a dark orb (an eclipse).

At a solar eclipse, without light produced at some parts of our upper atmosphere, the moon does not 'see' this light, and therefore, its surface is not 'lit' up, and is observed as a dark orb. This explains why an eclipse is seen only at certain areas on earth. As discussed earlier, if sunlight reached the moon, somebody somewhere on earth would have seen the dark orb of the moon, or an eclipse every day, as sunlight-if it reached it-would have lit up the side facing the sun.

We see a lunar eclipse only when the moon comes closer to the earth, as no part of its surface receives our upper atmospheric light from either side of the horizon. This explains why a lunar eclipse is observed everywhere on earth.

If sunlight propagated through interplanetary space reaching the planets, this light would have been intense, lighting up the entire solar system. We would not be able to see the moon, as it is a very small body close to the earth, only 3476 km in diameter, and 384400 km distance from the earth. If sunlight reached the moon, the dark orb would have been seen from everywhere on the planet. The solar eclipse is visible only along a narrow strip on earth's surface.

During a total solar eclipse all the light at the sun is blocked, and the sky is seen dark, and planets and brighter stars are visible. If sunlight reached the moon, or the earth, this would not have been the case. During this eclipse, as sunlight does not reach the earth we can also see the sun's corona.

Lunar eclipses

Since the sun and moon have the same size, as viewed from the earth, solar eclipses are rare. As the earth is relatively close to, and large compared to the moon, there are more lunar eclipses.

We are informed that there is a lunar eclipse, as the moon comes into earth's shadow, as the earth blocks the sunlight reaching it. This is incorrect. Normally, the moon's surface shines from the light propagated from our upper atmosphere, from one or both sides of the horizon.

We see a lunar eclipse, as the moon comes nearer to the earth than at other times. At this time its surface does not shine, as there is no light (from our upper atmosphere) from either side of our horizon. If sunlight propagated toward earth casting a shadow on the moon, observers somewhere on earth should have been seeing a lunar eclipse every month, as the moon orbits the earth once every 27-28 days.

The shadow during a lunar eclipse resembles a cloud of soot with a smoothly curved edge, being drawn over the face of the moon. After some time, the color will shift to a coppery glow, as it passes just about the mid point of the earth's shadow. The moon appears to take on this color, as at this point, its surface is only starting to receive a little light from our upper atmosphere, from one side of the horizon.

Light propagates, and there is no such thing as one beam, or one ray of light, bending around the surface of the earth to reach the moon, as we have been incorrectly informed. As there is less light received by the moon, as it comes closer to earth, and as there is darkness around, the lunar eclipse moon appears red. As the earth rotates, and as the lunar eclipse moon 'begins' to receive our horizon atmosphere light, we see the moon through these changing colors. In reality, the color at/of the moon is the same anytime, always, and it is our atmospheric colors that are changing.

Every lunar eclipse is different, as it depends on how close the moon is to earth, where it is viewed from, how much of a different atmosphere was generated with volcanic eruptions, and the amount of extraction.

We can see a lunar eclipse from all over the earth (unlike the solar eclipse seen from only certain areas), as the moon comes closer to the earth, and as its surface cannot light up without the moon receiving light at our upper horizon atmosphere.

A solar eclipse lasts a few minutes whereas the lunar eclipse can last all night

Although the moon orbits the earth once every 28 days, the earth rotates around its axis once every 24 hours, explaining why solar eclipses are seen only for short durations. During a lunar eclipse the moon comes closer to the earth, and its surface does not receive earth's upper atmospheric light from either side of the horizon. However, as earth rotates, parts of the moon's surface begin receiving our atmospheric light (this is why we see through different changing hues observing the lunar eclipse moon). This process takes longer, and we see the lunar eclipse moon for longer durations.

As the moon comes closer to the earth, a lunar eclipse can be seen everywhere on the planet unlike a solar eclipse, which is not only seen for shorter durations, but seen only at/from certain areas of earth.

Mercury is less brighter than Venus even though it is closer to the sun

Venus still holds on to its carbon dioxide atmosphere. With lesser interactions than earth’s atmospheric nuclei disintegrations, it shines bright. Mercury, being closest to the sun should shine brighter (receiving sunlight) than Venus, but without an energy transmuting cycle, or an atmosphere; it is now an impotent astronomical body, like some asteroids.

Why do we see different phases of the moon? Why is the waxing crescent the opposite shape than the waning crescent, and the waxing gibbons the opposite shape of the waning gibbons, the first quarter shape different from the last quarter, when the sun is in the same direction?

The phases

The moon has very little atmosphere, which has been depleted over the years, and is not interacted by the sun's radiated heat to produce light. The moon is not lit up from sunlight, but from the propagated light produced at our upper atmosphere.

During each lunar orbit, we see the moon's circular disk surface appearance change, from not visibly illuminated, through partially illuminated, to fully illuminated, then back through partially illuminated to not illuminated again.

Although this cycle is a continuous process, there are eight distinct traditionally recognized stages called phases. The phases designate both the degree to which the moon is illuminated, and the geometric appearance of the illuminated part. The phases of the moon in the sequence of their occurrence is new moon (its localization between the sun and the Earth, not visible except during a solar eclipse), the waxing crescent, the first quarter, the waxing gibbons, the full moon (its position where the earth is between the moon and the sun), the waning gibbons, last quarter, and the waning crescent.

New Moon occurs when the sun and moon are close together. Full Moon occurs when the sun and moon are at nearly opposite positions, which is why a full moon rises about the time of sunset, and sets about the time of sunrise for most places on Earth.

We do not see the new moon, as it is between the earth and the sun. Being close to the earth, and as the moon's surface shining light is not intense, as the glow at the sun, we do not, and cannot see it's lit up surface through our lit up atmosphere. If light propagated from the sun, it would light up the surface of the moon facing the sun, and some parts on our earth should be seeing a solar eclipse every month.

As the moon continues its revolution around the earth, more and more of its surface are visible. About a week after the new moon, half the moon is visible, and during the next week, more than half of the moon is seen, and finally, two weeks after the new moon, the moon and sun are on opposite sides of the earth. This is the full moon, as the side of the moon facing the sun is also facing the darkened earth, and the entire moon's 'illuminated' side is seen as a full moon. In the next two weeks the moon goes through the same phase, but in reverse.

We see different phases, as we can see the 'illuminated' surface facing us. As the moon is in the waxing crescent, only a part of it is observed. As it comes to ninety degrees in respect to the earth and the sun, more of its surface is 'lit' up. As it continues its orbit to the gibbons phase, yet more of its surface is lit up. As it localizes where the earth comes between it and the sun, it is fully 'illuminated' (the side facing us at darkness). Actually one half is fully lit making it appear as a full moon, and hence its name full moon phase.

At the crescent moon, part of the moons' surface shines (observed to us) with our upper atmospheric light. We see a crescent moon, as we are on the other side of our lit up atmosphere, which is lit up with radiated heat interactions (earth's upper atmosphere facing the sun). As the moon orbits the earth over some days, we then see a larger surface. We see the full moon, as the surface of the face (the complete half disk) of the moon facing us shines. Then at the waning side, as the moons surface being lit up from the other side of our upper atmosphere, we see the opposite shapes.

The moon never goes away, or changes shape-we just see a different fraction of the same moon's surface 'lit' up, with earth's upper atmosphere's light.

The waxing crescent, the first quarter, the waning crescent, and the last quarter shapes of the moon, as illustrated in astronomy books are incorrect. The waxing crescent is shown directly perpendicular, postulating that it is lit up with sunlight. The same is the case with the first quarter. In reality, this is not the case. The waxing crescent, the first quarter, and the waxing gibbons lit up areas are deviated toward the earth, as the surface of the moon lights up, from the light produced at our upper atmosphere, and not by sunlight.

We see opposite crescent, quarters, and gibbons shapes, as the moon is lit up from light of our upper atmosphere, as we observe it at different horizons. Observers in the northern hemisphere, and southern hemispheres, see the phases of the moon in opposite directions, evidence that the surface of the moon receives light from our upper atmosphere, and not from the sun.

Why does the moon's disk appear to be full for several nights in a row if it is clear, and the illuminated disk illuminated percentage changes very slowly at full moon?

At full moon, half of the moon's surface is lit up, receiving light from our upper atmosphere from both the sides of the horizon. As the earth revolves, the moon's surface at full moon receives light from earth's upper atmosphere from one horizon side.

This is also the case with the new moon, as the entire lit up upper atmosphere 'lights' up the surface of the moon facing us, but we cannot see it during daytime, as the glow at the sun is greater, and overpowers our observation.

The disk 'illuminated' percentage changes very slowly, as the full moon's surface receives light from both sides of the upper atmosphere horizon.

Why is the moon not seen in the sky during some nights?

You can never see the moon at nights when the moon is new, or when it is localized between the sun and the earth. The moon rises, and sets, every day appearing on the horizon just like the sun, as the times depend on the phase of the moon.

It rises about 30 to 70 minutes later each day than the previous day, so the moon is out during daytime, as often as it is out at night. At the time of the new moon the moon rises, and sets at about the same time the sun does.

As the days go by (as it orbits to becomes a crescent moon, a half moon, and a gibbous moon, on the way to a full moon), the moon rises during daytime (after the sun rises), rising later each day, and sets during the night, setting later and later each night. At the full moon, the timings of moon rise, and moon set, have advanced, so that moonrise is the same time the sun sets, and the moon sets, at about the same time the sun rises. As the moon orbits (becoming a half moon, and a crescent moon, on the way to a new moon), the moon rises during the night after sunset, and rises later each night. It then sets in the daytime after the sun rises. Eventually, the moon rises so late at night that it is actually rising around sunrise, and it is setting around sunset. That is when it is a new moon once again. This is when the moon is not seen in the sky.

Concluding:-

It is incorrect to state, that it takes about eight minutes for light to travel 150 million kilometers from the sun to the earth. Sunlight never reaches earth, and also never reaches Mercury (60 million kilometers away) the closest planet to the sun.

We see light at the stars, and light at bodies with atmospheres (that are being interacted with solar radiated heat and pressures) within their star fields. We do not 'receive' ' light' from the stars.

No form of energy including light can propagate through dark energy. We can see the light of quasars, and supernovae, at their place of occurrence, not their light, which is theorized to 'travel' to us, or anywhere at a constant speed.

TEMPERATURES

Temperature is a degree of hotness, or coldness, and therefore, heat and temperature is not the same thing. Heat, cold, and/or temperatures are forms/states or structures of energy.

As atmospheric space continuously propagates temperatures and pressures, and space confines every nucleus, nuclei configurations in elements, molecules, and compounds continuously alter. Therefore, if we lower or remove the heat from a thing, nuclei move slowly. When heat/pressure/force is added, the nuclei move and vibrate faster. Disintegration, or fusion of nuclei in high momentum, also engenders heat and pressure/force.

Vibrating nuclei generate momentum, and heat (high temperatures). The degree of temperatures at integrated nuclei, or an object, or an area, is proportionate to the vibrating intensity, and momentum of the nuclei. Higher vibration or momentum manifest high temperatures, and force/pressure.

When nuclei vibrate at very high frequencies 'energy' (pressure or force) is more intense, and things, or areas, get hotter, and hotter, and things are cooler, when they vibrate and move less. Depending on temperatures, momentum/force/pressures, nuclei tend to vibrate at low, or high frequencies.

Air or atmospheric nuclei, or air molecules move about freely, and if they move rapidly, we consider the temperature of the air is high, or the air is hot, and if they move slowly, as on a cold day we feel the air to be cold. Nuclei, and molecules, cannot move about as freely in liquids and solids, but there is motion-energy particles constituting nuclei never stop vibrating, or spinning.

Heat raises temperatures, as it motions in between nuclei configurations (space transmuting propagating heat around nuclei) increasing momentum. Heat raises temperatures, as temperatures are an integral part of force or pressure-both go together. Adding heat is in other words, adding pressure/force, explaining the rise in temperatures. Heat and momentum/motion, or pressure/force and temperatures, are forms of transmuting energy. Nuclei configurations forced to motion at high frequencies in a microwave oven generate heat.

The degree of temperature is proportionate to its pressure/force intensities. Molecules of air, or water, bounce into each other like ping-pong balls in a lottery number machine. The collisions transfer energy (heat and pressure/force) between the nuclei configurations of the molecules, and changes the speeds of individual molecules, but the distribution of speed remains the same, as energy only transmutes into yet other forms states, structures or processes. The higher the temperature, the faster on average the nuclei configurations of the molecules, propel into one another. At cooler temperatures, the molecules, or their nuclei configurations, move more languidly.

As heat motion nuclei configurations in molecules, gases and liquids expand. In other words, as energy (space) in the form of heat, and pressure/force, intertwine nuclei configurations of molecules they expand, or alter their state from solids to liquids, and liquids to gases. This explains why gases expand when heated, and contract when cooled.

Without propagating temperatures and pressure/force, nuclei configurations come to an almost complete stop. This is the coldest possible temperature known as absolute zero. Every existence can be altered, disintegrated, or terminated, with certain degrees of heat and pressure/force.

At certain high temperatures, and pressure/force, transmuting/propagating energy will manifest as light. In other words, only certain degrees of high temperatures and pressure/force intensities will produce light.

Thermal equilibrium

Temperatures can be described as that which determines the sensation of warmth, or coldness, felt from contact with it. When two objects/liquids are placed/mixed together (thermal contact) the object/liquid with the higher temperature cools, while the cooler object/liquid becomes warmer, until a point is reached after which no more changes occurs, and to our senses they feel the same. As the thermal changes have stopped, the two objects/liquids/gases (system) are considered in a thermal equilibrium-the temperatures of all are the same.

Heat or thermal energy flows from one nuclei space to another. Thermal energy is carried by the nuclei space of a molecule to another nuclei space of the other molecule, as they collide, or are put in contact. As the nuclei configurations alter, so do their confining space, explaining how heat/cold is transmuted or transferred, generating thermal equilibrium.

This mechanism for transferring thermal energy by contact is called conduction. Convection is a second mechanism of heat 'transport' involving more energetic molecules in a liquid or gas. For example, water in a pot boiling on a stove; hotter water closest to the flame will rise to mix with cooler water near the top of the pot. In this process, as the molecules are disintegrated by heat, the nuclei are susceptible to the energy transmuting cycle, and rise. As they collide, or mix up with the cooler nuclei of molecules above, heat is transferred. If the heat below is continued, all the molecules will heat up, and expand. Eventually, with continuous heat, all the molecules are disintegrated, and nuclei/plasma extracted into the air, amalgamate/into as atmospheric space.

Important note:

Thermal equilibrium is possible only with/at atmospheres, which are continuously integrating, and disintegrating. These atmospheric processes, is what keeps living things in our biosphere from experiencing extreme temperatures and pressures. Bodies not contained by, or, within continuously engendering, terminating, or engendering atmospheres would be heated up, and disintegrated. The process of termination, or disintegration, would release the heat, and thermal equilibrium would not be possible.

Within our solar system, and other star fields, heat energy from the sun propagated to gaseous astronomical bodies, are regulated by continuously disintegrating, or integrating, or engendering atmospheres.

Why do boiling and freezing points of water vary with altitude?

The multi-layered atmosphere of the earth continuously alters to different temperatures, pressures and densities.

Why do hot objects 'transfer'/'radiate' heat and not cold objects?

As the particle/nuclei configurations of cold gases, liquids or solids, are not in appropriate momentum to engender certain degree of temperatures (this is why they are cold), heat is not produced and therefore, cold objects do not transfer, or radiate heat. For propagation to take place, cold objects must produce higher than prevailing atmospheric temperatures.

If you put a hot object next to a cold one, or if you mix hot water and cold water, the hot object, or water, will alter their nuclei configurations to motion less and less, as their confining space mix, or amalgamates. The cold object, or the cold water particles/nuclei configurations will increase their vibration and momentum, as heat energy is transmuted the other way-thermal equilibrium.

Cold manifests as particle/nuclei configurations are in very low momentum. Heat or high temperatures translate into pressure or force, and this pressure/force within nuclei configurations increase, manifesting heat. Heat/temperatures, and certain pressure/force intensities, are intrinsically linked.

Fire and heat

Fire happens as oxygen in the atmosphere reacts with some sort of fuel (wood or gasoline, for example). Of course, wood and gasoline do not spontaneously catch on fire with oxygen around. For the combustion reaction to happen, the fuel must be heated to its ignition temperature. In other words, heat must motion nuclei configurations of the fuel.

Heat comes from different things such as a match, focused light, friction, lightning, or something else that is burning. These existences must exist at those temperatures, or their combination must transmute to ignition temperatures to begin a fire. For example, as wood is heated to a certain temperature its cellulose material decomposes into volatile gases. We know these gases as smoke (compounds of hydrogen, oxygen and carbon). The rest of the material of wood is calcium, potassium and so on.

As the compound molecules (nuclei configurations) break apart, the nuclei recombine with atmospheric oxygen (nuclei) forming water, carbon dioxide and other products, or in other words, this reaction is called burning.

These chemical reactions produce a lot of new heat, and it is this heat (within nuclei configurations) sustaining the fire. So wood will burn, as long as it continues to decompose, and oxygen continues to be available. Many fuels burn in one step-gasoline is an example. Heat vaporizes gasoline, and it burns, as a volatile gas.

As a candle burns, the rising disintegrating carbon nuclei emit light. This heat and pressure/force producing the light effect is called incandescence, and it is the same process that creates light in an incandescent light bulb.

The candle flame color depends on what material is burning, and how hot it is. Color variation in a flame is manifested by uneven temperatures-the hottest part-the top-glows blue, and the cooler parts below glow orange/yellow. As the disintegrating nuclei in those gases are extracted by/within the energy cycle, the top of the flame is hotter, and the flame is always pointed at the top. As fire burns, the material is disintegrated, and as the particles now are less dense, and lighter, they are subjected to extraction by the energy cycle and therefore, move upwards as smoke. This also explains why flames are always pointed at the top.

Fire produces heat, and keeps the burning material at the ignition temperature, so it continues, as long as there is material and oxygen around it. Fire heats any surrounding material, and transforms the gases as well, and if they ignite, the fire spreads. In short, fire is the result of a chemical reaction between two gases, typically oxygen, and a fuel gas. The fuel gas has to be generated by heat. In other words, with heat providing the necessary energy, nuclei in one gaseous compound break their bonds with each other, and recombine with available oxygen atomic nuclei in the air to form new compounds, plus lots more heat. Only some compounds will readily break apart, and recombine in this way.

When you boil water it transmutes into the gaseous form of steam, but this gas does not react with oxygen in the air, as hydrogen released is extracted at super speeds by the energy cycle. With the right temperatures, and pressure/force, oxygen combines with hydrogen in the atmosphere into/as water vapor.

Most flammable compounds contain carbon and hydrogen. Flammable fuels catch fire at different temperatures. It takes a certain amount of heat to change any particular material into a gas, and even more heat to trigger the reactions with oxygen. The necessary heat level varies, depending on the nature of the nuclei configuration patterns in the molecules making up the fuel. A fuel must have a certain heat level to form a gas, which will ignite when exposed to a spark, and the temperature must be much higher for the fuel to ignite without a spark. The fuel size also effects how easily it will catch fire (a block of wood or a toothpick). The fuel's shape also affects burning speed (block of wood or splinters). A fuel heat production depends on how much heat the gases release in the combustion reaction, and how quickly the fuel burns-both factors depending on the fuel's composition.

LIGHT

Light is a form of transmuting energy. As energy continues to condense as/at a star, and as energy is in continuously momentum, heat/temperatures, and pressure/force, concurrently continuously engender. These high degrees of temperatures and pressure/force intensities propagating within/as space, as it waves, manifests as light.

As space is an integral part of any particle/plasma, proton, nuclei, object, or an astronomical body, this generated pressure/force, or momentum, and temperatures, transmute/propagate as/within space, as space also transmutes to these forms of energy, as it waves. Space is a form of energy, and waves manifesting forms of energy. The intensity, frequency and wavelengths of space waves will manifest different forms of radiation/heat/light, color, pressure/force and sounds.

As these forms of energy dwindle with distance, and as temperatures and pressure/force alter, light is not apparent, and these forms of energy transmute/propagate, as other forms of radiation, ultraviolet, infrared, gamma and other rays.

All forms/states/processes or structures of energy exist (as humans ideate existences), as energy transmutes/propagates. Propagation is momentum-a form of energy. In other words, energy in/as motion transmutes, or these existences are observed, and experienced, as energy transmutes. In reality, as energy is everything everywhere, energy does not 'move' from there to here, or vice versa. As energy transmutes, momentum is apparent, but other forms of energy manifest.

Even though processes at stars produce extreme temperatures and pressure/force, without space, these forms of energy would not propagate/transmute. In other words, without space temperatures and pressure/force would not engender in the first place.

As mentioned earlier, energy in momentum engenders heat, and pressure/force. Electricity is a form of energy, or energy in/as momentum, or energy as pressure/force. As force/pressures are forced into a filament, the increase in force/pressure, and the concurrent heat produced, generates light.

There is no such thing as a photon, or a particle of light, as light manifests only as space transmutes, waving as light (in frequencies and wavelength proportionate to those degrees of heat and pressure intensities). Light is not an entity-nothing is. Every thing (all existences) is a state/form/structure or process of transmuting energy, as/within the fabric of existential energy.

As our upper atmosphere is interacted, and disintegrating nuclei produce light, this light propagates within/as atmospheric space, as it waves, explaining why we experience our visible light.

All forms of energy including light do not propagate from any 'source', and does not 'reach' any target'. Energy in the form of light is apparent, as energy transmutes to light propagating through/as space, as space waving transmutes to light. Within the same pathways, as light is propagating from its 'source', other forms, states or structures of energy are concurrently propagating the other way, as energy shuttles to-and-fro. Energy continuously transmutes from one state/process/structure or forms into yet another state/process/structure or form. There is light at a bulb, as electricity-a form of energy- is in momentum. Electricity as energy in momentum is generated by yet another form/state/process or structure of energy, as it transmutes.

Energy manifests as light, as space waves in wavelengths in the range of 400 to 700 billionths of a meter. But the full range of radiation wavelength extends from one billionth of a meter, as in gamma rays to centimeters and meters. Light waves also come in many frequencies. The amount of energy (as pressure/force) in a light wave is proportionally related to its frequency: high frequency light has high energy (high pressure force and temperatures); low frequency light has low energy (low pressure/force and temperatures). Light space waves not only wave at different frequencies; it also propagates at different speeds (depending on the intensity at the 'source'), dwindling with distance, and slowing down propagating through density-air, water or glass.

In reality, light does not propagate through space (as it waves), but the space per se waving at light frequencies is the light apparent. In other words, space transmutes into light, and as space is everywhere, there is light.

How does light and heat emerge from a light bulb?

Electricity as energy in momentum is generated by yet another form/state/process or structure of energy, as it transmutes. Energy in momentum, or electricity, transmutes the light bulb filament, as the force/pressure within/as electricity, or energy in momentum heats it. This transmutation produces both heat and light.

Measuring energy 'propagating'/transmuting speeds

To measure the speed of light, you must first produce light. The speed of light was measured in an experiment, as light continued to propagate (the 'source' continued to produce light energy). Light propagates through/as space, as spaces waves. Propagating (spreading out) light, or any form of energy dwindles (in intensities) with distance, as everything is a form, state, process or structure of energy. Momentum and/or propagation are also forms of energy.

The speed of light measured between two points depends on the light producing intensities at the 'source'. These intensities must be constant, and must be continuous. Moreover, as light, or any propagating form of energy, dwindles with distance, the speed of light nearer the source, and the propagating dwindling intensities must be known to reach accurate measurements.

As everything everywhere is energy; there is no void, or vacuum anywhere. The postulated 'vacuum' supposed to have been created in that experiment contained space-a form of energy. Light propagated as space (the supposed vacuum) waved, or transmuted to light. The vacuum created was void of all atmospheric particles/nuclei, molecules or compounds.

Light or sound can be measured at very small distances, as any form of energy propagating within/as space, dwindles with density, and also with distance. Forms of energy, as space waves propagating from a 'source' diverges dwindling intensities.

In reality, forms of energy, or existences, alter their characteristics continuously, and also with propagation. Light is not light propagating through certain distance (depending on the intensity at the source), but transmutes to other forms of radiation.

Therefore, we cannot accurately calculate the speed of light, or sound, or any form of energy. Space per se transmutes/transforms into those forms of energy. And as space is everywhere this exercise is futile. How can we measure the degrees of temperatures and pressure/force intensities of/at the 'source', or energy transmuting manifesting/propagating light? Moreover, all forms of energy depend on the intensities (producing them) at the 'source', which should be continuous, and constant. Furthermore, light or any form of energy does not propagate one way, as energy propagates/transmutes shuttling to-and-fro within the same pathways. There are no entities, and there is no causation.

It is assumed, that a force known as gravity exists, and influences more or less the speed of light. The next revelation details, why there is no such thing as gravity. Scientists have recently been studying the speed of gravity, and found out that it more or less matches the speed of light. The speed of light has been found to more or less match the theorized force of gravity. Earth's atmosphere continuously experiences radiated heat and pressure/force from the sun. This force is postulated as gravity, Light from our upper atmosphere propagates within or as atmospheric space, and the postulated force gravity more or less matches the speed of light.

There is no such thing as causation. Theories considering causation are completely incorrect. Energy cannot be created or destroyed, yet we hypothesize light travels indefinitely at great speeds. If energy is everything everywhere, another form or forms of energy should concurrently propagate within the same pathways as light-there is no such thing as a one way phenomenon. This explains why we cannot explain our one way conceptualized force of gravity.

Does light slow as it passes through water or glass, and how does it return to light-speed on its own once it exits?

Light or space waving as light, imparts a small momentum punch when it bounces off objects, and also by continuously shining light through a glass block will heat it, as light-a form of energy-is constituted of heat and pressure/force. Light as space waves interweaves through the nuclei configurations of water and glass, explaining why it slows down a bit, but as it exits these mediums-as space is everywhere continuously propagating these forms of energy-it continues its normal speed.

Is light an electro magnetic radiation?

We are informed, that light has both electric and magnetic fields referred to as electromagnetic radiation. This is incorrect.

Electricity is energy in/as momentum. Electricity forced through a filament transmutes to light, as heat is produced with pressure/force. Light is also energy in momentum. Electricity is a form of energy, as energy in continuous momentum. Light is produced or manifested, as energy in momentum, as high heat and pressure/force propagate through/as space, as it waves. Without space, temperatures or pressure/force will not engender and therefore, not propagate. Light is not an electro-magnetic radiation. Importantly, light is not light propagating through a certain distance, as heat and pressure/force (constituting light) dwindle in degrees and intensities.

We are informed, that light is an electromagnetic radiation existing as packets of energy called photons, and acting like wave-the energy of a photon is precisely related to its wavelength. It is stated that solid objects, or matter, consisting of particles (nuclei and molecules) include electrons. Electrons only exist at certain energy states, and a photon, which makes contact with a particle, may excite an electron or molecule to a higher energy state. This is incorrect.

The atom is hypothesized. Nuclei are confined by their space, which is further confined by atmospheric or interplanetary space. Space is a form of energy. The postulated electrons do not exist and therefore, do not orbit a nucleus, as we cannot explain their energy source. What makes an electron to continuously orbit that very same nucleus?

Light is not a particle, as a free particle is confined by space, and are integral parts of each other. Every particle, plasma, proton or nucleus comprising objects or astronomical bodies, and even galaxies are confined by their space.

Is light a particle or a wave?

In what form (particle or wave) does light travel-if it does-when it is supposed to have reached us from distant stars, after millions of years, or light years through space?

Light is not a wave, but is apparent, as space transmutes into light, and waves in frequencies, and wavelengths, manifesting the degrees of temperatures and pressure/force intensities (of light).

Light is not a photon, nor a particle. A particle, proton or nucleus has their confining space, and as these existences are in high momentum heat and pressure/force manifests. These forms of energy propagate through their confining space, and further within/as atmospheric or interplanetary space, manifesting light (with certain degrees of heat and pressure/force intensities).

The amplitude of a space wave tells us about the intensity, or brightness of the light, relative to other radiation waves. Radio waves, x-rays, ultraviolet light, infrared have their own characteristics wavelengths.

As light propagates, it is not light (as an assumed 'entity'/existence) per se that propagates, but in reality, energy/space is transmuting in the form of light (heat and pressure/force). As space is everywhere, light is apparent.

When you move your hand through a filled bathtub a wave is generated, as a result of transmuting energy to the water. Your hand is a structure of energy, and your hand movement (momentum) alters the fabric of water/energy. Energy in momentum, or as force/pressure, is transmuting in the form of a wave.

Light is not a particle. As light is apparent with the waving of space, we cannot state that light is a wave, as it is space (also a form of energy) that waves.

It is the waving of space at different wavelengths, frequencies or amplitude, or intensities, manifesting momentum, or pressure, or force, light, or heat, or temperatures and color, sound and other forms/states/processes and structures of energy.

What about the two slit and interference experiments?

The wavelike interference pattern in this experiment is always equated to water wave interfering. Our atmospheric space is filled with protons and nuclei. As space waves light, or heat, and/or pressure/force, it carries these particles/protons, or nuclei, which strike a detector or a screen. The double slit experiment is evidence that light propagates within/as space, as space waves, and particles and/or nuclei existing within atmospheric space are carried along within those waves. There are therefore, no particles of light or photons.

Does light travel in a straight line?

Light is not a particle and therefore, does not travel like a bullet. Light as space waves cannot separate into beams, or rays of single light lines. Light as space waves propagate from a 'source'. As space is everywhere, light is apparent.

Light as space waves will continue propagating to a certain distance as light, as long as the intensity is continuous, and constant, at its 'source'. However, those intensities dwindle with distance, and light will alter, or transmute to other forms of radiation. This explains why a candle, or a street lamp, lights up only a definite area. This also explains why sunlight does not reach earth, and why inter-planetary space is dark. Therefore, all theories about the speed of any of energy transmutation forms, state, processes or structures are preposterous.

What about light years and light traveling to us from distant stars for billions of light years? These are preposterous postulations, or theories, or stories, or science fiction, tickling the layman's illusory minds.

We can see the incandescent at the sun, or at a lit bulb, but we cannot see light. We see within, or by light. Our visual systems have evolved mechanisms to see objects within light. Our eyes are insignificant during darkness, or without light. We would not be able to see anything, if there was no scenery, or no objects. Objects and sceneries are observed only with light. We would not be able to hear anything if there was no sound, even with our auditory system.

What are shadows?

There is no such thing as a ray of light, or a beam of light, as we would be 'individualizing' light. There are no entities. Light is usually modeled as straight lines in drawings, which are very confusing.

Objects are structures of energy, existing within/as space. Therefore, light space waves unable to wave through the objects' particle configurations are stopped, generating a shadow. Light space waves can wave through certain particle configurations of objects, explaining why we can see through some materials, and why they are called translucent, and also explaining why these materials do not cast a shadow, as light interweaves through.

An object stops light space waving and therefore, stops light propagation. If there is no wall, or a surface, behind that object, light would continue as light space waves, as it is reinforced by space waving light behind the object.

A shadow only forms on a wall, or surface, immediately following the object traversing light. Space as a form of energy- if it is not waving-does not propagate/transmute/manifest any form, state, structure or process of energy. The size and other characteristics of a shadow, depend on the distance of the traversing object in front of the light from the source. The closer the traversing object is to the 'source', the larger the shadow. The greater the intensity of the 'source', the sharper is the shadow. A shadow is usually black. There is darkness, or black, with the absence of light. Shadows are not evidence that light travels in straight line.

Reflection of light

As light or space waves strikes an object, light energy, or space waving as light, transmutes into other states, forms or structures of energy.

We see objects out there, as they stand out in the landscape within light (as space light waves continuously strikes them), and not as light reflects from the object, as we have been wrongly informed. Light does not reflect from those objects as light. If it did, we would not evolve visual system to 'see' those objects, as reflected light would be continuously striking our eyes. We would have evolved receptors, or light sensing mechanisms to sense light, and not see objects within light. As light waves space, and vice versa, and as space is everywhere, there is light.

Refraction of light

Space light waves propagate three dimensionally, and do not 'travel' as a straight line, or lines, or a beam, or a ray. It is therefore, incorrect to state that light bends, as it refracts into another medium.

Some solids like transparent glass are comprised of certain particle configurations, where space light waves through their confining space, explaining why they are translucent. There are also some liquids as water, where space light waves through the molecules' nuclei configurations. An object such as a pencil in a liquid will appear bent, not as a result of light bending. We are observing the pencil through two mediums, the glass that contains the liquid, and also the pencil within the transparent liquid. Each medium is comprised of different particle configurations, and this makes the pencil to appear bent. Light as a propagating space wave does not, and cannot bend as postulated.

Light cannot be scattered or absorbed

Light is a space wave, and cannot be absorbed, or scattered as light, as it is not a particle. When light comes into contact with the surface of an object, it interacts with its particle configurations, and their confining space constituting that object. We cannot state light is absorbed, scattered, partly scattered, partly absorbed, or reflected off that object. Everything is energy in a state, processes, structure or form, and every existence continuously alters its characteristics.

As light strikes an object it is no longer light. Light is not distinct, as there are no entities, and so nuclei of that object will transmute into another state, process, structure or form of energy, and therefore, will not 'absorb' light, but heat energy. Light cannot also be 'scattered', as with interactions it will change its characteristics. However, if light strikes an object with certain nuclei configurations, as a mirror or still water, light space waves weave through these configurations, and propagate out with lesser intensities. Propagating back within light, space light waves reinforces these intensities as before.

Why and how does light go through transparent media and stops at opaque ones?

Objects are called transparent, or translucent, as they are comprised of certain nuclei configurations, wherein space/light waves through their confining space. Some objects do not have their nuclei configured in ways, facilitating light space waves to wave through, and are called opaque. Water and some types of glass are translucent materials. Crystals are also good examples of translucent substances, as they are comprised of regular arrays of nuclei, and the distances between them easily allow space waves through. Nuclei configurations in transparent/translucent materials are much spaced out, allowing certain frequencies of light space waves to pass through. Most materials are transparent to some frequencies, but not to others. For example, high frequency light such as gamma rays, and x-rays, will pass through ordinary glass, but lower ultraviolet, and infrared light, will not (as space wavelengths are larger). Some translucent materials like grease proof paper, and frosted glass let some, but not all light through transmuting, or diffusing the light.

Opaque objects do not have nuclei configurations allowing light space to wave through. Wood constituted of certain types of integrated nuclei makes it dark, or opaque, but light does alter its characteristics. High intensities of light can burn wood. Some materials are opaque to some frequencies of light, but transparent to others. Light propagates within/as space, so if you pull down a window shutter, you cut off light space waves, although there is space in your room-there is no light.

Some material transmutes light, or energy (heat and pressure/force), as the incoming light intensity, is at or near the vibrating frequencies of the particles/nuclei in the material. The energy transmutes the particles to vibrate at new frequencies, and new characteristics. The material may get hotter, emit colored light, or integrate with other particles, or disintegrate. It all depends on the amplitude, and frequency of the waving space light.

How does a camera work, if light is not reflected?

We are incorrectly informed, that light reflects from objects resulting in the changes in the chemical of the photographic film inside the camera. Light does not reflect the image of the object. Light propagates within/as space, and as it strikes the object an image of the object is generated. This image is captured by the chemical configurations of the film, as it was exposed to atmospheric light space waves by opening the shutter.

When you click the shutter you have frozen a moment in time, by recording the visible light exhibiting objects in the camera's field of view. In the same way that our visual systems have evolved to 'see' objects, or the scenery, which stand out by light, the chemical configurations of the film alter in ways, which can be developed-treated and modified-to produce a representation (a print) of that moment you clicked the shutter. When you click together with the flash, light propagates making the object apparent, and it is this exhibition that is captured by the film (particles/nuclei configurations in the film transmute with light space wave interactions).

When we open the shutter to take a picture, as light propagates everywhere, the chemicals in the film interact with only what they 'see'. This is why the shutter is designed to snap shut quickly. If the film is exposed before it is developed, light space waves will impose many images, spoiling the picture. Atmospheric 'lit' up space continuously waves light to-and-fro, from object to object.

Visual systems see depth, wide angles, stereo-type vision, height, distances, shadows, curvatures and other things having evolved mechanisms with foraging and survival requirements, or necessities, or considerations. In the same way, photographic film chemistry is created by our visual system, explaining why we can see prints in a film the same way we see objects or scenery. Other mammals not having evolved our type of visual systems will not see objects/landscape, as we see it.

Photo films are carefully engineered to chemically alter to light space waves, generating an image. Photographic films are made with organic molecules known as spectral sensitizers, and are added to the surface of the grains to make them more 'sensitive' to blue, green and red light.

Light does not reflect from an object at the same intensity, as it strikes it. We see landscapes or an object, as they stand out within light (being energy structures within/as the fabric of space/energy).

Our eyes have liquid as the external membrane, which is similar to the particle/nuclei configurations at the surface of still water, or a mirror. If we put in an eye drop, we are unable to see, in the same way disturbed water, or a broken mirror cannot 'see'. We do not see our reflection on other surfaces, as those surfaces are not configured to mirror surface type nuclei configurations. The film in a camera also has the same nuclei configurations chemically engineered that 'transmute', as the shutter is open, instantaneously capturing the object/landscape. Light does not reflect to our eyes from the mirror, or from the surface of the water, or to the camera film.

More on light

Bright light on a piece of mirror or glass (smooth surface) will form a blotch of light on the opposite wall or ceiling. With a mirror and some smooth surfaces, space light waves through their particle configurations, propagating back within/as space light. However, the intensity of this 'reflected' space light wave alters. This distorted light space waves, then create a blotched up light image on the opposite wall, or ceiling.

Light does not reflect from the mirror, but the mirror now becomes the 'source' of another form (intensity) of light.

Space light waves propagates through the nuclei configurations of translucent glass (this is why it is translucent), and will not create a blotched up light on the opposite wall, or ceiling-there is no reflection.

How we see ourselves in the mirror if light at that same intensity is not reflected, and why does the mirror point at you when you point at it is discussed within revelation 14 (light and sounds).

Why does light through a magnifying glass start a fire?

The magnification mechanisms (nuclei configurations so patterned) in a magnifying glass, converge light space waves to a point on the other side. As the intensity of light waves together at a point increases, so does heat and pressure/force, which heating up the object at the other end, generates a fire.

How do solar cells work?

Solar cells are made of special materials called semiconductors such as silicon, currently the most commonly used. As these cells are exposed to atmospheric light, the nuclei configurations transmute light, or heat, or thermal energy. Forms of energy transmute to yet other forms of energy. These cells now hold 'current' or electricity. For example, this current can now power a calculator. This is a basic process, but there is really much more to it.

COLOR

Light is a form of transmuting energy. Certain high degree temperatures, and pressure/force intensities, propagating within/as space-as it waves-manifest light. Lesser and higher temperatures and pressure/force intensities, propagating within/as space manifest other forms of radiation, x-rays, gamma rays and radio waves. Light does not contain colors, as we have been informed, and therefore, lesser degrees of temperatures, or heat, and pressure/force intensities do not manifest colors.

As light, forms of radiation, heat and pressure/force interact with nuclei increasing their vibration/intensities, different colors manifest. Color is therefore, linked to temperatures/heat.

Light is apparent as temperatures and pressure/force propagate as/through space, as space alters or transmutes to light, and waves. As space is everywhere, light propagates until it dwindles with distance, depending on the intensities at 'source', explaining light everywhere.

Color is apparent, as nuclei vibrate/rotate, generating certain temperatures/heat intensities. However, as nuclei exist within atmospheric space, they cannot propagate as space waves (as light does), and therefore, the generated colors are apparent at/as nuclei, and/or areas of vibrating/rotating nuclei. We can see colors only where there are nuclei for example, objects constituted of nuclei, nuclei within our atmospheric space, atmospheres of planets, and nuclei within the interstellar medium. This explains why we see things in color (things constituted of nuclei), and why we do not see propagating colors, or colored light.

Particles/plasma/protons, and nuclei are integral parts of their confining space, and their confining space exhibits the colors generated by vibrating nuclei. We see streaks of meteor showers of different colors, as nuclei are interacted with heat transmuting to colors. As all forms of energy dwindle propagating through/with distance, these streaking across colored nuclei are observed for short duration.

Different intensities or vibratory/rotary levels of nuclei will manifest different colors, and correspondingly varying degrees of temperatures. Each degree of heat manifests a different color, or nuclei vibrating/rotating at varying intensities, display the corresponding colors. As each color is generated with each degree of temperatures, color is linked to temperature/heat.

Color is only apparent with/at nebula, interstellar medium or atmospheres with/of nuclei. Objects and landscapes comprising nuclei will also be apparent with color, if there is light (heat and pressure/force).

With light (as atmospheric space propagates heat through nuclei configuration space), nuclei constituting that object vibrate/rotate, manifesting their colors within their confining space. Particles/nuclei vibrating/rotating to different intensities manifest varying colors, and correspondingly degrees of heat/temperatures and pressure/force. The faster nuclei vibrate/rotate the more heat they generate, and the color manifested is closer to the blue end of the color spectrum.

Colors determine the temperatures of a body. Things glow with different colors when heated. All objects, no matter what their chemical nature, or composition, send out a certain color if these objects are heated to the same temperature, evidence that a certain color is linked to a certain degree of heat. This also explains that nuclei configurations were patterned, and/or structured with the same degree of heat and pressure/force.

We cannot see colors without heat and pressure/force. In other words, we cannot see colors without light, as light comprises temperatures and pressure/force. This is why in/at objects of color (where their nuclei configurations are nearly at a stand still) light must activate them to vibrate/rotate, and they manifest their color.

The color of an object changes, as it is slowly heated. A cool object gives off no apparent color (as their nuclei have nearly stopped vibrations), but a hot poker glows red, then orange, yellow and then blue. Red is the lower end of the color spectrum where temperatures and pressure/force are less.

A candle or a campfire does not give out white light, but an incandescent. As a candle burns, or with a campfire, atmospheric nuclei are being disintegrated and therefore, there is no white light produced. As there is no white light generated around a campfire all objects have the same hue, and there is no color.

Just as you mix hot and cold water you get a different temperature of water, so also if you mix two colors you get a different color with light, as nuclei configurations alter.

If color is linked to temperatures why are red colored objects not hot?

Each nucleus engendered, and endures, as a result of certain prevailing temperatures and pressure/force. As nuclei configurations conjoin to form objects, they loose their 'individuality' within space, and are not easily influenced with altering temperatures and force/pressure. However, constituting the object they display their colors within light. Light (heat and pressure/force) activates nuclei on the surface of the object, explaining why those colors manifest. This explains why we cannot see colored objects in the dark.

Atmospheric colors

Our atmosphere consists mostly of nitrogen, oxygen and hydrogen. Hydrogen is also abundant (50% more than oxygen in the atmosphere at any given time), but it is extracted at super speeds, as replenishment material for the sun, and its amount in the atmosphere cannot be determined.

Each nuclei, or nuclei configuration, or element, or molecule interacted with light, or heat and/or pressure/force, increase their vibrations/rotations, generating yet more heat, and or pressure/force, manifested as colors, and these colors will be different corresponding to the heat and pressure/force. Interacted with radiated heat and pressure/force from the sun, atmospheric nuclei configurations vibrate/rotate to their color intensities.

Nitrogen manifests the red color (the abundant nitrogen is why we have colored sunrises/sunsets). Oxygen manifests the green color, and most plants appear green, as they are continuously processing oxygen (during the day). The interaction of radiated heat on the departing hydrogen/protons gives out a blue color, explaining why our sky is blue (more blue when skies are clear-more departing hydrogen with more extraction).

The oceans also look blue from above, as the water molecule consisting of two parts of hydrogen is continuously disintegrated. Observing it closer, the ocean water looks greenish-blue, as one part of disintegrating oxygen also makes up the water molecule.

Plants/trees are mostly green, as oxygen is being produced. At night without light space waving they take in oxygen, and produce carbon dioxide. At night plants breathe in oxygen, and it is only with light that they can be seen. With light they begin processing oxygen, and are observed green.

Our troposphere is very dense. With radiated heat and/or pressure/force, vibrating/rotating nuclei manifest colors. However, we do not see our troposphere full of color, as our atmospheric light generated at our upper atmosphere propagates through/as atmospheric space to the surface of the earth, and we experience light. Moreover, atmospheric nuclei transforming into colors collectively produces white light, enhanced by the upper atmospheric light, and also by the light seen around the sun (not from the sun).

Our visible light is not light from the sun, or sunlight

Energy in the form of radiated heat, pressure/force and protons propagate outwards from the sun, interacting bodies/atmospheres within interplanetary space. These forms of energy interact with our upper atmospheric nuclei.

Protons extracted from our oceans, lakes, surface and the atmosphere leave at super speeds, interacting with other protons, and forms of energy from the sun. These interactions disintegrate our upper atmosphere nuclei, generating heat and pressure/force.

Neutron plasma and protons from disintegrations of tritium, deuterium and helium, generate heat and pressure/force with their release. Intense heat transmutes nuclei into plasma (our ionosphere is filled with plasma) which is easily extracted to the sun. This is how, and where our visible light is produced. As light propagates within/as space, atmospheric space waves as light, and we experience our visible light.

Hydrogen or a proton with its confining space does not glow with heat or pressure/force. The space confining a proton must be removed in order for naked plasma to glow. This explains that a star/sun does not contain hydrogen. The sun or a star glows, as plasma produced at the photosphere is expelled via the corona. With equal intensities of incoming heat and pressure/force interactions, the corona of departing and incoming plasma enhances the glow produced at the photosphere.

During the day, our upper atmosphere is continuously being interacted, disintegrated, and integrated again, with radiated heat and pressure/force from the sun, explaining our day light.

We have our visible light for many reasons.

1. The light at the sun (not from the sun) dominates over our atmosphere.

2. There is no nitrogen at our upper atmosphere, and therefore, no colored red/orange light.

3. The interactions from/by radiated heat, and pressure/force from the sun, activate upper atmospheric nuclei, which are ionized producing light. This light is propagated within/as atmospheric space, explaining why we experience light even at the surface of the earth.

4. As the light at the sun (not from the sun) completely dominates over our atmosphere (during the day), and these three combination colors of red, green and blue collectively alter to a single 'white' color, or white light, atmospheric space waving white light is enhanced, propagating light.

5. At nights, without direct radiated heat and pressure/force, our upper atmospheric nuclei are not interacted, and there is no light (we are in darkness). At nights, we do not see our dense troposphere of colors, as/and there is darkness, as there is no light, as colors (with heat and pressure interactions on nuclei) manifest only in/within light. In other words, our troposphere colors are not observed both during daylight, and at nights.

We assume that light propagates from the sun, but this is incorrect. We do not receive any sunlight, but only radiation in the form of temperatures/heat, ultraviolet, infrared and gamma rays. Light and other forms of energy dwindle, propagating through distance. This explains why we do not receive those extreme temperatures and force/pressures, generated at the sun. It is within/through this visible light (generated at our upper atmosphere) that we observe the glow at the sun through the radiated heat/pressure/force pathways. In reality, the glow at the sun enhances our upper atmospheric light, otherwise our atmosphere would probably be multi-colored, and during the day we would observe the sun, as we see it during sunsets and sunrises.

The three abundant gases of our atmosphere are nitrogen, oxygen and the departing hydrogen; therefore, the dominant colors of the atmosphere are red, green and blue. The three primary colors are red, blue and yellow. They are called primary colors, as there are no two 'lesser' colors that make them up. Mixing red and yellow gives you orange, and yellow and blue give you green. Blue and red give you violet. Purple is not a primary color, as it can be created through combining equal parts of blue and red. White is a combination of the colors green, red and blue.

Mechanical things that make color like a television, or a computer screen are created through the mixing of three basic colors to make other colors, a process known as 'additive color'. The three colors used to start additive color mixing are red, blue and green, as starting with one composite color green you can still create yellow, as it is contained in the green. Actually, you are actually starting with four colors red, green, blue and yellow.

Subtractive color is the concept of combining colors to make another, like mixing red and blue paint to get purple. That may sound like additive color, but in reality colors are made by subtracting a hue out of the color scheme by adding more of another. Adding more white to black makes it more silver, subtracting more black, as more white is mixed in.

The difference between additive and subtractive color is what you get when you add them all together. If you add all the colors together in a subtractive color method you get black. This results, as you added them all together, and all those colors subtracted from all the others leaves no set color or black.

Within our atmosphere or even a computer, the colors are mixed with light as the key component. Without light there is no color. Mix a computer's additive colors red, green and blue together, and you get white. Red, green and blue light of equal brightness mixes to make more or less white light.

The atmosphere of Venus is constituted mainly of carbon dioxide, and yet it shines bright. Venus' atmosphere is interacted with radiated heat and pressure/force from the sun, and the disintegrating nuclei into protons and plasma generates an incandescent. Mercury and Mars have lost their atmosphere, and so there is very little incandescent.

Is there any such thing as white light?

We can see the incandescent at the sun, the stars and lightning. We consider this white light, as it is seen at a distant. However, we cannot see light when it is around us. Our visual system comprises atmospheric nuclei. As atmospheric nuclei light up to radiated heat and pressure/force, the nuclei in our eyes concurrently light up-we then 'see'.

We would not be able to 'see' if there were no objects/landscapes, but only sense light. We can see objects/landscapes within, and by light, but we cannot see our atmospheric light, as our visual mechanisms that 'see' are light per se. As light/color is linked to temperatures, we see different types of light.

Why is the color of the sun white?

White light cannot be described, as it is not white color, as humans describe it. We conceptualize existences, and it is therefore, difficult to explain/describe them.

We see the incandescent/glow at the sun through our upper atmospheric light, and also through our colored troposphere. We see the incandescent at the sun, as a golden yellow or silvery white. The color at the sun seems to alter, depending on clear or cloudy skies. With clearer skies more hydrogen is extracted (the sky is deep blue), enhancing the silvery white light at the sun. As we see the glow at the sun, it appears yellow, as yellow equals white minus blue. This mix up of colors makes the sun to appear yellow. However, the color at/of the sun stays the same overhead, or at the horizon.

Why is lightning white light?

Light is not white, and light cannot be explained correctly. Clouds of nuclei collide with each other resulting in light, and thunder (sound).

Lightning should have appeared colorful, as our troposphere is dense, and only colors, not white light should have been seen. However, lightning is mostly seen during dark skies, and intensities are propagated within/as space as light. The pressure or force generated at these collisions, force nuclei configurations asunder. Vibrating nuclei at those intensities produce light (the lightning) as atmospheric colors collectively manifest as white light, and as atmospheric space is not waving white light during dark skies.

The steaks of lightning observed are chains/floods of vibrating nuclei scattering haywire, trying to find pockets within the dense troposphere. We do not see continued light with lightning, as the source of heat (collisions), and pressure/force, must be continuous.

Why do we see mostly white light from stars?

We see light at star field bodies, and not only from a star, although the star has a greater incandescent, explaining why most star fields appear white. Some stars have hues/colors, as we see this light through their outskirts of star fields filled with nuclei, elements, molecules, and compounds.

As our hominid ancestors migrated from their habitats, developing their visual systems gradually over generations, only humans can see stars. As we evolved to see distances over generations, looking afar, we evolved visual systems to first see the moon, and eventually the stars. If we stare at the stars over many generations the stars will get clearer and brighter, and we will be able to see more distant stars, as our visual systems will get progressively better.

How can you distinguish planets from stars?

Color is found in areas where nuclei are present, such as the air in/as our atmosphere, nebula, and the outskirts of star fields. All stars radiate white light, as we see their incandescent through the outskirts of their star fields, filled with nuclei and molecules. Collectively they manifest as white light. Some outskirts of star fields have different types, or an abundance of one type of element or molecules, explaining why some stars are of a certain hue.

Some planets like Mercury and Mars have depleted their atmospheres over the years. Some like Venus and the Jovian planets have dense atmospheres, and shine bright with radiated heat/pressure/force interactions of/at their atmospheres.

Earlier it was difficult to distinguish between stars, and some planets, but now it is much easier with sophisticated instruments, and advances in calculation/measuring methods and systems. Even today, there may yet be some planets, which have not yet been discovered.

Why do stars twinkle? Who do planets not twinkle?

The twinkle observed at stars is evidence that there is a functioning energy transmuting cycle between the star, and its nebula/bodies within its star field. Light does not 'arrive' from any star. We see light at a star, and not light from a star. As we see this light through their interstellar medium filled with nuclei/elements, molecules and compounds, and as the star field is in continuous rotation the light at the star twinkles.

As the star rotates at super speeds, star field space rotates the other way producing spiral arms. Each spiral arm ephemerally blinks the light at the star producing those twinkles. Horizon stars twinkle less than those observed overhead, as there is more atmospheric density between us and the horizon.

Planets have space/atmospheres around them, which as interacted produce some kind of incandescent, and their interiors do not glow.

Why is a burning candle flame, or fire not white?

What is the color in fire?

Fire produces heat. As a candle burns, the wick and the wax is disintegrating, and different temperatures and pressure/force are generated, thus giving out varying colors.

As fire or a candle burns, their constituents are disintegrating. Furthermore, these processes also utilize atmospheric nuclei, which are also disintegrating. If you heat a horse shoe, it will turn red in color, further heating will turn it into white color. Adding heat alters the degree of color seen, as nuclei configurations alter.

With a fire, or a candle flame, the red color is observed as atmospheric oxygen is used to incite the combustion, subsequently disintegrating into nitrogen. With nitrogen being dominant in the atmosphere the flame appears red. Oxygen and nitrogen then break down into carbon and hydrogen. Carbon immediately mixes with atmospheric oxygen giving out smoke, explaining why with burning oxygen, and the atmospheric oxygen, the color green is not easily observed. However, we do see some amount of green in a candle flame. The tip of a candle appears blue, as the oxygen and other material broken down into hydrogen is now seen leaving the flame, or the fire. Above the blue flame of the candle, and the tip of the fire there is smoke, as hydrogen further breaks down amalgamating (disappearing) into the atmosphere or air.

Does white light alter to color light when its intensity is decreased?

Light and color are both forms of energy. Light is constituted of heat and pressure/force. As light energy decreases with distance and densities, light alters to another form of energy, or forms of radiation. White light does not alter to color, as light propagates as a space wave, and color is manifested, or is apparent within space confining altering nuclei (heat and pressure/force increasing intensities).

Colored sunrises and sunsets

Why does the moon not change its color overhead, and at the horizons as the sun?

It is incorrect to theorize, that the sun changes in color at sunrises and sunsets, as we see the sun through a greater part of our atmosphere when it is at the horizon. If this was the case, the moon should have also altered its color at the horizon, and also altered its luminosity.

As our visual systems are comprised of troposphere nuclei, we can only see (concurrently as atmospheric nuclei light up) during the day, as they transmute/transform to light. As there is no light at night, we cannot see through darkness.

Seeing through our troposphere or our biosphere, the upper atmospheric light is more intense than our troposphere colored atmosphere, and therefore, we experience our so called visible light, and the light at the sun is observed at its normal incandescent.

Seeing the sun at the horizon, we observe the sun through more of our troposphere colored atmosphere, and so we see it orange/red, as nitrogen is abundant (standing at the horizon, the troposphere distance increases-earth's diameter is 12000 km-overhead around 600 km of atmosphere). We do not see the color at the sun (it has not changed color), but we see through the colors of our atmosphere, explaining why the clouds and surrounding landscape are the same color, as those of the sun.

Sometimes you may see orange/red at your window or wall during sunrise or sunsets. At those times, a little of our upper atmosphere at the horizon is 'lit' up. Light from the upper atmosphere, enhanced by the glow at the horizon sun passes through our troposphere colored nuclei, giving us those beautiful moments.

Light propagates within/as space, and not through the atmosphere or atmospheric colored nuclei. We see the horizon sun through atmospheric nuclei, which are of color, which are 'lit' up, as solar radiated heat interact some portions of the upper atmosphere horizon. In the case with the moon, when the moon is overhead, or at the horizon, our troposphere nuclei are not heat radiated, and as there is no intense light around, there are no troposphere colors, explaining why we see the moon at its same color.

Why does the moon sometimes look larger at the horizon than overhead?

At the horizon, during those days when the moon comes closer to the earth, more of the moon's basalt surface is 'lit' from the light propagated from our upper atmosphere, and as a larger area of the moon is seen lit up to the observer, the moons looks larger and closer. When it is overhead the moon's surface 'sees' less of the earths lit up upper atmosphere, and also receives less intensities of light from our upper atmosphere, and as a smaller area of the moon is seen lit up, the moon looks smaller, and further away.

A sphere's surface lit up area will be seen larger with a high intensity of light, and/or closer to the source of light, and smaller areas will light up, or a smaller surface will be apparent with less light, and being further away from the source of light.

This is evidence that there is no light from the sun or sunlight, but our visible light (that the moon 'sees'-as we see it) is generated at/with interactions, and disintegrations of nuclei at our upper atmosphere.

Why is the sky blue?

As water molecules and other material consisting of nuclei in the air, and the oceans, are disintegrated, hydrogen is swept away at super speeds. Departing protons interact with the incoming protons, and disintegrating nuclei at the upper atmosphere produces our visible light. With departing protons the blue color manifests, explaining blue skies. Clear days give us a deeper blue, as extraction increases.

Why is the sea blue?

The water molecule is continuously disintegrated, and integrated, within the energy cycle, as hydrogen is extracted. The water molecule is comprised of 2 parts of hydrogen, and 1 part of oxygen. With interactions (heat and/or pressure/force) hydrogen gives out a blue hue, and oxygen green. From above the water, the sea looks blue-the dominant color of the water molecule (more hydrogen) but in reality, as oxygen also constitutes the water molecule, at near, or above surface levels, the sea look bluish-green, or cyan.

Are black and white colors?

Most creatures would not move away from their habitats if foraging was easy, and there were no predators. Moving away from your habitat necessitates the evolution of better eyesight. Visual and other sensory systems of creatures evolve according to foraging, and/or survival necessities/requirements. There are creatures today existing with visual systems sensing, or seeing the environment in darkness, or in light, in black or white, or in colors. Some mammals may still be seeing nature in black and white. Hominids moved out of their millions and millions years habitat, and migrating into open grasslands they evolved better visual mechanisms to see further, wider, higher, and in color.

White or black are human concepts. It is difficult to explain the different whites in different 'lights'. It is also difficult to explain the different shades of black in/as shadows, and darkness. The white light from a fluorescent is different from an ordinary lamp, and also daylight.

By mixing some colors, and shutting off their nuclei configuration capabilities to light space waves, you obtain the color black. With no light there is darkness. We did not evolve mechanisms to see by darkness; otherwise we would be able to see in the dark. There are creatures that see in the dark, and cannot see by daylight.

Black is a mixture of colors. As many colors mix, manifesting black, a black object will have more nuclei in a given square inch than any other color. We also get black when things are burnt (black smoke), as things are disintegrated with fire, or heat nuclei configurations comprising that object disintegrates. Disintegrated particles rise up with the energy extraction part of the cycle, explaining rising black smoke. No color will be apparent without light. A half full red paint tin will appear black, as the tin sides prevents light on the paint. If you mix some colors in different proportions, you get a black color. You mix red, blue and green in equal proportions, you get white.

Light is not scattered or absorbed

Light is a form of energy, and changes its characteristics into another form, or state, or structure, propagating with distance, or on collision with an object, or another energy structure. Therefore, light as 'light' cannot be scattered or absorbed.

Colors or colored objects are seen only within light. As light (heat and pressure/force) activates those nuclei constituting those colored objects to vibrate, they manifest their colors.

Light is not reflected off a black glove, explaining why a black glove looks just as black in light, or in a shadow. Light does not reflect the color, but light enables us to see the color (mixed up colored particles constituting the black glove), as it interacts with the mixed colored nuclei to vibrate manifesting their color. As black is a mixture of colors, light makes black apparent either in a shadow or in light.

White light through a red light filter would give you red on the other side, not 'because' red absorbs, and allows only red light through, but as the red filter is comprised of red color nuclei configurations. Therefore, if you put two colored filters you get only black, and no color, not 'because' they 'allow', or 'absorb' only red or green light, but the mixed nuclei configurations (allowing propagation of light) have together cancelled (mixed up) out their colors.

Black clothes becomes hotter on a sunny afternoon than white ones, not as a result of 'absorbing' heat, but nuclei configurations generating the color black are patterned in ways, which retain heat. As black is a mix up of colors, there are more nuclei in a square inch of black cloth than any other color, explaining why heat is retained.

Colors through a rainbow and a prism

Nuclei configurations and molecules constitute the atmosphere, or exist within atmospheric space. Light space waves propagating, and interweaving through nuclei configurations of the prism, alters its intensities. However, these lessened intensities are reinforced, as light space waves exit the prism (as space is everywhere propagating light-heat and pressure/force). If light constituted colors, we would have seen different colors of light immediately as it left the prism, before it hit the screen. We do not see colors behind the prism, as dwindling intensities of light is reinforced with atmospheric space waving light. We only notice the sequence colors, as those in a rainbow on the screen.

Light space waves interacts nuclei configurations within air, or the atmosphere. This results in nuclei to increase their momentum, thereby increasing heat and pressure/force intensities manifesting colors. However, within our atmosphere these colors collectively combine to enhance our visible or white light, propagating within/as space.

With decreased intensities leaving the prism, light space waves interact with nuclei configurations within atmospheric space, between the prism and the screen. These interactions of nuclei manifest different types of color (as nuclei are subjected to different intensities of heat, as the prism is so shaped, altering the intensities of light space waves). However, we do not see these nuclei altering colors between the prism, and the screen. As light space waves propagate, these colored nuclei are forced on to the screen, and with continuous collisions at the screen (as light space waves are continuously propagating), these nuclei manifest their colors. If there is no screen placed behind the prism, those altered nuclei would propagate through space, but as light is more dominant within/as space, we will not see their color. If there was a glass placed instead of a prism, all the light space waves leaving the glass would be of the same intensity, and there would be no colors on the screen, as nuclei would not alter to colors.

Light does not contain colors, and/or colors are not apparent with decreased light intensities. As light dwindles with distance, light energy transmutes to other forms of radiation, not colors.

The rainbow sequence of colors observed at the screen is evidence that light does not contain colors. Light propagates through/as space, as it waves. There are no single rays, or beams of light, and light does not 'travel' in a straight line. There is no single light ray striking the prism.

The rainbow colors observed at the screen are in sequence, as the prism is so shaped, layer by layer, altering light space waving intensities. Each decreased light space wave intensity will activate nuclei (between the prism and the screen) to vibrate at different degrees of intensities, explaining different colors at the screen. We see distinct colors in sequence one over the other, and not scattered or overlapping colors (as light space wave propagates), as a result of the shape of the prism.

If light 'traveled' in a straight line, or a beam, or as rays, different colors of light would smudge the colors on the screen. Moreover, many light rays would enter the prism from different angles presenting a multicolored, or a black screen (black is a mixture of colors).

Light through a crystal, soap bubble, oil drop and a rain drop (ice crystal)

A crystal has many surfaces at different angles. Space light waves, propagating through the many angles of crystal surfaces, activate different nuclei configurations into varying intensities, displaying different colors seen within the crystal.

The soap bubble, the oil drop, and the rain drop are spheres. We observe myriad colors at oil drops, and soap bubbles, as they comprise different nuclei configurations allowing light space waves to weave through.

Why are there rainbow fringes when we change the angle we are looking at?

You can observe changing colors at transparent spheres like droplets or bubbles, as light space waves propagates within every nuclei configuration, altering degrees of heat and pressure/force, which manifest different colors. Their front surface is curved, and not flat. Light does not reflect from the surface facing us, as colors are produced, as light space waves activate nuclei configurations. If light reflects, and also if light contained all the colors, we would observe scattered multi color, and not distinct colors, as seen in sequence at rainbows, and through a prism.

Why are rainbow colors in sequence, and not overlapping each other?

This question is evidence that light does not travel in a straight line in beams, or in rays, but propagates within/as space, as it waves.

Rainbows are seen only when rain drops are ice crystals. As ice crystals form, they begin their descent from the cloud. Clouds can hold only so much of ice crystals for certain duration. By the time other clouds form, and produce ice crystals, the sun and your angle of observation alters. This is why rainbows last for only a short while.

Millions of drops like tiny prisms are falling close to each other during a rain storm. Different colors are observed, as drops fall (observed at an angle). Each layer of the ice crystals on their descent will display only a certain color. As each layer of crystals is observed at a certain angle, and altitude, there is a particular color. As ice crystals fall we see red, and as it further descends the red changes into yellow, then green, blue and finally violet. We see all the colors, as there are many layers of ice crystals falling down at the same time.

Light is not made up of colors. As explained earlier, each nucleus vibrates to certain intensities manifesting only a certain color. The same nuclei configurations can manifest different colors, depending on the light (heat/temperatures and pressure/force), and the angle corresponding to its vibrating intensities, but only one color at a time is manifested.

As the ice crystal falls, it alters our angle of view (colors are manifested viewed only at a certain angle), and a different color is seen. This explains why rainbow colors are in sequence, and not overlapping, and we observe a distinct colored arc. Each ice droplet alters its characteristics, as it descends through different temperatures, densities and pressures. We see the colors within the ice crystals, and not behind them. We see the different colors of the rainbow, as layers of ice crystal keep falling down, and at each angle, a different color is observed. If light contained all the colors we would not have a distinct sequenced different colored rainbow. We would have overlapping colors.

The sky is darker above the rainbow, and brighter directly underneath the rainbow, and it grows dim toward the centre of the rainbow. The sky is darker, as water vapor is transmuting, or condensing into nuclei configurations forming clouds. As light cannot pass through these nuclei configurations the area is darkened. As ice crystals form, and as light passes through, the area starts lighting up, and as the ice crystals disintegrate on their descent the area below the rainbow is brightened.

Why is a rainbow in a half circle?

The higher the sun behind us the less we see rainbows. However, sunlight does not reach us. Our upper atmospheric light is enhanced by the glow at the sun (not from the sun).

Rainbows are mostly seen when the sun is low at the horizon nearing sunsets, and immediately after sunrises. The rainbow is always semi-circle, as the horizon is in the way. If you look at the rainbow from an air plane or high up on hills and mountains, you will see it more of a circle.

Rainbows are always forming, but there is rainbow only when we see it at a 42 degree angle. We see the rainbow when the sun is always behind our back, and the rain drops in front of us. A perfect half circle rainbow is when the sun is exactly at the opposite horizon, rising or setting. The line from the centre of the sun to the centre of the rainbow arc runs through the head of the observer. The highest point of the rainbow arc is at 42 degrees. As the sun rise the rainbow arc sinks lower. When the sun reaches a height of 42 degrees in the sky, the rainbow arc disappears at the opposite horizon.

Why and how is our atmosphere/biosphere full of colors?

We learnt how earth obtained its new atmosphere. During the maelstrom, as clouds formed, there were continuous thunderstorms and lightning. With tremendous force and high temperatures, the myriad types of interacting, disintegrating and integrating atmospheric nuclei were forced down on to the surface of the earth, and its oceans. A large number of atmospheric nuclei conjoined at certain degrees of temperatures and pressure/force intensities, produce a particular color. Each type of nuclei was engendered at a certain degree of temperatures and pressure/force intensity. With these processes, the surface of the earth (as the mantle and crust formed, concurrently confined, or engendered the oceans) including its oceans were filled with nuclei/elements, molecules and compounds, as nuclei interacted, disintegrated and integrated.

Combinations of nuclei configurations produce different types of molecules, and as our atmosphere is dominant with nitrogen, oxygen and the departing hydrogen, our biosphere is full of colors-objects and landscape.

As our atmosphere is dominant with nitrogen (red), oxygen (green) and the departing hydrogen (blue) our interacted atmosphere facing the sun during the day, is filled with color by/with radiated heat and pressure/force.

From certain molecules/material in the oceans, cells engendered (revelation 13) originating living things. As cells contain these colored nuclei configurations, living things are made up of myriad structures of colors.

As light space waves strike nuclei configurations of living and non-living things, we observe everything in color, or black or white. As living things evolved, they processed atmospheric and surface nuclei, randomly evolving their colors, explaining why there are myriad type organisms with different colors. Trees, leaves, fruits, and flowers of different colors, creatures of different color combinations, and other colorful objects, and things in the biosphere are all consequent to the atmospheric colored nuclei configurations that patterned, and/or were structured.

Why do things have color?

We learnt, why and how objects/landscapes of the biosphere are filled with living, and non-living things with color. Light (heat and pressure/force) activate nuclei configurations in living and non-living things, manifesting their color. We can see objects, and or landscape in colors, black and white, within/with light, and not in darkness.

As protons or hydrogen is interacted with heat and pressure/force, blue color manifests. Hydrogen or protons is continuously extracted from the disintegrating atmospheric nuclei, and the water molecule in the oceans, lakes and rivers, from the surface of the earth, and from organisms. These protons are extracted, departing at super speeds away to the sun. This explains why the color blue in animals, birds, fruits, flowers or leaves is rare. Blue is observed in nature, as a combination of the other dominant colors red, green with other colors. Certain degrees of temperatures and pressure/force turn nitrogen into red, oxygen green, and hydrogen blue.

How do we see colored things?

As light strikes an object of color it activates the nuclei configurations constituting that object, and as these nuclei are activated to certain intensities they manifest their colors. Each type of nuclei was engendered with a certain degree of heat and pressure/force, and manifest that color, as color is linked to these forms of energy.

If you heat up a horseshoe with a blow torch it will get red hot, and if you heat it enough it gets white hot. Red is the color exhibited, as heat energy is low, so in a red-hot object the nuclei configurations are just getting enough heat in between their nuclei, emitting that color As the nuclei configurations are heated up, nuclei configurations concurrently alter to higher degrees of heat and pressure intensities, thereby exhibiting the white color Materials/object must be comprised of certain integrated nuclei configurations, enabling them to transmute heat energy as different colors.

We have been incorrectly informed that light contains colors, and objects 'absorb' those colors, and this is the reason we see colored objects. If light contained colors, an object would 'absorb' any, or many of the various colors of light. During daylight there is continuous space waving light, and we see the objects/landscape continuously, as those nuclei configurations exposed to light (heat and pressures) are activated, manifesting their colors. Those colors do not reflect from the object to our eyes, as we have been incorrectly informed. Colored objects therefore, are apparent only with light, and cannot be seen during darkness.

It is incorrect to state, that any object or living thing absorbs, or reflects, white, or colored light. Light and color are forms of energy, and energy cannot be 'absorbed' or 'reflected' in/as the same state, structure, process or state of energy. We see objects or a landscape, as they are constituted of nuclei where space waving light cannot refract through it. Otherwise they would have been translucent. They therefore, stand out within light. It is therefore wrong to postulate, that we can see colors, as an object directly emits light waves in the frequency of the observed color, or an object absorbs all other frequencies reflecting back to your eye only the light wave, or the combination of light waves that appears as the observed color.

Light does not reflect from objects (as light). If it did, we would not evolve visual system to 'see' those objects, as light would be continuously striking our eyes. We would evolve receptors or light sensing mechanisms to sense light, and not see objects within light. Although space waves light, and light is reinforced after it strikes an object, light cannot reflect to our eyes from that object as an image. There are images in a scene, and if light reflects all the images to our eyes, the scenery would be distorted.

How did living things evolve their color?

This answer will be found under the topic 'color in ecology'- revelation 17.

Color in outer space

Astronomical images are taken through a distinct wavelength band showing different intensities, but no real color.

Nuclei/elements and molecule configurations are only found in nebula, atmospheres of planets, and the outskirts of star fields. If these configurations are interacted with radiated heat and pressure/force, we observe color.

There is no color anywhere within interplanetary (other than atmosphere of planets), interstellar, or galactic space. Although the outskirt of our solar system is filled with nuclei, molecule and compounds, radiated heat from the sun dwindles in intensities, and without interactions those areas are observed dark.

It is shocking to comprehend, that if our atmosphere was not constituted of its present quantities of the present dominant gases, most creatures, and we would not have evolved. Moreover, there would have been no visible light, but a colored atmosphere. Without our present atmospheric nuclei combinations there would also be no visible light, the type through which we observe, and experience our world. Why do we intend to live elsewhere on the moon, or some other planet? Imagine a life without color.

SOUNDS

Light, and heat, or temperatures, and pressure/force propagate, or transmute, or manifest, as space transmutes to these forms of energy, and waves. Color as a form of energy manifests, as nuclei vibrates/rotates to/at certain degrees of temperatures and pressure/force. As nuclei exist within atmospheric space, color is apparent, within their confining space. There is too much of space waving light, and this explains why we do not see propagating colored nuclei light.

Sound is produced, as nuclei increase their vibration/rotation with propagating temperatures and pressure/force, and also as nuclei, or objects, or matter, collide with each other. Nuclei, objects and/or matter exist within atmospheric space and therefore, sound does not propagate/transmute within/as space, as light does. Sound is relayed nuclei to nuclei, as space between each nuclei waves conveying those intensities, or sounds, from the 'source' to the listener.

The different wavelengths, and frequencies of waving space, generate light, and other forms of radiation. In the same way, the different wavelengths and frequencies of space, waving between sound producing nuclei, generate different types of sounds. Different sounds will manifest, proportionate to the different vibrating intensities of nuclei.

A sound engenders, as nuclei collide, or nuclei configurations comprising two or more objects collide. Therefore, a sound can only occur within air, nebula, atmospheres, objects, or mediums (water) consisting of nuclei, or nuclei configurations. Color can be seen, and sound can be heard around nuclei in momentum.

Within our atmosphere, the collisions of nuclei interacting, disintegrating and integrating continuously generate tremendous sounds. Creatures living within earth's atmosphere have fine tuned auditory mechanisms/systems avoiding being attentive to these sounds, as these sounds are not required for foraging and survival. Without inheriting these mechanisms, creatures would have gone insane the minute they were born.

The vibrating object which generates the disturbance of air, or nuclei (sound) could be the vocal chords of a person, a vibrating string, the sound board of a guitar, a violin, the vibrating tines of a tuning fork, or the vibrating diaphragm of a radio speaker. The sound could be musical, or it could be noisy; but regardless of its quality, the sound was created by a vibrating object, or the collision of nuclei, or nuclei configurations. All nuclei comprising objects when hit, or struck, or plucked, or disturbed will increase their vibration. If you drop a pencil, or strum a guitar string, their nuclei configurations will vibrate to a particular frequency, or a set of frequencies. If the amplitude of the vibrations is large enough, sound will be heard.

Colliding nuclei, or nuclei configurations, producing sound intensities activate space to wave, which relays, or conveys those intensities to adjacent nuclei. This is how sound is propagated.

Pitch and frequency

A strummed guitar string has its nuclei configurations vibrating at 500 Hz, and that intensity are conveyed by space, waving from/between nuclei to nuclei, which propagate the sound to the receptors in the ear of a listener, detected as a 500Hz sound.

As the guitar string vibrates less in time, we should be hearing a different type of sound with the dwindling of intensity (as different intensities produce different sounds). With a greater intensity more nuclei are activated to the sound, and the pitch is high. As the string vibrates less in time, a lesser amount of nuclei are sound activated, and the pitch is low.

Our auditory mechanisms have been fine tuned, over millions of years, to continuously hear the first sound striking a receptor. Mechanisms are also in place to hear the high, and the low pitch. Auditory systems have been fine tuned taking foraging, predation, and/or survival considerations. Every step on a dried leaf will produce a different type of sound, and at a different pitch, warning the creature of an approaching, or departing predator/prey.

Intensity

The greater the amplitude of the vibrating string, a greater number of nuclei involved with the relaying of the sound, and louder or more intense is the sound.

As a sound propagates (nuclei to nuclei, with space waving in between) through a three dimensional medium, the intensity of the sound decreases with increasing distance from the 'source'. This results, as the sound is spreading, and the energy/intensity of the sound is being distributed over a greater surface area. If the distance is increased, the intensity decreases, and vice versa. This explains how creatures hear loud and low pitch, and can calculate distance from source.

The same sound will not be perceived to have the same loudness to all individuals. Age is one factor which affects the human ear's response to a sound. Two sounds with the same intensity, but different frequencies, will not be perceived to have the same loudness. The human ear has a tendency to amplify sounds having frequencies in the range from 1000 Hz to 5000 Hz. Sounds with these intensities seem louder to the human ear (Mammalian survival mechanisms). Despite the distinction between intensities and loudness, more intense sounds will be perceived to be the loudest sounds.

Sound in a vacuum

There is no such thing as a void, or a vacuum; and/as it cannot be generated. We may remove the atmospheric nuclei from an enclosure, but we will not be able to remove space-a form of energy, as everything everywhere is energy.

When a ringing bell was placed in a jar in a postulated 'vacuum', there was no sound. What was removed was air, not space. The sound of the bell could not propagate nuclei to nuclei, without air in the postulated 'vacuum'. Without air, or nuclei configurations existing within space, sound cannot be propagated, or is not heard. In another experiment to measure the speed of light, atmospheric nuclei were removed, and a postulated 'vacuum' was not created. Light was supposed to have been measured, or there was light, as it propagated within/as space, as it waves.

The speed of sound

The speed of sound in an experiment was calculated to be about (335 meters per second) 750 miles per hours at 0 degrees C. In reality, with the myriad forms, intensities, frequencies, magnitudes of sounds, and the dwindling of all forms of energy through distance, and densities, it is difficult to set a standard.

A nuclei disturbed or activated, generates a force, waving space, and this is conveyed to the next adjacent nuclei disturbing it, and transporting the energy/sound. The speed refers to how fast the disturbance is passed from nuclei to nuclei.

All forms of energy 'propagate' within/as space, or through space, and it is impossible to measure or calculate speed. We cannot calculate the quantities of speed (how fast) or frequency (how often). We should know the intensities of the processes at 'source', and these processes, and or intensities must be continuous, and be constant. Furthermore, forms of energy ‘propagating’ dwindle with distance and densities. With energy continuously transmuting, everything is flux, and as energy is everything everywhere, experiments (within three dimensions) to calculate the speed of forms of energy are futile.

As sound propagates within space, density is significant. However, as nuclei comprise gases, liquids and solids, the structure of nuclei configurations and integrations (density) will dictate the speed of all forms of energy. Depending on temperatures and pressures, a sound will propagate faster in a denser material than a less dense material, as there is less space waving in between. Solids have the strongest interactions between nuclei, followed by liquids and then gases. For this reason, sound travels faster in solids, than they do in liquids, than they do in gases (depending on temperatures and pressures).

Speed, velocity and related phenomena are human constructs. These are required for human planning, and understanding phenomenal reality.

What is a shock wave? What is a sonic boom?

Sonic booms can be heard with super speed aircraft. An aircraft at super speeds, thrusts through atmospheric nuclei compressing them. This concurrently produces air pressure, resulting in sound waves, similar to the water waves that result by/from a ship's bow. These pressure waves combine, and form shock waves, which travel forward from the generation or 'release' point. As an aircraft flies at supersonic speeds, it is continually generating shock waves, dropping sonic booms along its flight path. The sonic boom which is heard on the ground is the sudden onset, and release of pressure, after the build-up by the shock wave, or 'peak' overpressure. When nuclei are compressed, the resulting pressure generated from their release generates a sound called the sonic boom. The boom is merely a loud noise resulting from the pressure generated (released), as nuclei supplant the displaced area with an equal tremendous force.

Reflection, Diffraction, Transmission and Refraction of sound

Sound as a form of energy is unlike light. Light propagates (as space waves as light) as heat and pressure/force. As light propagates, it dwindles with distance, altering into other forms of radiation. Heat or momentum, or pressure, or force, as an integral/intrinsic part of light also dwindles in degrees, and intensities with distance. Sound continues dwindling in audibility with distance. Light therefore, cannot in reality, reflect, transmit or refract, as with temperature and pressure changes with distance and collisions, it is no longer light, but other forms of energy.

Although heat and pressure/force are intimately linked, as increasing heat will also increase pressure force, this is not the case with nuclei. Increasing heat, or activating nuclei to vibrate to heat intensities will manifest color, and this color will be apparent within the nuclei's confining space. However, as nuclei are activated with force/pressure (as in collisions), sound manifests. With pressure/force, sound is conveyed by waving space, relaying those intensities from nuclei to nuclei.

Sound (as pressure/force intensities) can bounce off a surface called reflection. They can bend around an obstacle called diffraction. They can cross the boundary into the new material or obstacle called transmission, and if sound changes the 'speed' and direction during transmission, it is called refraction. Sound reflects, diffracts, transmits and refracts, as space waves convey sounds between atmospheric nuclei.

Reflection

Reaching a boundary a sound wave undergoes reflection. Space waves not only convey sound from nuclei to nuclei, but as space waves with momentum, nuclei are also set in motion.

The amount of reflection is dependant upon the density of the obstacle. Smooth walls have a tendency to direct sound wave nuclei in a specific direction. Rough walls tend to diffuse sound waves nuclei, reflecting it in a variety of directions allowing a listener to perceive sounds from every direction of the room making it seem lively and full. For this reason, auditorium and concert hall designers prefer construction materials, which are rough than smooth.

Reflection of sound waves nuclei off surfaces is also effected by the shape of the surface. Curved surfaces with a parabolic shape focus sound wave nuclei to a point. Parabolic shaped disk collect large amount of sound wave nuclei, and focus it a focal point. At that point you can hear even the faintest whisper of someone talking across the room. Bull Moose utilizes his antlers as a satellite disk to focus sound. Owls have evolved spherically shaped facial disks, which can be maneuvered in order to gather, and reflect sound waves nuclei toward their ears.

Diffraction

Sound waves nuclei change in direction (as air is everywhere), as they pass through an opening, or around a barrier in their path. We notice sound waves diffracting around corners, or through door openings allowing us to hear others speaking to us in adjacent rooms. Owls are able to communicate across long distances, due to the fact that their long (high intensity) hoots are able to diffract around forest trees, and carry further than the short tweets (low intensity) of song birds.

As elephants migrate in large heard, and can sometimes get separated, they emit certain sound of low frequencies to communicate over long distances. As the matriarch at the front of the heard makes a turn to the right, all the other elephants in the herd will do the same thing. These synchronized movements occur, despite the fact that the elephant's vision of each other is blocked by dense vegetation, but is possible as these movements are preceded by certain type of communications.

Refraction

Sound waves refract when 'traveling' over water. Water has a moderating effect upon the temperature of air; the air directly above the water tends to be cooler than the air far above the water. Sound travels faster in cooler air than they do in warmer air. With cooler air, nuclei are configured more closely than with warmer air, so space sound waves convey the disturbance easily.

Reverberations

Reverberations occur in small rooms, or areas within 17 meters. Mammals have evolved mechanisms, where the initial sound is kept in memory for up to 0.1 second. This mechanism evolved to make double sure, that a sudden sound was actually heard. We have inherited these mechanisms, and can hear reverberations when talking in an empty room, honking the horn while driving through a tunnel, or underpass, or singing in the shower. As these mechanisms evolved in open areas, reverberations occasionally occur in concert halls, and lead to displeasing garbled sounds. The initial sound would reflect back from trees, or objects, and is heard prolonged. It is consequent to creatures evolving these types of mechanisms, that they could proliferate, as staying in groups, every individual could hear the growl, or roar of the leader, avoiding predation or dangers.

Echolocation

Echoes occur when a reflected sound reaches the ear, more than 0.1 seconds after the original sound was heard. Eventually, reflecting from wall to wall the first sound dies out. With echoes the arrival of the second sound is perceived, as a second sound, rather than the prolonging of the first sound, as with a reverberation.

If you have ever screamed within a canyon, perhaps you have heard an echo of your scream off a distant canyon wall. The time delay between the scream, and the echo, corresponds to the time for the holler to travel the round-trip distance to the canyon wall and back.

Sound travels from nuclei to nuclei, as space waves in between, conveying the disturbance. If a particular sound propagated through/as space (as light) the intensity of a particular sound would break up at an obstacle. The returning sound wave would be completely distorted, or never returned. As all forms of energy dwindle with distance, echoes are not as loud, as the initial one, getting fainter every time around.

We have been informed that bats have evolved organs to expel particles (pulses of sound) through space, and to instantaneously discern objects with the reflecting pulses. Echolocation is the ability to emit pulses of sound, and create an image of the surroundings from the returning echoes. Bats can track down a fast flying insect, and can distinguish different types of prey by their acoustic shape and texture. This information is doubtful, and refutable. If this theory was true, why have bats evolved eyes? They should have evolved distinct unique organs that expel, and also analyze pulses of sound.

Our visual systems have evolved from air and light up, as the atmosphere is lit up, and we 'see'. We see objects/landscape standing out within light. Our visual systems to see within light evolved, as we live above the surface of the earth in open areas.

Bats have evolved their visual system from/as atmospheric nuclei. As they have always lived in darkness they have evolved their eyes to see structures of atmospheric nuclei within darkness. There are many creatures that 'see' in the dark.

The Doppler effects

The Doppler effects can be described, as the effect produced by a moving cloud of sound nuclei. We are most familiar with the Doppler effects, as a result of our experiences with sound. As the police car approaches you with its siren blasting, the pitch of the siren sound is high; and then suddenly after the car passed by, the pitch of the siren was low. That was the Doppler effects. Also, as you move toward a sound 'source'-if the sound is continuous, and at the same frequency-you will hear the sound getting louder. It is just the opposite when you are moving away. If the sound is continuous, and is at the same frequency, and moves toward you, it will be heard louder, and louder, and the opposite if it moves away.

Sound nuclei in a chain reaction relays, and or activates adjacent nuclei, as space in between waves those intensities, explaining how the sound reaches the auditory receptor of the listener. The audibility of a sound, depends on the closeness of the listener to the 'source' of the sound, the intensity at the 'source', and whether the 'source' is approaching, or receding from the listener. However, the same siren (tone or sound) will be heard.

A sound should normally change in audibility (as intensities decrease); as all forms of energy dwindle with distance. This is not the case with sounds. Space waves in between sound generating nuclei, conveying those intensities to adjacent nuclei. As more nuclei vibrate to sound the audibility is greater.

As a police car siren is approaching you the siren is activating more and more nuclei. So there are more sound waves of the same intensity approaching the listener, at a more frequent rate (high pitch). As the siren passes by, the sound and/or sound nuclei are dispersed, and the siren is less audible. Audibility increases nearer the 'source', as the sound nuclei are less propagated (and denser) than at further distance. The effect is not a result of an actual change in the frequency at/of the 'source'. The source puts out the same frequency, explaining why the sound does not alter, but remains the same, and only audibility increases, or decreases.

Clouds, Lightning and Thunder

Evaporation takes place, as heat is generated with atmospheric interactions, and this heat changes water into vapor. As water or liquids are subjected to this heat generated by atmospheric reactions, nuclei configurations move around faster, and moving quickly they break away from the surface of the water/liquid, and with heat-that initiated the process, as well as heat from interactions-they are susceptible to be extracted away as vapor, or gas, rising into the atmosphere-evaporation. As the vapor, or gases rise up, the heat that they carry with them decreases, as heat is lost with atmospheric interactions. As it continues to rise to higher altitudes, and lower temperatures, eventually enough heat is lost, and the vapor condenses, returning to a liquid state.

Some of the water vapor condenses (moisture accumulates) into clouds (containing millions and millions of water droplets and ice particles suspended in the air). As the process of evaporation and condensation continues, these droplets encounter many collisions with other moisture that is in the process of condensing. As the rising moisture encounters colder temperatures in the upper cloud regions, some regions begin to freeze, but some regions do not. Rising air currents take the unfrozen droplets to the top of the cloud, whereas the 'heavier' frozen droplets fall to the lower portion of the cloud, or continue on to the ground-precipitation. Rising moisture also collides with ice, or sleet, that is in the process of falling to the earth, or located in the lower portion of the cloud.

These regions of the atmosphere with all the ongoing processes of temperature and pressure changes from heat to freezing, condensation, upward air movement and so on, generates atmospheric density variation/fluctuations resulting in pressure pockets. These pockets sometimes extend from the surface to the clouds in many directions. These pressure pockets with adjacent less dense areas, quickly altering densities and pressures, generate cloud collisions, resulting in lightning and thunder.

Thunder happens concurrently with lightning. Lightning rapidly heats up, and expands nuclei to one thousand times, forming a powerful compression that is heard as thunder. Thunder is usually not a one-time, but a rumbling sound, as the sound echoes and reverberates. This happens as the air in those pockets, or tubes, expand and contract. However, the first loud sound originates from the collision.

Thunder and lightning concurrently engender, and are not 'caused' by the other. The explosion has those 'lit' up nuclei being thrown away as lightning, within those tubes or pockets (streaks). Other nuclei rush in those tubes and pockets, and vibrate to those same intensities, as those nuclei activated in those collisions.

Lightning comes from striking nuclei. The first thing you see is a flash from clouds colliding, followed by branches which flash subsequently after the main strike. We observe lightning in long streaks filling up those pockets, normally approaching the surface.

Light from lightning propagates within/as space, as it waves, and therefore, thunder (or vibrating nuclei) intensities are conveyed nuclei to nuclei after space waves light. If lightning or light at these collisions, and also the collisions were continuous, we would not have heard thunder.

Processes that generated the alteration in atmospheric densities, resulting in altering pressures originate from the surface, and lighting streaks fill up those pockets-action and reaction. There are different types of lightning strikes, cloud to ground, ground to cloud, and cloud to cloud, observed as lightning fills the pockets.

Why do secondary strikes occur when the flash from the main strike is still visible? After the main strike there are many dimmer flickers a few more times. This happens as when clouds collide they bounce back in a reaction, and come back, but not in the same intensity-as with a bouncing ball. Just as we hear thunder rumble we also see secondary lightning strikes. Both occur concurrently, and as a result of each other. Light from the collision propagates through/as space, but sound nuclei utilize space to convey sound to adjacent nuclei. This can only happen if space is not waving (as) light. As light and sound nuclei alternately utilize space in sequence, thunder is heard in a rumble, and secondary lightning strikes are observed. As space continuously waves light during the day, we do not observe bright lightning, and loud thunder.

As clouds collide, the nuclei around the strike become extremely hot that it actually explodes, as the heat generates nuclei to expand rapidly. This explosion is thunder, which is a shock 'wave' (sound nuclei) radiating away from the strike path. This explosion also generates space to wave as lightning (heat and pressure/force). When atmospheric nuclei heats up they expands rapidly, generating a compression 'wave' that propagates through the surrounding air. This compression ‘wave’ is the thunder. Thunder propagates much slower than light, as sound is propagated from nuclei to nuclei with space waving in between, whereas light propagates within/as atmospheric space.

The audiovisual spectacle of thunder and lightning is a combination of the vibration of nuclei, and their disturbance by sudden forces. The collisions of clouds, or large areas of nuclei send them fleeting, and as nuclei vibrate to ‘light’ frequencies, we see the light around these collisions.

Lightning does not last long, as collisions are instant, and not continuous. Cloud collisions produce both lightning and sound. Lightning is observed first, as floods of nuclei at the place of occurrence transmute vibrating into intensities exhibiting ‘light’.

We see the lightning first, as we can see it at the 'source', and it propagates faster within/as space, and also it is apparent as light. We hear the thunder later, as sound nuclei utilizes space alternatively, as space waves first as light. We do not hear the thunder at the same time, as we see the lightning, as we cannot 'see' the thunder. Moreover, a sound relayed from nuclei to nuclei takes more time to reach the listener.

Lightning and thunder are not continuous, as the 'source' of light and/or thunder is not continuous, explaining why the sky is not continuously lit up. As all forms of propagating energy dwindle with distance, thunder is not heard more than 16 kilometers away. Thunder and lightning are often experienced during precipitation and monsoons.

Why can radio waves go where light cannot?

Light travels within/as space, and if there is an object traversing space, light cannot propagate through it, this explains a shadow.

Radio 'waves' are sound 'waves'. Sound travels from nuclei to nuclei, as space in between waves conveying sounds. So sound nuclei can reflect, bend around, refract and diffract, as nuclei exist everywhere within atmospheric space.

The atmospheric density determines how fast sound will travel. Sounds at night, under certain conditions, may be heard a long distance away, as the air nearer the ground is cooler than the air higher up. Cool air is denser than warm air, so sound waves between nuclei are shorter, and relay easily and faster, and so sound can travel for long distances. Sound maintains its intensity longer in a denser medium than a tenuous one. Sound is heard clearer, and travels faster in solids, and liquids, than air.

Why does sound travel faster in deeper water?

Water as a liquid has nuclei configurations closer together than atmospheric nuclei, or water vapor. The density, temperatures and pressures of water plays a significant role in sound propagation. Solids have nuclei configurations (very close together) where sound propagates ten times faster than air. There is more pressure in deeper water, as radiated heat and pressure/force delve down to the oceans to disintegrate the water molecule. With pressure nuclei configurations are closer together (denser), and as space waves are shorter, sound travels faster in deep water.

Concluding Note:

Propagating light decreases in varying intensities through different sections of the prism, as it is so shaped. We observe those rainbow colors behind the prism, not as a result of altering wavelengths of light. Each atmospheric nucleus when heated manifest different corresponding colors, to various intensities of light (heat and pressure/force).

Astronomers state they can tell us a star's specific distance, age, and chemical components by deciphering the components of a star's light using a spectroscope. A spectrograph splits up light into the spectrum of colors. This spectrum is not like the rainbow, but it is interrupted by dark lines. From the relative number, thickness and position of these lines, astronomers calculate temperatures, compositions, speed, directions, rate of motion and other things of a star. If light dwindles with distance, and as atmospheric nuclei emit color corresponding to the intensities of space waving light, how can astronomers tell us how far a specific star is, it age, its chemical constituents or composition, its temperatures, speed, direction and rate of motion?

REVELATION TEN

NO SUCH THING AS GRAVITY OR FORCE

Nothing is a one way phenomenon-Energy transmutation as momentum and weights

ELUCIDATING FORCE/PRESSURES, OR ENERGY IN MOMENTUM, OR ENERGY TRANSMUTATIONS

Physicists have been trying to unify the four forces, as the Theory of everything. The four postulated forces are explained following: A force called gravity, which is supposed to act between all mass, and bodies having an infinite range. The electromagnetic force acting between electrically charged particles, for examples electricity, magnetism, and light. The strong force binding neutrons, and protons, in the cores of nuclei, and the weak force resulting in beta decay.

Force or Pressure is a form of energy

Existential energy exists continuously in momentum, or is 'energetic' (human concepts), as energy continuously transmutes. Force is a form of transmuting energy, as the process of transmuting is momentum per se. You cannot action, or force, anything without transmuting a form, state, structure or process of energy. A form, state, process or structure of energy does not engender without force/momentum, or in other words, as energy transmutes.

With force or pressure, heat concurrently engenders, and vice versa. Heat also is a form of energy. Heat and force/pressure are intrinsically linked, or are integral parts of each other. Different degrees of heat manifest different intensities of pressure/force. It is preposterous theorizing the force of gravity, without considering temperatures as its integral part.

Certain degrees of heat and pressure force intensities manifest as light, or forms of radiation, gamma and ultraviolet rays, x-rays and other phenomena. All these forms of energy are apparent, or observed, and/or experienced propagating within/as space, as it waves. Trying to figure out how the four forces can be united, without taking heat/temperatures into consideration are exercises in futility.

Energy is everything, everywhere. Energy continuously transmutes into state, structures, processes or forms, virtual/latent or observable. Humans call these existences, but as everything is energy there are no 'entities'.

All forms of energy as temperatures/heat/light, forms of radiation, sound, and pressure/force/momentum are manifested/engender, as energy transmutes/propagates through, or within/as space, as it waves. The different waving frequencies and wavelengths of space corresponds/manifests the different intensities of/or the forms of energy. Space waving per se is motion, or pressure/force, and therefore, the different frequencies and wavelengths of space waves, determine the intensity of the force/pressure, and degrees of heat.

As we create existences, we have problems explaining them. Energy is explained, as vigor, or power, shown in action, exertion, or performance, and force is explained, as the capacity to do work, or to cause physical change, strength or power. Energy, force, and everything, is the same thing and therefore, all descriptions, or explanations, or meanings are of the same thing. Force engenders, as energy propagates, or transmutes, interacts, integrates, or engenders, and disintegrates, as everything is energy everywhere.

Energy propagation is not a one way phenomenon. As heat/temperatures, all forms of radiation, pressure, or force, or momentum, propagate from the star, other forms, states, processes or structures of energy propagate (transmuting) toward it (as replenishment for the expended material/energy). This explains the energy transmuting cycle between the sun and the planets/bodies. This also explains why there is no such thing as gravity-a one way postulated force.

As dark energy confines galaxies, there is an equal continuous opposite force confining dark energy, as energy transmutes (as star fields). Within star fields, energy as forms, structures, processes and states, propagate, or transmute through, or within/as space, as it waves, and as energy transmutes/propagates/shuttles to-and fro, there is a continuous force between existences.

As free energy transmutes into a particle, there is a force containing it, and an equal force confining the space that confines it. In other words, as energy condenses into a particle, force as spin and momentum engender, as a particle, and its containing space exists.

The strong and the weak nuclear forces, the non-existent electromagnetic force, and gravity, are all theorized to explain, or describe, energy transmutation phenomena.

NO SUCH THING AS THE STRONG NUCLEAR FORCE

It is postulated a strong nuclear fundamental force, acts between elementary particles, mainly quarks. It is assumed, that a strong force binds quarks together in clusters to form protons, neutrons, and heavier short lived particles. The strong force is assumed to hold together the atomic nucleus, and underlies interactions among all particles containing quarks. The strong force, or the color force, is the dominant force inside the nucleus, which acts between the quarks, inside the protons, and the neutrons.

Continuously engendering plasma is continuously expelled, together with heat and high pressure/force from a star, and as space is an integral part of the dense region, it confines plasma expelled from the star, which immediately transform into protons. Nuclei engender, as neutron plasma and protons are further confined by space (revelation 4).

Force as everything else is 'individualized' in physics. As energy condenses/transmutes, and as energy is continuously in momentum, the transmutations per se, is the momentum or force manifested. There is no such thing as a strong nuclear force.

NO SUCH THING AS THE WEAK NUCLEAR FORCE

It is postulated there is a weak nuclear fundamental force, which underlies some forms of radioactivity, and certain interactions between subatomic particles. The weak nuclear force is now believed to have intrinsically the same strength as the electromagnetic force.

As discussed earlier, there is forces/pressure as energy transmutes. As energy transmutes into different forms, states, processes or structures, or existences, there are different intensities of force/pressure.

As energy is continuously transmuting, and everything is interconnected and interdependent, everything everywhere is in flux. Every particle/plasma/proton/nuclei transmutes/transforms into yet different processes, structures, forms, or states, continuously altering its constituents and characteristics. No existence exists/endures with the same characteristics for any duration. At stars, energy is continuously condensing into particles, and particles into plasma, and plasma expelled transmute into protons.

Our solar system engendered, as fragments of neutron plasma and protons from an exploding neutron star darting through, were confined by space, generating deuterium, tritium and helium. Eventually, as the inner planets atmospheres were instantaneously depleted, helium with deuterium and tritium were integrated into many different types of elements. As protons and nuclei are subjected to the energy transmuting cycle, between the sun and the earth, protons are extracted from elements, molecules and compounds, which have formed the myriad forms of matter. As a result, over time, nuclei gradually disintegrate depending on the integration processes (temperatures and pressures/force) that engendered them. Nuclei as deuterium, tritium and helium, 'individually', or constituting elements, molecules, compounds, or matter, are continuously disintegrating, are radioactive, decay, emit gamma, other rays, and other activity, as a result of the energy transmuting cycle.

Earth's atmosphere, oceans, the surface of the earth, and its interiors are continuously altering their characteristics, consequent to the energy transmuting cycle, between the sun and our atmosphere, and the many energy transmuting cycles within the atmosphere, and earth's surface and interiors. A process, form, structure, or state of energy, or an existence, can transmute or transform, alter, decay or terminate instantaneously, or after a moment, a few seconds, or last for years, or millions of years. There is no such thing as a weak nuclear force.

NO SUCH THING AS THE ELECTROMAGNETIC FORCE/RADIATION

The creation of electromagnetic radiation and the electromagnetic force

Many years ago, humans compared the magnetic needle to the earth being a giant spherical load stone, and the power that attracted the magnetic needle was thought to be in the earth, and not the skies. Earth's rotation in some ways was connected with magnetism. People were fascinated by the compass, as/and magnets repelled and attracted, and postulated that magnetism was governed by a more mysterious force than gravity (which was taken for granted) and therefore, this assumption gave birth to universal gravity. Magnetism therefore, was not associated with gravity, but with electricity. Rubbing amber (the Greek word for electrons) flimsy objects attract each other, and these properties are also shared by glass, sulphur, crystals, resin, and other substances.

With the magnet's attraction, and repulsion, and continuous electricity, conductors and insulators followed. Electricity through wires (cannot flow across glass or air) over distance, gave us current, or liquid fire. Mathematics entered electricity, and magnetism, but nobody could explain their physical existence. The parallel synthesis of electricity, and chemistry, and the battery, gave us electric current-chemical interactions produce electricity. Electric current through certain mediums showed decomposition, explaining that chemical affinity has an electrical basis. A certain electrical discharge called the cathode ray consisted of sub-atomic particles, and these having derived from different source such as hydrogen and oxygen, taught us that sub-atomic particles, and the postulated atoms built elements.

The realization that an electric current produced a postulated magnetic field unified electricity, and magnetism-the electromagnet was born. Yet nobody could explain what created the current. Then magnetism and electricity was reduced to a common source-the electron. Where is the electron? It is there somewhere 'because' mathematical models worked?

We then produced the dynamo, and the electric motor-electrical energy. Yet nobody can explain electrical energy. It was found that all changes in electric, and magnetic forces, sent waves spreading through space; and that these waves had the same transverse character, and the same speed of light. As light undulations are similar to electricity, with the influence of magnetism, it is considered an electromagnetic radiation. Magnetism, electricity and light are now united.

Electro-magnetic radiations came to be regarded as rapid alternations of electrical and magnetic stresses in space, where each change in the electric stress gives rise to a magnetic stress, which again give rise to an electric stress, and so on. Soon the range of these radiations was shown to comprise not only the visible spectrum between the ultra-violet, and the infra-red of radiant heat, but to extend to the ultra-short gamma rays of radioactivity, and to the kilometer long waves used in radio-communication. So with electricity, magnetism, light, heat, electromagnetic radiations, science finally realized that the various powers of nature were merely different forms of energy. Natural phenomena were convertible, as mechanical energy could produce heat, heat into motion, motion into electricity, electricity into magnetism, and so on.

Physics, chemistry, biology and most disciplines of science are the same thing. Everything is one and the same thing-energy transmuting into different states/structures/processes or forms. We created phenomena as different existences, and theorize to explain our creations. How can we ever come up with correct and complete explanations?

The postulated electromagnetic radiation

It is postulated, that energy is propagated within free space, or through a material medium in the form of electromagnetic waves. Examples include radio waves, infrared radiation, visible light, ultraviolet radiation, x-rays, and gamma rays.

The postulated electromagnetic force

It is postulated, that the electromagnetic force is one of the four known basic forces in the postulated 'universe'. It seems electromagnetism is responsible for interactions between charged particles, which occur because of their charge, and for the emission and absorption of photons (electromagnetic radiation). The phenomena of electricity and magnetism are consequences of this force. The electromagnetic force is postulated to be much stronger than the gravitational force, but significantly weaker than both the weak force, and the strong force.

Like all physical phenomena, magnetism and electricity are human constructs/concepts. Before this revelation, we did not know what they are, how they occur, and how they are generated, yet we had associated these phenomena to energy radiation, as the electromagnetic force. There is no such thing as an electromagnetic radiation, or electromagnetic force. The phenomena electricity and magnetism are elucidated following:

Electricity

The concept or the word for the phenomenon electricity has several meanings. These meanings contradict each other, and the contradictions like hundreds of our constructs confuse everyone. Teachers, engineers and scientists have a hard time understanding it. Only one of the meanings can be right in the following conflicting definitions. Electricity means: electric charge, flowing motion of electric charge, electrical energy, amount of imbalance between quantities of particles and space, a class of phenomena involving electric charges, an electric field, the nitrogen/oxygen plasma (sparks of electricity), a field of science, or as discussed an energy transmutation.

Electricity is a form of energy occurring in nature, and is observable in natural phenomena such as lightning. It can also be produced by friction, or a chemical reaction of mechanical effort. Electricity is also electric current.

As some substances are rubbed together, their nuclei configurations disintegrate, or alter, expending energy, or force/pressure and heat, or electricity. In chemical and mechanical reactions, the nuclei configurations in the postulated atoms, or nuclei in molecules are agitated, or 'energized' thus producing thermal energy, or heat and pressure/force.

Energy 'expended' or energy 'energized' is in other words, force or pressure. In reality, as everything is energy, a part of energy cannot be 'expended'. Energy is 'energetic', as a form, state, process or structure of/as energy transmutes. This transmutation per se is momentum, or force, or pressure. Nuclei are integrated and disintegrated, as energy transmutes and therefore, as the configurations are 'disturbed' energy is expended. In reality, energy is not expended, but transmutes into yet other states, processes, structures or forms.

As space propagates temperatures and pressure/force, and as space confines nuclei, these forms of energy propagate within/as space confining nuclei, explaining how force/pressure, or energy in momentum, 'traveling' within a conductive material is termed electricity.

Electricity and light are forms of energy (in momentum). Light propagates within/as space, and also through the nuclei configuration's space in nuclei, and molecules. In reality, light (energy) as temperatures and pressure/force propagates/transmutes within/as space, as it waves, as space transmutes to light-light is apparent. This explains why electricity through conductive materials generates light (with a filament), or different degrees of heat/temperatures and pressure/force intensities.

As space confines nuclei, and is everywhere, electricity-as a form of energy in momentum-can propagates through conductive materials. These materials are conductive, as they are constituted of certain nuclei structural patterns. Electricity is energy 'forced' through conductive materials like a copper wire. As electricity propagates, or transmutes, within nuclei configurations' space, electricity, or energy as/in momentum, is not apparent. However, electricity as a form of energy transmutes to light, and or heat when forced through a filament, or materials offering resistance (heat).

It is energy in momentum, within/as a conductive material, which we postulate as electricity. The intensity of the pressure, or force, in motion (energy in momentum) through conductive materials, will manifest the corresponding heat/light, or forms of radiation at certain materials. The greater the resistance, the greater is the heat generated. This effect is used in electric heating devices.

Energy cannot be stored, created, or destroyed. Energy or electricity cannot be 'stored' in a battery. Energy is everything everywhere. To generate electricity, or energy in momentum, another form, state, structure or process of energy transmutes. To ‘activate’ energy, or to make the battery work, the terminals must be connected to a filament, or other mechanisms to activate the chemicals. In a simple electric circuit, electricity (pressure/force) propagates through a wire, as a cycle in a complete circle. Light bulbs utilize electricity, or energy in momentum, or force or pressure, transmuting the filament into light/heat energy. Space inside, and outside the bulb will transmute/propagate, as light, and heat.

It is incorrect to state, that electric current in a wire results with a flow of electrons dislodged from atoms. There are no such things as electrons, as space as a form of energy, confines nuclei of the postulated atom. The pressure/force (electricity) or energy in motion is measured, as it passes any point of wire per unit time. There is also no such thing as an electric force. Both electricity and force are forms of energy, and not distinct of each other. Electricity in other words is pressure or force, or energy in momentum through a conductive material.

There is no such thing as an electric field, and an electric charge. Electricity and 'charge' per se, are 'forces' of different intensities, and force per se is a form of energy. There is repulsion and attraction, or a stronger or weaker force, as energy is continuously transmuting into yet other forms, states, processes and structures of energy.

Humans have created existences, or physical phenomena, and therefore, cannot explain/describe physical reality. Electricity or electric current is a form of energy in momentum, and as energy is in momentum there is pressure/force. In reality, energy cannot 'travel' from one point to another. Energy is in momentum, as it is continuously transmuting into yet other states, processes, structures, or forms, or existences.

In an electric circuit, the 'source', the load, the propagation, the force/pressure, light or heat generated, space that propagates light or heat, are all forms, states, processes or structures of energy transmutations. Simply, electricity is energy in momentum, or force or pressure within/as a conductive material. There is no action or momentum, without transmuting a state, structure, process or form of energy. Everything everywhere is the same thing-energy.

Magnetism

Magnetism is explained as a power to attract iron, or the ability of something to attract (as a magnet), or the science that deals with magnetic phenomena. Applications of magnetism range from the simple magnetic door catch to medical imaging devices, and electromagnets used in high energy particle accelerators. A magnet is a body able to attract, and repel, especially iron, and some ferro-magnetic metals.

Magnetic fields

We have been informed, that magnetism is a phenomenon associated with magnetic fields, the effect of such fields, and the motion of electric charges. Magnetic fields exert forces on moving electric charges.

A magnetic field is the region around a magnet, electric current or changing electric field in which magnetic forces are noticed. The field around a permanent magnet, or wire carrying steady direct current is stationary, while that around an alternating current or changing direct current is continuously changing. Magnetic fields are commonly represented by continuous lines of force, or magnetic flux, that emerge from north-seeking magnetic poles, and enter south-seeking poles. The density of the lines indicates the magnitude of the field, the lines being crowded together where the magnetic field is strong. Magnetic lines of force are a way of representing a 'magnetic' field. They point from north to south outside a magnet, and from south to north inside a magnet; magnetic lines of force form complete loops, and never cross.

What really is magnetism?

We learnt, as the earth's original atmosphere was depleted, the interiors of nuclei were exposed to radiated heat and pressure/force from the sun. As helium (being heavier and constituting most of the interiors) disintegrated into deuterium and protons, the engendering increases of pressure or force, and heat in a reverse process, concurrently 'began' integrating helium, tritium and deuterium, into elements heavier then helium. At the iron stage, heat and pressure/force intensities were at the maximum level, which is possible anywhere. Concurrently, the new forming atmosphere with disintegrations engendered oxygen, and eventually water vapor. The formation of clouds, and the precipitation on the interiors, disintegrated/solidified the elements, as temperatures and pressure or force decreased. However, precipitation generated iron to disintegrate explosively, and with the sudden blasting away of heat with tremendous pressure/force, heavier elements were integrated. As iron solidified (with the sudden release of heat and pressure/force), its nuclei integrated tightly, explaining why it has the most tightly bound nuclei among all the elements.

As iron nuclei are tightly bound, there is little or no space surrounding its nuclei configurations. The nuclei of iron are patterned in certain arrays with a direction (as pressure or force, and heat were 'released' during iron nuclei structuring), and space cannot propagate temperatures and pressure/force around iron nuclei configurations. There is space in between nuclei configurations in other elements, but this is not the case with iron, and some ferro-magnetic materials. When iron is magnetized, pressure/force and heat not only 'forces' the nuclei apart, but also confines its nuclei. As a result, when magnetized iron is placed with similar nuclei configurations of another material, a loop forms, and/as space waving temperatures and pressure or force, activate the nuclei configurations. If the nuclei configurations of both the units brought together face the same direction, there is attraction, and if they face each other in opposite directions there is repulsion. Therefore, like poles will repel each other, and unlike poles with attract each other. As our atmospheric space is continuously propagating temperatures and force/pressure, these forms of energy propagate through a magnet's nuclei configurations, explaining why magnetic fields display lines of force from north to south outside a magnet, and from south to north inside a magnet, forming loops, but never cross. This is magnetism. How then can we postulate light as an electro magnetic radiation?

No such thing as the electromagnetic force

Light is a form of energy, and manifests, as energy in the form of heat and pressure/force, transmutes/propagates within/as space, as it waves. Electricity is a form of energy, and manifests as energy in momentum or as pressure/force, transmutes/propagates, within/as nuclei configurations in conductive materials. Magnetism is experienced only with ferro-magnetic materials. Magnetism is the force experienced/manifested, as a magnet attracts/repels other similar material.

Nuclei are continuously interacted within the energy transmuting cycle, between the sun and the earth. Heat and pressure/force propagate within/as space (as it waves these forms of energy), altering nuclei configurations, and as a result, nuclei disintegrate, integrate, are radioactive, decay and expel forms of radiation, heat and pressure/force.

As explained earlier, as iron structured during the release, or decreasing heat, and pressure/force, its nuclei structured in certain arrays with certain directions. The continuous pressure/force of the energy transmuting cycle propagates, interweaving through magnetized iron's nuclei arrays leaving it at one end, explaining the postulated/theorized magnetic or vector fields, and/or its attractive and repulsive behavior.

Electricity is force/pressure, and is a form of energy, and magnetism generates a force/pressure in one direction in certain materials. As light is a form of energy, or heat and pressure/force, and above all, as forms of energy transmutes/propagates shuttling to-and-fro from 'source' to 'target', light cannot be an electromagnetic radiation. There is also no such thing as an electromagnetic force.

Temperatures/heat/light and forms of radiation, pressure or force or momentum, electricity and magnetism are forms, or states, or structures, or processes, or existences, as energy transmutes.

Why do properties of material change when their structure is reduced to the nano-level, and why electric current passed through silicon nano wires cause them to luminescent when it does not in a silicon wafer?

From micro-structures researchers have delved into nano-structures trying to create nano materials, which can self-organize as they think nature does-like DNA and proteins do. However, we will learn that there is no such thing as self organization. When a material is restructured to nano-levels, nuclei configurations are altered changing their properties.

Electricity is energy in momentum through certain conductive material. Through nuclei configurations in certain material this energy in momentum, or electricity, transmutes to heat, and/or illuminates.

Why do magnets have poles? Why does a magnetic needle face the North Pole?

Why is iron one of the materials that magnetizes? Why do magnets have this inner force to attract or repel another magnet or attract itself to another metallic element?

How can a fridge magnet cling to the side of a refrigerator without any power source?

Why does endless magnetic energy emanate from permanent magnets?

We learnt how our new atmosphere formed with water vapor condensing into clouds and precipitation. With continuous precipitations iron disintegrated explosively, generating heavier elements than iron, as heat and pressure/force decreased, and increased, in incremental steps. However, as these elements disintegrated, they were susceptible to/within the radiated heat and pressure/force of the energy transmuting cycle. Disintegrating nuclei which were concurrently solidifying were extracted more from one side of the earth facing the sun. As a result, as these elements solidified, there was more land mass within what is now our northern hemisphere, explaining why there are more oceans in the southern hemisphere. During this time the northern hemisphere faced the sun, as earth orbited the sun. However, without an energy transmuting cycle, the same side of earth faced the sun, and earth was not rotating around its imaginary axis. Eventually, as the new atmosphere formed, the energy transmuting cycle took hold, as concurrently the mantle and crust formed, and precipitation was contained in/as the lakes and oceans. As the oceans sustained the depleting atmosphere it gradually depleted, exposing the land in the northern hemisphere. As a result, earth tilted slightly.

As the southern seas are interacted, disintegrated and extracted, the earth comes closer to the sun, and as the northern surface, land and oceans material are extracted, earth goes further away (revelation 8-seasons). With more and easier extraction from the southern oceans, earth is dragged closer to the sun (during the southern hemisphere's summer).

During that time, as elements was solidifying, and concurrently being extracted, iron nuclei structured in particular arrays giving them a direction. This explains why a magnetic needle faces the North. Most navigational compasses are mounted on a vertical pin, and can only move on a horizontal plane. A compass needle is a magnet attracted toward the northern hemisphere, where earth's liquid inner core is continuously being subjected to the energy radiation cycle-the postulated earth's magnetic field. There is more iron in the northern hemisphere, as it contains more land mass.

At different periods of the Earth's history, the magnetic needle faced different directions. This is consequent to the continents drifting apart. A magnetic needle does not exactly face the North Pole of the geographically described earth.

Iron nuclei are the most tightly bound, and without much space confining their nuclei, heat and pressure/force quickly disturb their magnetized nuclei configurations when in close proximity to other similar nuclei. As their nuclei are structured in particular arrays giving them a direction, they possess their attractive and repulsive behaviors-explained earlier.

There is more land mass exposed to the energy transmuting cycle in the northern hemisphere, and with more iron in the interior there is possibly more activity. As iron with other nuclei continuously disintegrate in the interiors, a compass needle (suspended on a thread) points downwards toward the interior of the earth in the northern hemisphere. As there is more extraction from the southern oceans, this same needle will point upwards in the southern hemisphere.

As iron engendered, heat and pressure/force were at extremes. With sudden precipitation, and the release of heat and pressure/force, iron engendered as the most tightly bound nucleus with explosive disintegration. As iron nuclei are tightly bound (no room for space to propagate heat and pressure/force) it cannot 'hold' pressure/force. In other words, space continuously waving heat and pressure/force does not easily disturb its nuclei configurations. This explains the malleability, tensile strength, and other characteristics of metals

Iron has a really tricky chemical property, as fusing iron is actually 'absorbing' ‘energy’ (force/pressure) instead of giving it off. To alter iron nuclei in the direction of more complex nuclei, or of less complex nuclei requires an 'input' of energy, or more force. At certain degrees of heat and pressure or force, iron and some other materials (ferro magnets) alter their tight nuclei configurations, resulting in them being magnetized. Once iron is magnetized, temperatures and pressure/force propagate through its nuclei configurations. These nuclei configurations are then activated when similar nuclei configurations are brought in close proximity. However, as these nuclei have a direction, they will ‘attract’ other iron nuclei facing the same direction. In other words, the propagating atmospheric temperatures and pressure or force, through these nuclei configurations then forms a continued pathway or a loop. This explains why magnets attract and repel.

A refrigerator has its outer body constituted of iron nuclei configurations, and as magnets are brought in close proximity it will stick to the refrigerator, as temperatures and pressure/force can now propagate continuously through the loop which has formed. As explained above, this piece of iron when magnetized transforms its nuclei configurations facilitating propagation of continuously existing temperatures and pressure/force. This explains the attracting and repelling behavior, and why it maintains its strength indefinitely (attached to the fridge)

It is incorrect to state that a magnet stuck to a fridge is held by magnetic energy. A magnet does not draw its power source from an external source-there are no entities. However, a magnet as a structure of energy existing within atmospheric space ‘allows’ the propagation of heat and pressure or force, making it ‘energetic’.

All nuclei engender/exist, and endure with the same degrees of temperatures and pressure/force intensities, within the energy transmuting cycle, between the sun and the earth, possessing their characteristics under these conditions. Altering heat and pressure/force, will alter, and/or disintegrate nuclei. Ferromagnetic materials heated above certain temperatures completely lose their ferromagnetic behavior. As these materials are integrated together with particular nuclei configurations, certain temperatures, and or pressures will disturb these configuration patterns, and alter their magnetic characteristics.

It is stated that some materials have a feature known as Ferro-magnetism. The prefix 'Ferro' refers to iron which is one such material. We have been informed that these materials have the ability to 'remember' the direction they have been subjected to- this is nonsensical physics.

Does earth have a magnetic field?

It is postulated that earth has a magnetic field, as a compass needle points approximately north at most places on the earth. In reality, as only energy exists, there are no entities, as everything exists within/as the fabric of space/existential energy.

Nothing is a one way phenomenon. Energy is shuffling to-and-fro within the energy transmuting cycle, between the earth and the sun, explaining why earth cannot possess a magnetic field. As radiated heat and pressure/force of the energy transmuting cycle interacts with earth's atmosphere, it creates a bow shock similar to water being deflected around the bow of a ship. This phenomenon deflecting solar winds, believed to create the postulated magnetic field is incorrect.

NO SUCH THING AS GRAVITY OR THE GRAVITATIONAL FORCE

Gravity is/as a human concept, postulated as a universal force, which acts between all bodies that have mass. It is also explained, as the force which attracts a body to the centre of the earth. It is assumed, that although gravity is the weakest of the postulated known forces, it shapes the structure, and evolution of stars, galaxies, and the entire postulated 'universe'. It is also assumed, that the laws of gravity determine the trajectories of bodies in the solar system, and the motion on earth, where all bodies experience a downward gravitational force exerted by the earth's mass, the force experiences as weight.

Nobody can present an explanation why gravity attracts, and not repels. How does gravity 'cause' falling objects, and orbiting planets, without drawing on any known power source? We are informed, that gravitation is the tendency of masses to move together. Weight is a force caused by gravity equal to the product of the acceleration of gravity, and the mass of the object. Exactly, why two bodies of mass, separated in space, have a gravitational attraction to one another remains largely unknown, despite much research, and various theories.

Gravity clashes with quantum theory, as it does not fit in the standard model. Nobody has spotted the particle (graviton) responsible for it. Newton's apple now contains a whole can of worms.

Aristotle, Newton and Einstein have all theorized gravity as a force, which attracts bodies depending on the mass and distances. Newton developed a quantitative theory of gravitation postulating a force of 'universal gravitational attraction', which was proved only mathematically accurate. However, this has helped in sending rockets to the moon, or around the solar system.

Einstein proposed a whole new concept of gravity involving the four-dimensional continuum of space-time, which is curved by the presence of matter. Why should the mere presence of matter warp it, and also where does this unlimited power source for it all come from?

Both these theories are incorrect, as there is no such thing as gravity, or a gravitational force, which is refuted following. We talk about gravity that keeps our feet on the ground, keeps the moon in orbit around the earth, and keeps the planets in orbit around the sun. Yet nobody knows what 'causes' it.

In 1918, Einstein proposed that gravity would spread like ripples moving at the speed of light, through his four dimensional universe of space and time. Even today after many decades, and after an expenditure of hundreds of millions of dollars on hair triggering detection systems, no one has actually recorded unambiguous evidence of their supposedly omnipresent gravitational waves. This is explained following.

Combining the postulated force of gravity with these forces of quantum mechanics, to create a theory of quantum gravity is currently an important topic of research among physicists. General relativity is essentially a geometric theory of gravity. Quantum mechanics relies on interactions between particles, but general relativity requires no particles in its explanation of gravity.

Scientists are very mystified by gravity knowing nothing about its true nature, and yet also claim to explain it. For years, scientists have theorized about the graviton (a particle that transmits the force gravity), but are frustrated in their attempts to find a consistent quantum theory for it. Scientists are still searching for the graviton, hoping for some sort of physical verification, and we can be pretty sure there will soon be another theory. New theories such as quantum gravity, and the string theories, are being invented, and modified, or shelved. More will follow to explain the unexplained previous ones.

Why Newton's theories of gravity are incorrect. Why Newton's dynamics are wrong.

With Newton's dynamics, star fields should all fall together at some point, or simply be attracted to each other. Newton had explained this would happen, if there were a finite number of stars distributed over a finite region of space. On the other hand, if there were an infinite number of stars, distributed more or less uniformly over infinite space, this would not happen, as there would not be any central point for them to fall into. So how can we explain why star fields are not attracted to each other in a galaxy? Why do they not fall onto the centre of their galaxy?

Making gravitational forces repulsive at large distances, with the attractive forces between nearby stars, balanced by the repulsive force from those that were further away, would create an equilibrium where everything would be unstable, if stars got near, or far away. What about the interactions between star fields conjoined to form galaxies?

Newton believed the infinity of the universe keeps the space bodies from falling onto themselves, due to their mutual gravitational attraction. He thought that evenly distributed matter throughout an infinite space, will put itself together in an infinite number of big masses, instead of collapsing in one mass. The origin of the space bodies is uncertain. There is no prospect of identifying the mediator of gravity. The motion of the space bodies cannot be accounted by the visible matter in the postulated universe. The postulated gravitational pull exerted by the matter found in stars, is not enough to keep the cluster of galaxies stable.

Gravity cannot exist at the surface or the centre of a body, and if it does, it would attract all particles, nuclei/elements and molecules that form its atmosphere, and bodies would have continued increasing in size. All the gaseous elements in/of/as the air would crash down onto the surface. An astronomical body would have pulled its atmosphere onto the surface, and kept on increasing its size by gulping smaller bodies, eventually growing into one large astronomical body. There would have been only one immensely large orb. All star fields would also fall onto the centre of a galaxy. There would be no other existences, and no momentum, force/pressures or temperatures. Energy as matter would be inert, and inactive without any transmutations-nothingness.

According to Newton's dynamics, my body and the earth are attracted to each other with equal force-that means my body should accelerate a lot faster toward the earth than the earth toward me. What about my head, as only my feet stands on the ground? Every part of my body should lay flat over the surface, and so should every living organism. All plants and animals would have been attracted flat on the surface of the earth. All plants and animals should cover the surface of the earth. Seeds should germinate, and grow downwards, or should have evolved to stay underground. Tall trees would not have been able to grow straight up, and lift tons of water upwards.

Newton’s dynamics do not account or explain; why some planets/bodies have elliptical orbits-it explains circular orbits-small ones around larger ones. It does not explain the eccentric orbits of Pluto, and the comets around the sun. If gravity or the gravitational force exists, why do planets orbit the sun, and not crash into it? It also does not explain why all astronomical bodies spin or rotate. Why should different planets have different rotation speeds? Why should Venus and Uranus rotate opposite or retrograde ways? Why do Mars suddenly stop rotating, and start rotating the opposite ways for some months? Why are some planets tilted? With Newton’s dynamics, every moon and satellite should orbit the sun independently. Every planet should orbit each other, every star should orbit another, and every galaxy should orbit another galaxy. Why should some planets have discs, and why have they not formed solid bodies? Why have atmospheres of planets not integrated with their interiors, some with partly solid bodies? Why does our atmosphere not integrate with our oceans? All these questions are answered within revelation 7.

A force is an energy transmutation, or a form of energy. Nothing is a one way phenomenon. All forms of energy propagate through or within/as space, as it waves. Space waves as forms, states, processes or structures of energy are propagating, or transmuting, or shuttling to-and-fro. Therefore, the one-way postulated gravity, or this force cannot exist. Every thing exists/endures within/as the fabric of space/energy. There are therefore, no entities, and without entities there can be no causation. Everything everywhere is energy, or energy transmutations.

Why Einstein's theories of gravity are also incorrect

Einstein was dissatisfied with no understanding of gravity, so he presented another theory known as general relativity, some parts of which are even more mysterious, and unexplained. Einstein improved on an already incorrect theory of gravity by Newton. Theorizing relativity and space-time, he took into accounts extremely large gravitational fields, and objects moving at velocities close to the speed of light.

Einstein's theory of general relativity states, that the presence of mass, energy and momentum 'causes' space time to become curved. Because of this curvature, the paths that objects in inertial motion follow can 'deviate', or change direction over time. This deviation appears to us as acceleration toward massive objects, which Newton characterized as being gravity. Einstein explained that 'gravity' works, as the mass of an object 'causes' space to curve around it. He states, that as a less massive object (like a planet) approaches a more massive object in space (like the sun), the less massive object follows the lines of curved space, which draws it near to the more massive object, and into orbit. In general relativity, however, this acceleration or free fall is actually inertial motion. Objects in a gravitational field appear to fall at the same rate, as they are in inertial motion, while the observer is the one being accelerated.

There is no such thing as causation, as there are no entities. Existences exist/endure as parts of a cycle, sustaining each other. Einstein fails to explain, why and how a less massive object approaches a more massive object. What is the force behind this approach? What generates the momentum? Or in other words, why do planets orbit the sun, and why moons and satellites orbit the planets, and together also orbit the sun?

What is 'four-dimensional space-time' and why would the mere presence of matter wrap it?

Is light bent and is space curved?

Einstein argued that light should be affected by curved space, the same way that matter is, as light is also matter. He argued that light bends, as it passes through a strong gravitational field. This was confirmed when a solar eclipse gave scientists a rare opportunity to study stars while the sun was up. Normally sunlight outshines the light of all stars, but during an eclipse the sun is obscured, and other stars are visible.

Scientists carefully measured the positions of several stars, which appeared close to the sun. Six months later, as the Earth moved along its orbit to a point on the opposite side of the sun, so that it was once again in line with the stars studied during the eclipse, the position of those stars were once again recorded, and were found to differ slightly from their positions during the eclipse. The only explanation for this is that the star's light had been bent by the sun's gravity.

Revelation nine details that light as a form of energy propagate within/as space, as it waves. In reality, light from those stars could not have propagated out of their star fields. Moreover, it is impossible for light from those stars to propagate through ours. We see light at the sun, and at the stars. Light dwindles with distance, and we do not receive the light generated from the sun processes, and not from any star. Those scientists confirming Einstein’s' light-bending theory observed light at those stars, not light from them. From a distance light at those stars would appear to be bending around the sun, as they would have been seeing light at those stars through the sun's atmosphere, which is millions of kilometers thick, and full of outgoing and incoming plasma.

There is also another explanation why the stars differed slightly from their position. Every six months the earth gets closer or further away from the sun, and those stars would have changed their position observing them from earth.

Einstein's general theory of relativity states that gravity is not so much a force, as it is an inherent property of space and time. Accordingly, earth orbits the sun, not because it is attracted by gravity, but because it has been caught in a big dimple in space-time caused by the sun, and spins around this dimple like a fast moving marble caught in a large bowl. This is incorrect.

There is no such thing as gravity, as time is a human construct for human living, and space is a form of energy-not a void, or a vacuum. Time for humans began at the postulated Big Bang, and is independent of space-a form of energy. There are no entities and everything is interdependent, and interconnected. Existences are forms, states, processes or structures of transmuting energy, existing within/as the fabric of space/energy. Revelation 7 details why planets orbit the sun, and why they rotate around their imaginary axis.

Einstein's theories do not pay any attention to quantum mechanics-the realm of the extremely small-not because gravitational forces are negligible at small scales, but because gravity does not exist. All these unsolved problems prove there is no such thing as gravity, closing Newton's dynamics and Einstein's theories.

THE POSTULATED GRAVITATIONAL FORCE OUTSIDE EARTH

If there is no such thing as gravity how are planets or astronomical bodies kept at certain distances from the sun, and why do they orbit the sun? Why do they not crash into the sun?

How the postulated gravity drives the dynamics of astronomical bodies? How do planets orbit without any known power source? What brings the space bodies together and what keeps them apart? What keeps the space between bodies together, and what maintains the distance between them? These questions are answered within revelation 7.

If there is no gravity in space why do shooting stars fall?

There is no such thing, as falling down, or going up, within existential energy's expanse. Shooting stars are debris of disintegrating comets. As these interact with our upper atmospheric nuclei, they disintegrate, and 'light' up, and are observed as 'shooting stars'.

Earth's atmospheric nuclei are comprised of many elements, and with particular degrees of heat and pressure/force they are activated, and 'light' up in colors. They are observed shooting across, as they are captured, and transported by one of earth's many layered rotating atmospheres.

The force on various planets in our solar system

Incoming forms of energy from the sun comprises radiated heat/temperatures, pressure force and protons. Protons interact and disintegrate atmospheres, and surfaces of bodies without atmospheres. However, as bodies/planets possess an atmosphere, the incoming radiated heat and pressure forces down the integrated nuclei/air toward the surface. This explains why things can be weighed on earth. However, this is not gravity, as there are equal outgoing different forms of energy toward the sun.

Mars and Mercury and our moon (without atmospheres) have more or less heat, and the same intensity of force from the sun, which is less compared to that of Venus and Earth with atmospheres (similar incoming force intensities).

The force at Jupiter is 2 ¼ times than that at the earth. Saturn, Uranus and Neptune have a slightly more intense force than earth. The force on the Jovian planets is calculated on top of their atmospheres, as they are gas giants. The force is stronger at the Jovian planets (even though forms of energy dwindle with distance), as the amount of extracted material/energy is less than that at Earth and Venus. The force at Pluto is negligible, as it is distant from the sun.

THE POSTULATED GRAVITY AROUND EARTH

No such thing as gravity

The concept/word/phenomenon gravity is taken from the Latin word 'gravitas' meaning weight. We postulate there is gravity everywhere, as things can be weighed-weighed down toward the surface of the earth. Things weigh, and fall to the ground, as a result of the energy radiation/transmutation cycle between the sun and planet earth.

As everything everywhere is energy, energy only transmutes into different forms, states, processes or structures. Nothing is a one-way phenomenon.

As the atmosphere continuously disintegrates and integrates, temperatures, densities and pressure/force are manifested. With interactions, disintegrations and extraction processes within/as the atmosphere and the oceans, there is a slightly greater force toward the surface of the earth. As a result, everything is weighed down, or falls to the ground, and/or is accelerated toward the surface of the earth.

Things and objects are weighed down, if they are constituted of nuclei configurations heavier than those of air. If nuclei were lighter than air (as protons) they would have been extracted toward the sun, and if there were of the same weight as air, or atmospheric nuclei configurations, they would float as air.

We are unaware of other forms of energy being extracted from the earth, as we have not evolved mechanisms to do so, as these were/are not significant requirements necessary for foraging and survival. We cannot feel certain degrees of temperatures, densities or intensities of pressures/force, as we are comprised of atmospheric nuclei. We can feel the warmth or coolness, sense humidity, and feel the breeze/winds, as the atmosphere alters.

Nuclei have integrated to form objects/bodies. Subjected by/to and within the pressures and temperatures of the energy transmuting cycle, objects/bodies are 'weighed' down, or fall toward the surface. This explains what keeps us to the ground, why objects fall down, and what provides the acceleration for an apple to fall to the ground.

If there was such a thing as gravity there would be force as a one-way phenomenon. Nothing is a one way phenomenon, as everything everywhere is energy, or continuously transmuting energy in states, processes, structures or forms. Every existence is part of a cycle sustained by another existence, and vice versa, and endures as long as the cycle continues. There are no 'sources' or beginnings, or originations, and no targets, or ends or terminations, only energy transmutations between existences, as parts of their cycle.

As things are weighed down, or fall to the ground, and as we cannot observe the extraction of protons, we postulate there is a continuous incoming existing force we call gravity. Newton's dynamics cannot explain why things are held forcefully on the ground, or accelerate downward when dropped, or why the postulated force called gravity only attracts, and not repels.

Why the postulated force called Gravity is not the same everywhere, but weakens rapidly as we move from the surface of the earth?

We are informed that the planet Earth is not a homogeneous sphere, and the force of gravity actually fluctuates in surprising ways as you go deeper underground. As so much of the planet's mass is concentrated toward the centre, the acceleration of gravity actually increases, as you begin to probe downward probably rising to as much as 5 percent above surface, and accelerating near the core boundary. From here, force and acceleration gradually decreases to zero at the centre. If gravity existed, and Newton's dynamics were correct, this would not be the case.

There is more land mass in the northern hemisphere, but with more oceans in the southern hemisphere there is more extraction there, as the water molecules disintegrate. There is incoming energy or force, and equal outgoing energy in different forms. As earth has different types of surface all around, and different atmospheric layers are continuously altering temperatures, pressure/force and densities, the incoming force is not the same everywhere.

Going down into the oceans, or the crust/mantle/outer core, the pressure increases, and/as there is an increase in density. This density increases as we go deeper, both in the oceans and the crust and outer core, explaining why force/pressure weakens approaching the centre of the earth.

Why should gravity weaken as one travels south?

There is more extraction in the southern hemisphere, as there are more oceans and therefore, a lesser force experienced. However, the incoming and outgoing energy transmutation is the same. There are other processes, structures or forms of energy, shuttling through-and-fro the energy cycling pathway.

What makes water gather in the lowest places? What keeps milk in a glass?

Nuclei configurations comprising the postulated nuclei/elements, molecules or compounds, which are continuously interacted, disintegrated and integrated, remain as air, or atmospheric nuclei, or the atmosphere. Gaseous elements/molecules are lighter, and easily expand, and exist as air.

Some nuclei/molecules forming substances or compounds are confined at the surface, or in the crust of the earth, as they are weighed down, being heavier. Molecules constitute water. Liquid water is heavier than water vapor and therefore, weighed down at the surface. However, extraction processes of the energy cycle continuously breaks down the water molecule, and water in our lakes, rivers and oceans are continuously being interacted, disintegrated and extracted. Milk is constituted of molecules, and the glass contains it, as it is being weighed down.

Air is constantly being disintegrated and integrated. Liquids if exposed, dry up with evaporation (as temperatures and pressures disintegrate the molecules), and solids decay with radioactivity (as temperatures and pressures gradually alter their nuclei configurations). Everything in the troposphere, subjected to the atmosphere-surface-ocean cycle, and the incoming pressures or force, is weighed down, and kept toward the surface of the earth.

Water, milk, or liquids, and solids being composed of integrated materials are heavier than air (depending on temperatures and pressure/force), and remain contained, unless temperatures and pressure/force alter, or the nuclei disintegrate.

What give the pendulum its steady beat? What makes a pin drop?

A pendulum is suspended at a certain point. At the beginning, the pendulum must be raised horizontally to the point, and released. It falls down subjected to the energy cycle, and would fall down if it was not suspended. Reaching the opposite side, it stops more or less at the same horizontal level where it began.

As it is suspended, and with the continuous incoming force, it swings to the other side, where it is again subjected to the same equal force of the energy radiation cycle. As the pressure/force at both ends is the same, this swing continues giving the pendulum its steady beat. The pendulum must be of a certain weight, and be suspended at an appropriate point to be effected by the same pressure, and continue its swing.

A pin drops as it is composed of nuclei configurations, collectively heavier than material that is extracted, and therefore it is weighed down-it falls down. If the nuclei comprising the pin are disintegrated they would float as air, and with further disintegration into particles they would be extracted, as they would become much lighter.

How do tall trees lift up hundreds of gallons of water?

If gravity existed, tall trees would not be able to lift up water. Revelation 20 details, that all organisms are structures of the environment. Evolved from environments, which are engendered by the energy incoming-outgoing transmutation cycle, all organisms conform with, and/or by the environments.

Plants have evolved mechanisms and exist, as they extract nutrients and/or water through their roots in the soil. Their mechanisms have evolved utilizing the extraction power, conforming to the processes of the energy cycle. In reality, plants grow as this extraction process pulls in nutrients/water through their bark, stem and leaves. The extraction process of the energy cycle further extracts this material from their new shoots, and this explains how plants grow, or exist, and endure.

Tall trees have evolved certain type of mechanisms, and can easily lift up water, as they use those same outgoing extraction pressures of the energy transmuting cycle. These processes are found in all trees and plants, explaining why some trees/plant stand tall, and how growth takes place. Most plants/trees have their branches growing outwards from their barks, facilitating their leaves to better interact with the incoming, and the outgoing energy transmuting cycle, processing the continuously altering atmosphere.

What makes a spinning top remain upright? Why does a spinning hard boil egg stay upright? Why does a hard-boiled egg spun on a tabletop rises on one end and whirls like a top? Why does a moving bicycle stay upright?

Any rotating system generates torque-a turning moment-a type of force. But from where does torque come from? The faster the bicycle or the top turns/motions, the greater is the torque, and the greater is the stability of the bike, and the rider on it. Why is there a turning moment and why is the direction of the force straight up? Why does a top gradually begin to stabilize when it slows down?

If you stand a bicycle, or a top, and let it go it will fall down. Now this does not happen because there is a force we call gravity- there is no such force. Mathematicians have explained the phenomenon to friction. But how does friction occur? Surface friction has nothing to do with it. We are told that gravity wins. Winning and losing are human words/constructs for human minds, and planning.

In reality, as energy is everything everywhere, nothing is lost, or gained, or won. As you spin a body, or an object you are agitating its nuclei configurations. A moving or spinning object/body agitates the space confining its nuclei and, as the nuclei expand, the object/body gets lighter; and/as the pressure/force upon it decreases. There is a centripetal and centrifugal force, within a spinning body. Spin per se is momentum-a form of energy. As the top spins, force (space propagating temperatures and pressures or force) propagates around its nuclei configurations, and as nuclei expand, the nuclei, and the top, are more susceptible to the extraction processes of the energy transmuting cycle. This explains why a spinning object/body gets lighter. The object/body gets lighter, as spin increases.

Increasing the spin, the nuclei configurations with increased momentum expand further, and the body/object will eventually disintegrate. This explains why, as a moving bicycle gets lighter, it is easier to ride it. This also explains why a spinning top getting lighter stands up, and with the decrease in spin wobbles, and lays flat. A moving bicycle stays upright with momentum, as it gets lighter. As the rider also gets lighter with momentum, cycling (as the wheels turn-momentum) becomes easier. It becomes easier to cycle with speed/momentum.

With a certain level, or degree of vibration, or agitations, or heat and pressure, or force, integrated nuclei expand, and subsequently disintegrate. With nuclei expanding, matter be it solids, liquids or gas becomes lighter, and more susceptible to be extracted within the energy transmuting cycle and therefore, stand or rise up. For example, liquid water will transmute to water vapor with heat and pressure/force (as space propagates heat and pressure/force interweaving its nuclei configurations).

Nuclei comprising a spinning egg, or a spinning top, expand with momentum, or spin, and rise up. The top will rise upwards on its bulkier side, as more nuclei configurations expand, and more of the surface is exposed to the energy transmuting cycle. The pointed tip facilitates, and abets the spin/momentum.

The egg tends to rise in just the opposite way, as the bulk of the mass will be lower, and although most of its mass is below (as the egg spins) the egg is so shaped, that the opposite end of the concave end point upwards-it is forced to rise up. It is less dense on the smaller end, and this area is more susceptible to be extracted (remember it is a hard boiled egg-the red yolk is heavier). Anything spun (nuclei in momentum expand) will overcome the pressure, or 'weight', and will rise. The shape of the egg plays a part, and also the fluid hardened by being hard-boiled. As the spin (or momentum) decreases, the nuclei configurations are getting back to their normal conditions.

Objects of the same volume with more matter/mass will be forced 'downwards', more than objects with less mass/matter. Objects with more nuclei, or nuclei configurations, which have engendered close together, will weigh more than objects with fewer nucleuses, or nuclei configured apart.

Why does a ball bounce?

A ball consisting of nuclei configurations heavier than air is subjected to incoming energy (temperatures and force or pressure). Reaching the ground with this force, air within the ball, or the nuclei configurations comprising the ball is displaced. As space confines nuclei configurations, this space displacement generates a reaction (nuclei expand), and the ensuing pressure rebounds the ball. This bounce is also enhanced by the outgoing processes within the energy transmuting cycle.

Mass and Weight

What is mass?

Energy transmutes into a particle concurrently with its confining space. As particles integrate together, it becomes observable by humans, and we call it mass. This is also matter. Visible energy in any form, state, process or structure is matter. In other words, invisible matter is energy. Humans create existences and name them, explaining why we cannot correctly describe our own existences of physical phenomena. Mass is how much matter a body contains, or an amount of substance. It tells you how much particles, nuclei or molecules a body has, not what they weigh. Mass is detailed in the next revelation.

What is weight?

Energy in the form of pressure/force, temperatures and protons are propagated from the sun. Our atmospheric nuclei and ocean/water molecules are continuously interacted, integrated and disintegrated. Lighter particles and protons from these interactions, and disintegrations are extracted.

Objects/bodies comprised of nuclei, which are collectively heavier than atmospheric nuclei configurations are therefore, subjected to the incoming pressure/force. This pressure/force weighs down these nuclei and objects, and this force is calculated, or measured as weight. The object-depending on its mass-is then explained to have a certain weight. This pressure force is also responsible for objects falling down to the surface of the earth.

Weight and mass are easily confused, especially since both are measured in pounds or kilograms. Weight is the force of incoming pressure on an object within the atmosphere of the earth. Weight varies according to where something is weighed, while an object's mass is unchanging. As each layer of our atmosphere has different densities, pressures and temperatures, weight correspondingly alters at different altitudes, and at different places.

Mass or 'inertial mass' refers to the amount of matter contained in something. Mass and weight are human words/concepts. Mass is different from weight. Depending on how nuclei integrated to form an object, some objects will weigh more or less than others, even though they are of the same volume or size. Dense objects (with more nuclei) are obviously heavier.

You can weigh things, and calculate how much they weigh in comparison to your own constant. But a thing weighs differently at different places on the earth, at various altitudes, and at every other place in the solar system.

Why some things weigh more than others even though they are the same volume or size?

It all depends how nuclei integrated to form objects/bodies. Nuclei constituting bodies/objects have space confining them. Dense objects with more nuclei configurations are obviously heavier, and are more subjected to the densities, pressure and temperatures of the changing atmosphere. The atmosphere is per se densities, pressures and temperatures.

As our atmosphere is continuously interacted, disintegrated and integrated, the radiated heat and pressure/force from the sun, pushes down on atmospheric nuclei. In turn these nuclei push down on objects/bodies to the surface of the earth. If the nuclei comprising an object/body are collectively heavier than atmospheric nuclei configurations they would weigh more, as they are forced down toward the surface of the earth, and can be weighed. If they are lighter (they are lighter as those nuclei are being subjected by extraction processes of the energy transmuting cycle) they would have no weight, and/or be extracted. If they weigh more or less as atmospheric nuclei they would constitute the atmosphere, floating freely like air.

Why does a body having the same mass on earth weigh less on the moon than on the earth?

Earth or any astronomical body does not exert any 'gravitational' force on another body, as there is no such thing as gravity. Within a star field, or our solar system, energy transmutes/propagates shuttling to-and-fro, to atmospheres of planets/bodies. As the moon has nearly depleted its atmosphere, it is an impotent body without any cycle. Without a cycle, energy is not transmuting and/or propagating, and there is negligible force on the moon.

Weight can only be calculated within an energy transmuting shuttling to-and-fro cycle pathway. Weight is in reality, the amount of 'pressure' on an object/body toward a surface. Temperatures, atmospheric density and other factors alter weights.

Without an atmosphere, a feather and a hammer dropped together would never fall to the surface of the moon, but float aimlessly. An atmosphere must be dense, interacted and disintegrated, and interacted again to generate changing temperatures, densities and pressure/force, facilitating things/bodies to fall down to the surface, or to have weight, or to stay at the surface.

Earth has a thick atmosphere, and there is large air resistance to a falling object. On the moon the atmosphere is so thin; it has effective zero air resistance. The hammer and a feather may, or not eventually fall on the surface of the moon, as the moon still has some atmosphere, but how long this will take is uncertain.

As planets/bodies are differently distanced from the sun, and have atmospheres of different densities, the pressure/force at each of them would not be the same. The same body/object will weigh differently on each of them.

At higher altitudes with different densities, pressure/force and temperatures, things/objects configured from the biosphere may weigh nothing at all. This explains why an astronaut out of his vehicle floats aimlessly.

How do heavy objects rest on a table without its molecules giving way, collapsing the table?

The table engendered from certain structural designed nuclei configurations. Any object exists/endures structured to the prevailing temperatures and pressures, propagating both around the object, and within its nuclei configurations. Depending on the weight placed, its nuclei configurations will hold, but as weight is increased there is an increase of propagating temperatures and pressures around, and within the table's nuclei configurations. This will result in the molecules breaking apart, giving away.

Temperatures, pressures and densities affecting weight and mass

Energy from the sun propagates/transmutes shuttling to-and-fro, to all planets/bodies within the solar system. As our visual systems comprise of atmospheric nuclei, we cannot observe particles/nuclei extracted from the surface of the earth, our oceans and the atmosphere. We are the environment/atmosphere/ocean of temperatures, pressures/force and densities and therefore, we also cannot sense these forms of energy. However, when the atmosphere alters to higher or lower temperatures, pressure/force and densities, we can sense these in the form of warmth or coolness, breeze/wind, and the humidity.

If we wet any part of our body we feel coolness there, as the water molecules are disintegrating, and hydrogen and parts of oxygen are being extracted-this is evaporation. As we perspire we feel the 'coolness', as this is transpiration-the same process. The coolness is felt as temperatures decrease, as nuclei disintegrate.

Organisms comprising the environment (of the same densities, temperatures and pressure/force) cannot exist outside the troposphere. Cells have evolved or engendered within these atmospheric conditions, and will disintegrate elsewhere. Astronauts in space, avoid this disintegration with their special space suits, and exist with the supply of oxygen etc. Space vehicles are shielded from different conditions, and have booster engines for thrust and transport. At certain altitudes everything engendered of/from the biosphere will drift/float aimlessly, and eventually disintegrate.

A balloon floats or rises upwards if it is filled with gases lighter than atmospheric air, or atmospheric nuclei configurations. Hydrogen and helium gases both have lower masses, and weigh less than air, or other gaseous elements, and are continuously extracted. Heating helium makes its nuclei more susceptible, and so it rises even faster.

An object floats in water or in a liquid, as it is lighter-less dense-than an equal volume of the fluid around it (or displacing)-buoyancy. An object's density is the amount of nuclei it has, or how tightly nuclei or molecules are packed together. A piece of cork will float in water, as it is lighter, and less dense than an equal volume of water. A balloon floats in the air, as it is lighter, and less dense than an equal volume of air surrounding it.

Molecules in air bounce around and bump into each other. When you heat air in the balloon the air molecules move faster, and bounce further away. That makes the hot air less dense than the air around it, and as hot air (its disintegrating nuclei configurations) is more susceptible to the energy cycle, it will rise.

Objects/bodies are engendered with certain temperatures, and/or pressure/force. Increasing or lowering the intensities will alter the nuclei configurations. As nuclei get lighter they are subjected to the extraction part of the energy cycle, and rise.

As you go up in altitude air pressure goes down, as the higher you go the less air (tenuous upper atmosphere) there is pressing down on you. Air density changes with altitude too. As air compresses under its own weight, its density is greatest at sea level, and decreases with altitude.

A rising balloon filled with lighter gases than air will rise. A balloon filled with air will also rise, as the air is heated, as the air or nuclei are disintegrating, and are subject to the extraction part of the energy cycle. But far up, as the pressure decreases, as there is no density it will stop rising.

Hot air rises, but cools with altitudes. Closer to the surface of the earth, with a denser atmosphere the pressure/force increases. Going deeper, the water gets denser with increasing pressures.

Every existence exists/endures, consequent to it being engendered by/with certain temperatures and/or pressures. If those pressures and/or temperatures alter, that existence will concurrently alter. With extremities of pressure, and/or temperatures that existence will disintegrate.

Why do popcorns pop?

Popcorns having tight hull structures allow kennels to explode with pressure and heat. A leaky hull prevents moisture build-up, and kernels do not pop. Kennels are like pressure vessels, exploding when pressure and heat reach a certain point.

Weightlessness

Typical orbital altitudes for human space flight vary between 192 to 576 kilometers above the surface of the earth. Orbiting spacecraft like the Space shuttle or Space Station are not kept in orbit around the Earth by gravity. Gravity does not keep satellites from falling toward the earth. At certain altitudes they are not subjected to the atmosphere-surface-ocean cycle. Sometimes they have to be boosted by engine power to straighten out, and keep their orbits. Shuttles at their altitudes are subjected to both the energy transmuting cycle from the sun, and the atmosphere-surface-earth cycle, and they have to travel at very high speeds to prevent them being constrained by these cycles.

Astronauts, the space station, and an apple are all bodies composed of nuclei/elements, molecules engendered within the biosphere. If you drop an apple within our biosphere it will fall down to the surface of the earth. The structure of the apple is designed by evolution, so that by falling to the earth its seeds germinate more trees. This is why apple trees are still around. The nuclei configurations of an apple are heavier than air.

An astronaut has his body structured by cells to help him action, and move about using the ground as leverage and support. Both the apple and the astronaut are structured of, from, and according to conditions in the environment. As both are the environments (the environment subjected to continuous pressures/force, air density and temperatures) they both continue their energy structural processes, or continue to survive and exist, as the environment does. In some upper layers of the atmosphere with different incoming and outgoing pressures, temperatures and densities, an apple and an astronaut will not fall down, but will float. In yet other extreme environments, both the apple and the astronaut will have their cells disintegrated into particles.

In space there is no such thing as weight, or falling down. Here within our biosphere as things have engendered of, and from the environment, we can calculate weight (with incoming pressures generating densities) and/as things 'fall' down. Up there nothing falls down, as although there are different temperatures and pressures/force, there is less density. Without density, nothing can be weighed. However, energy is also being extracted within the same pathways, so things should also have weight in space. This is not the case. As we go higher in altitude the atmosphere diverges explaining our tenuous upper atmosphere. Closer to the surface of the earth, as forms of energy converge on earth, and with our continuously altering atmospheres of temperatures and pressure or force, there is density, and things can be weighed.

In space, the apple, astronaut and the space station 'fall' together, but they are not falling toward earth-they are floating. Objects/bodies inside of the station appear to float in a state scientists call zero gravity, or more accurately micro gravity. But since there is no such thing as gravity, we may call this weightlessness. Micro gravity is what space scientists use to determine at what speed the space shuttle should orbit the earth, to avoid being pushed toward the earth, or away from it, with the incoming and outgoing pressure/force of the two energy cycles acting upon it.

Many amusement park rides create brief periods of free fall. Some rides that operate vertically are actually 'free fall rides'. Most roller coasters have a set of parabolic (rolling) hills that also create brief periods of weightlessness. At these rides, your nervous system cells alter their structural processes to maintain the body responding to being set in momentum. As this momentum terminates the system again sets to normal. However, before your system normalizes you experience a state of weightlessness.

We have been informed that at a certain distance when you are traveling from the earth toward the moon, the moon's gravitational pull will get stronger than the Earth. Gravity does not exist, and within your journey to the moon, there is no point you will feel pulled anywhere above Earth's atmospheres. Everything around the surface of the moon floats aimlessly, as it has no atmosphere, and no density. However, depending on the interactions between the energy transmuting cycle of the sun, and other planets you may be pulled (thrown away) toward any of them.

REVELATION ELEVEN

NO SUCH THING AS MATTER

Matter is duration visible/observable condensed energy.
This revelation discusses the formation, and characteristics of the myriad forms/states/processes/structures of energy/matter

ENERGY AS/IS MATTER

The formation of particles, quarks and plasma

As energy is in continuous momentum, and/or continuously transmutes, and as everything is flux, no existence maintains its characteristics for any moment of time.

At star processes, particles, quarks, and plasma, are progressive stages of continuously condensing energy. Plasma is condensed energy (the dense region, or a star). We can postulate that the dense region is comprised of energy particles. However, condensing energy, or energy particles, or quarks can all be called plasma. It all depends how we wish to explain, or describe our created existences.

As plasma is continuously expelled from the star, it is confined by space, transforming into protons. Therefore, particles, quarks or plasma constitute the proton, but these existences are virtual, and hypothetical. A naked proton or integrating particles is plasma, or proton plasma. Proton plasma 'individually' confined by space is hydrogen, or a proton. This space confining proton plasma, and a nucleon, has different properties than atmospheric or interplanetary space. At a dense region, or at star processes, as energy is continuously condensing into particles/plasma, and as the dense region is collectively confined by space, particles or plasma are not 'individualized', and cannot be described as protons.

Protons/plasma as/with radiated heat and pressure/force, interact with atmospheric, or nebula nuclei, within interplanetary space. Particles and plasma from disintegrating nuclei are extracted back to the star/sun, within the shuttling-to-and-fro energy transmuting pathway cycle, as replenishment energy. These processes endure the star, the bodies within the star field, the star field, and the energy transmuting cycles.

At particle accelerators, nuclei and protons (comprised of particles/quarks and plasma) are broken down into smaller units of energy, what physicists call energy particles. As nuclei, or a proton, or disintegrated particles can be observed with sophisticated instruments/detectors, they disappear very quickly, as they amalgamate into/as space, or as free energy again, as particles, quarks and plasma are forms of condensed energy.

Why cannot protons be engendered at star processes?

The dense region (a star/sun) of continuously condensing engendering particles, and/or plasma, is confined by star field space. Plasma continuously expelled from a star is immediately contained by space, transforming into protons.

Every existence is continuously changing. As plasma propagates from the star, it appears to us that the star has a fixed atmosphere. In reality, the continuous expulsion of plasma 'endures' the star's atmosphere.

Protons existing 'individually', as hydrogen do not possess the same characteristics, as when integrated with neutron plasma forming the nucleus. The nucleus with its confining space (with different properties) is a new existence with new characteristics.

Protons are only engendered at functioning star fields. Nobody will be able to create a single proton, or hydrogen, as we have to first source free energy, and then condense it. By disintegrating a nucleus, neutron plasma amalgamates/transmutes into/as space, and the proton may be contained.

The formation of Neutron Plasma

No such things as Neutrons

As some types of star fields terminate, the star explodes its atmosphere of plasma, which amalgamates with star field space, producing a nebula of protons. These protons do not disintegrate, as star processes have ceased, and replenishment material is not required.

Without confining space, the cold core rotates faster, and/as some parts of the core concurrently disintegrates into particles. These particles then form the periphery of the new core. Disintegrating particles/plasma would transmute into protons if there was space around. As there is no space within the terminated star field (utilized to engender protons) the periphery of particles transmute into neutron plasma. If space was not utilized, the disintegrating core of particles would amalgamate into/as space or free energy.

The super speed/rotation of the periphery (neutron plasma) attracts the nebula of protons into the core, and concurrently further condenses the core, conjoining neutron plasma with the protons- this is a neutron star.

The concentration continues, until all the neutron plasma and protons are tightly conjoined into a very dense sphere. With continuous rotation, and as concentration can go no further, the neutron star explodes.

Fragments from the neutron star within its star field disintegrate into protons, and the neutron plasma. Neutron plasma (as disintegrating condensed energy) disintegrates, amalgamating/into as free space/energy. Hydrogen/protons also disintegrate, and with/as free space, amalgamates/drowns into/as a black hole.

Neutron plasma engenders as there is no space to confine it. Therefore, there is no such thing as a neutron. Neutron plasma is forced to conjoin with protons, as a neutron star rotates rapidly. If this was not the case, free neutron plasma immediately transmutes/amalgamates into/as space.

Neutron plasma is/as disintegrating condensed energy, is forever trying to break free to transmute/amalgamate with space. This explains why deuterium is easily disintegrated, and why tritium (with two parts neutron plasma) is radioactive, and not easily found in nature. This also explains, why heavier elements with a greater amount of neutron plasma than protons are radioactive, and/or unstable.

Protons are condensed energy, or plasma confined by space. However, protons continue to be confined by space, until heat or pressure/force alters these configurations. In this case, proton plasma will disintegrate transmuting/amalgamating into/as (its containing space) space, or free energy. Neutron plasma is disintegrating condensed energy. At/as a neutron star, it is conjoined with protons. Within a nucleus, neutron plasma with protons is confined by space. If not, neutron plasma immediately amalgamates, or transmutes into/as space/energy. Therefore, there is no such thing as a neutron, as neutron plasma (disintegrating condensed energy) cannot be 'individually' confined by space, as in the case of a proton.

Neutron plasma is not engendered at any functioning star field, and only engenders at a neutron star and therefore, only hydrogen exists freely within functioning star fields, outside a star processes and its atmosphere. For a nucleus to engender, protons (condensing proton plasma confined by space) integrates with neutron plasma (disintegrating condensed energy), and these two existences are further confined by space.

At neutron stars, protons only conjoins (not fuse) with neutron plasma, as there is no space. As neutron plasma is disintegrating condensed energy, there is a tremendous outward force from a neutron star, as neutron plasma forever tries to break free. This will explain why a neutron star, continuing to expel force, is concurrently condensing with an equal inward force (as the neutron star concentrates), which is also enhanced by the continuing increase of its orbital rotation (without space confining it). This would also explain why a neutron star pulses.

Protons engender/exist only if plasma, or condensed energy, is confined by space. Neutron plasma engender/exists only at neutron stars. Neutron plasma also engender/exists as nuclei (together with a proton), as/if fragments of an exploding neutron star are instantaneously confined by space.

Free neutron plasma (outside a neutron star, or a disintegrating nucleus) immediately transmutes, or amalgamates into, or as space or energy. Free proton plasma (without confining space) also immediately transmutes/amalgamates into/as space/energy.

The formation of nuclei

How, when, why and where does the nucleus form

The formation of deuterium, tritium and helium

It is incorrect to postulate that the fusion of hydrogen to helium is a power source in the sun/stars. At/as a functioning star, energy is continuously condensing, and as a result, particles engender, and with further concentrations of particles/energy, plasma forms.

Space confines expelled plasma from the star, which transmutes into protons. Therefore, at functioning stars, protons exist outside of star processes.

Hydrogen is a proton, and not a basic nucleus, and cannot be called an element. A nucleon, or the basic unit of matter, is a functioning whole with two different existences of different characteristics. Deuterium, and Tritium, as elements are therefore, the basic units of matter.

The formation of nuclei

Neutron plasma is formed only at neutron stars. At neutron star processes protons are conjoined with neutron plasma. As a neutron star explodes, neutron plasma disintegrates into/as space, or free energy, and protons disintegrate into plasma, and subsequently into particles, and with space into/as free energy. However, if these exploding fragments dart through a nascent star field space, they are confined by space, engendering nuclei.

An exploding neutron star happened to be in close proximity to our nascent star field. If our star field was functional, the fragments of this exploding star would disintegrate on reaching the interstellar medium. As our nascent star field was not full fledged (no energy transmuting cycles in force), these fragments were able to dart through our nascent star field space. These fragments (protons conjoined with neutron plasma) were confined by space, engendering nuclei. Concurrently, these processes incited, and enhanced the processes of the nascent sun.

As the sun or the dense region was now operational, energy in the form of temperatures and pressure/force and plasma/protons expelled from the sun, interacted with the fragments-turned-nuclei. As interactions continued, concurrently generating atmospheres enveloped the fragments-turned-nuclei, engendering the planets.

As energy transmuting cycles took hold, our solar system was born. As energy transmuting cycles were in place, all these fragments transformed into full-fledged planets, as their continuously interacting/disintegrating/integrating outer layers, transformed into altering atmospheres, enveloping their interiors.

Nuclei or deuterium and tritium engendered, as space confined the fragments (consisting of protons conjoined with neutron plasma) of the exploding neutron star, concurrently with increases of temperatures and pressures.

Tritium which is also a basic nucleon also engenders the same way as deuterium, but with an extra amount of neutron plasma, it is radioactive and unstable, as neutron plasma (disintegrating energy) has innate tendencies to disintegrate. Therefore, any element that contains more of tritium is unstable, and radioactive.

Deuterium will not integrate with another deuterium and therefore, deuterium integrates with tritium, as tritium possesses extra amounts of neutron plasma. It is the release of the extra portion of neutron plasma with this integration, helium engenders. Moreover, the release of neutron plasma concurrently generates heat and pressure/force, and more space, which facilitates and abets the integration of deuterium and tritium into helium. In the same way, helium will not integrate with other helium, unless there is a release of neutron plasma, explained elsewhere.

This integration processes continued until all of the protons and neutron plasma integrated to engender deuterium, tritium and helium, as concurrently the energy transmuting cycles between the sun and the planets steadfastly set. However, as disintegrations and integrations continued, the 'release' of neutron plasma forming helium generated more space, concurrently increasing the temperatures and pressure/force. As the fragments continued to be interacted and disintegrated with radiated heat and pressure/force from the sun; disintegrating and integrating nuclei, and generated space, transformed into atmospheres enveloping the interiors-planets.

The original atmosphere of all the fragments turned planets, were comprised of protons, deuterium, tritium and helium. After the explosion of the planet between Mars and Jupiter, all the inner planets' atmospheres were completely depleted, exposing the interiors. This is when nuclei integrated into heavy elements, as explained within revelation eight.

Nuclei can only be engendered, if fragments of an exploding neutron star pass through a nascent star field space, as neutron plasma and protons are conjoined together with confining space.

As proton plasma is condensed energy, and neutron plasma is disintegrating forms of energy, freed neutron plasma, and freed plasma, or proton plasma, would transmute into/as space/energy, and would never conjoin to engender a nucleus. As a proton is confined by space, neutron plasma conjoins-but does not fuse-with it, engendering a nucleus, or deuterium, or tritium-the basic of all matter. As a proton and neutron plasma is conjoined, but not integrated or fused, a nucleon can disintegrate into a proton, and neutron plasma. Moreover, nuclei can conjoin to form elements, molecules, and other matter. This would not have been the case, if a proton and neutron plasma has integrated/fused together, as a functioning whole.

A deuterium contains equal amounts of proton plasma (proton) and neutron plasma, or equal amounts of condensed energy, and disintegrating condensed energy, explaining why a nucleus is the basic of all elements, or matter. Every existence engenders, exists and endures, concurrently with another existence of different characteristics sustaining one another, explaining why proton plasma, or condensed energy, and neutron plasma, or disintegrating condensed energy conjoins forming a new existence.

Tritium does not contain equal amounts of proton (condensed energy) and neutron plasma (disintegrating condensed energy) and therefore, tritium does not easily integrate with another tritium to form any element. With certain temperatures and pressure/force, as its extra neutron plasma is released, heat and pressure/force is generated, and it transmutes into a deuterium. This process facilitates and abets its integration with another deuterium to engender helium, and with helium to engender heavier elements. It is consequent to tritium possessing this extra amount of neutron plasma, that nuclei can integrate with each other to form elements, molecules, and the myriad forms of matter.

As a proton is an 'individual' entity of condensing, or compressing energy, you cannot describe a nucleus, as a fusion of protons and neutrons. There is no such thing as a neutron. Neutron plasma is disintegrating condensed energy, but is not confined by space to be understood by humans as an 'entity'.

We can state that a proton and neutron plasma conjoins (as a functioning whole) as a nucleus. We cannot use the words integrate or fuse, as this would mean that a nucleus is a new existence blended, or amalgamated into a new unit. This is not the case, as the nucleus can disintegrate into protons, and neutron plasma again.

We can disintegrate light and heavy elements (fission, fusion-A-Bombs, H-Bombs, or thermo-nuclear generation), but we cannot produce a nucleus, or an element. We also cannot produce a proton, or neutron plasma, neither can we conjoin them. Protons are only produced leaving a functioning star, and neutron plasma is only produced at a nascent neutron star.

How can deuterium or a nucleus of a proton and neutron plasma be confined by space, within space?

A proton is proton plasma confined by space, and is an 'individual' existence-hydrogen. A nucleon is a proton conjoined with neutron plasma confined by space, also an 'individual' existence. Space confining proton plasma, or as a proton, and space confining a nucleon, possess different properties than atmospheric and interplanetary space (continuously propagating temperatures and pressures/force). Both these existences exist 'individually' outside of a functioning star, and/or a neutron star, and within star field, or atmospheric space.

Atmospheric or interplanetary space continuously exist propagating different degrees of temperatures and pressure intensities, and every existence engenders, or exists, or endures, or alters, or terminates depending/subjected to continuously altering temperatures and pressures.

How do you take a nucleus apart?

If you want to pull a nucleus apart, we have to get rid of its space (continuously propagating certain degrees of heat and pressure) that binds it together. You can shine light on the nucleus, or expose it to radiation with different intensities, or you can hit it with other nuclei. In reality, everything is made up of energy, so shining light, or hitting it with another nucleus, is really just a special case of whacking it with other forms of energy. In reality, we use other forms, states, processes or forms of energy to transmute the nucleus into yet other forms, states, processes and structures of energy. Heat and or pressure/force are other forms of energy that will do the trick, as it will make nuclei in an element to move quickly, and hit each other.

To knock out the nucleus apart, will demand more energy than removing the surrounding space. You can of course, hit a complete nucleus with something hard enough breaking it into bits and pieces. The nucleus will disintegrate into a proton and neutron plasma. Neutron plasma will amalgamate into/as space/energy giving out heat and pressure/force. The proton will also eventually disintegrate, amalgamating into/as space/energy. However, as the proton is condensed energy, its different types of disintegrating particles will be visible by sophisticated detectors for a very short duration.

New techniques have been developed using super fast pulses of space (not electrons) to burrow insides molecules, and send back information about the arrangement of the nucleus. The nuclei/molecules are accelerated at high speeds producing x-rays, and other forms of intense radiation used to probe the inner workings of molecules.

Why does helium have the lowest melting point, and the only liquid that cannot be solidified by lowering temperatures? Why does it not easily integrate with other helium even at high temperatures?

A deuterium at earth's upper atmosphere is easily disintegrated by radiated heat and pressure/force from the sun. A tritium is a nucleus with extra amounts of neutron plasma (disintegrating condensed energy). As neutron plasma continuously seeks to transmute/amalgamate into/as space, tritium is radioactive or not stable. Tritium also easily disintegrates at earth's upper atmosphere, explaining why it is not found in nature. All nuclei of/as the atmosphere are continuously disintegrating.

Deuterium is the basic unit of all matter. With equal portions of condensed energy (proton plasma), and disintegrating condensed energy (neutron plasma) it will not integrate with another deuterium, or two deuterium will not integrate to form helium, which is in reality, two deuterium. A helium nucleus will only engender, as a deuterium integrates with a tritium, as the tritium's extra neutron plasma is 'released', generating heat and force or pressure, and space. In reality, it is the release of heat/force that integrates a deuterium and tritium into helium.

As helium engendered with a 'release' of force/pressure, it does not easily integrate with another helium, or a deuterium, unless there is a release of neutron plasma (generating heat and pressure/force and space), concurrently with integration. This explains why helium is a gas at certain temperatures and pressures. We can postulate that helium integrates with a tritium to form a heavier element-another reason why tritium is not easily found in nature.

As helium engendered with a certain degree of heat and pressure/force intensities, and with the release of space (pressure/force) it exists as a liquid at certain temperatures and pressures. The two nuclei of equal amount of condensing and disintegrating condensed energy, bound under these conditions prevent helium from easily turning into a solid at lowered temperatures.

As helium engendered with unique nuclei configurations under decreased temperatures and pressures (as its nuclei expanded), it alters its phase turning into a super fluid, or a liquid that runs. However, at very low temperatures, and under high pressure you can alter the state of helium.

The formation of elements heavier than helium

Why and how did heavier elements than helium form?

Scientists tell us that 14 billion years ago there was a huge explosion called the Big Bang that brought the postulated universe into existence. Before the 'universe' oozed out from a point there were no laws of physics, no time, and no particles- How ridiculous! There is more!

We are informed, that at the very beginning there was an immense amount of radiation-from where and how? It took hundreds of thousands of years after the Big bang for matter to form-Why so long and how?

At the very beginning the postulated universe was small-the size of a single atom. As the temperatures were at ten thousand million millions degrees atoms could not stay together, and broke together into many exotic particles-the primordial soup. Where did the atoms come from?

We are informed that this primordial soup lasted for a nanosecond, and then the postulated universe started growing for the last 14 billion years. As the universe expanded it cooled and new particles formed. Within what was it expanding? When it was ten billion of a second old, the temperatures had fallen to a million millions degrees, and now protons and neutrons begin to form. Why, how and from where did the force come?

When the universe was a tenth of a second old, protons and neutrons joined together to make helium, and the temperatures were cool enough to stop these nuclei breaking apart. The universe was 300,000 years old before atoms began to form and the postulated electrons could stay bound to the nucleus at decreased temperatures. However, there were still atoms of only two elements-hydrogen and helium. Between the ages of a hundred million years and a billion years the postulated universe began to get its first stars. How?

It is theorized that clouds of hydrogen and helium collapsed under their own gravity and reached incredibly high temperatures. How and why can and did these clouds collapse when the postulated universe is expanding?

It is stated, that hydrogen nuclei were moving fast enough to be able to collide and fuse together forming helium. The enormous amount of energy released in such fusion reactions is what lights up the stars and maintains their huge temperatures. Stars are still being born today, even 14 billion years after the Big Bang. However, they do not last forever and millions of stars have died since the Big Bang. Why and how?

It was in these dying stars that heavier elements such as carbon and iron were first formed. How? These elements make up the contents of any planets that orbit newer star like our sun. How? All the above is incorrect.

Following is the refutation:

After the planet exploded, the interiors of the inner planets were exposed. As radiated heat and pressure/force from the sun were in force, and as disintegrations and integration continued, there were increases in temperatures and pressure/force intensities. With the gradual increases in temperatures and pressure/force, deuterium, tritium and helium integrated with other deuterium, tritium and helium, or combinations of one another forming heavier elements than helium. All elements engendered with one or more helium, and/or a deuterium, and/or a tritium.

One by one, heavier elements integrated, concurrently as temperatures and pressures increased with integrations and release of 'space'. When heat and pressure/force were at extremes, the heaviest element generated was iron.

As elements are integrating below the element iron there is a release of energy, or outward pressure/force and temperatures. Therefore, the inward integrating pressure/force concurrently generated an equal outward pressure/force with temperatures. All this release of excess forms of energy or outward force becomes available to nearby nuclei, and one by one heavier element integrated to iron. Moreover, and importantly, as neutron plasma was 'released', more and more space was generated, and this facilitated and abetted further disintegration of nuclei, and also concurrently the integration of nuclei into heavier elements. Elements become heavier and heavier, as more nuclei can be integrated with the proportionate rise in temperatures and pressures. At the iron stage, extreme temperatures and pressures or force have been reached.

The formation of elements heavier than iron

Temperatures and pressure/force reached the extremes as iron engendered. At the same time iron was engendering, nuclei were disintegrating profusely at the outer regions (atmosphere) of the interior, and concurrently nuclei were also integrating, as oxygen and other gases, as/in, or engendering the new atmosphere.

With the prevalent temperatures and pressure/force, oxygen generated at the atmosphere was breaking down into other elements. As explained within revelation eight, there was an alteration in earth's atmosphere as ozone formed, and oxygen was able to combine with the departing hydrogen (from disintegrating nuclei) forming water vapor. Water vapor forming clouds brought about precipitation.

As rain fell on the fiery interiors iron disintegrated explosively. The powerful blasts generated heavier elements than iron. Intense sudden heat, and pressure/force, facilitated the integration of heavier elements than iron behind the outgoing shock waves, as nuclei collided/interacted and integrated. However, concurrently as temperatures and pressures suddenly decreased, only small amounts of heavier elements than iron engendered, some rare.

Heavier elements engendered from disintegrating nuclei, and quick variations in temperatures and pressures/force. As a result, heavier elements engendered with different mixes of nuclei (Deuterium, tritium and helium), and more amounts of neutron plasma was 'released' during these event. This explains why heavier elements than iron have more of neutron plasma, and why most of them are unstable, and radioactive. This also explains why fusion generates more force/power, or heat energy than fission. It can be postulated that over thousand of years, as heavier element comprised more of a mix of neutron plasma they disintegrated easily, as they are radioactive, also explaining why some of them are rare.

Elements

The nucleus is tiny, and it is only when millions and millions of them join together that we can see visible lumps of elements. A nucleon can combine with the nucleon of the same element, or they can make compounds by combining with nuclei of different elements. The number of different combinations of nuclei (making compounds) is vast.

In an element the nuclei are of the same kind, but the varying strength, and interactions, between the properties of surrounding space (differences between the degrees of temperatures and force intensities on each nucleon) on different types of nuclei, means that the elements have very different physical and chemical properties. There are more than 109 different types of nuclei-one for each element.

A single nucleus exhibits all the chemical properties of the substance we call gold. Gold has 79 parts of proton plasma or protons, and roughly 117 parts of neutron plasma. If we break the gold nucleus separately into protons and neutron plasma, these pieces do not exhibit the properties of gold. We can state that gold has 79 protons.

By 2001 there were 115 known elements. However, those above 109 are highly unstable, and have been made in only tiny quantities in the laboratory. They exist for a short time, as both neutron plasma and protons without confining space (with certain degree temperatures and force intensities) transmute into/as space/energy.

From about 109 elements, we get millions of different materials we see around us every day. All matter is same, as all matter is made up of nuclei. But matter can be different because objects can be made up of different kinds of nuclei. Gold is made up of the gold nuclei, whereas salt (sodium chloride) is made up of two different kinds of nuclei-sodium nuclei and chloride nuclei.

The elements nitrogen, oxygen, fluorine and chlorine form strong bonds with two identical nuclei linked together. They therefore, exist at room temperatures, as gases, with pairs of linked nuclei moving chaotically in space. Two or more nuclei linked in this fashion are described as molecules. Most elements do not form discrete molecular entities such as those described, but have structures that are held together by chemical bonds. Mercury, sulphur, iron, tin, lead, copper, silver, gold, phosphorus, carbon, arsenic and salt can be broken down into simpler substances. As long as a substance could not be broken down by chemical techniques, it was an element. Different substances are composed of different nuclei, or combinations of nuclei, and any one substance could be converted into another by rearranging the nuclei.

As elements were discovered, and each of them had different properties, scientists arranged the elements of increasing nucleus weight in a tabular form, culminating in a periodic table. Although there are only 109 known elements, there are millions of chemical substances found in nature, or made artificially. These substances are not simply mixtures of two or more elements: they are specifically determined chemical compounds, formed by the combining of two or more elements together in a chemical reaction, and the 'glue' that holds these compounds together is actually space (propagating temperatures and pressure/force), and is known as chemical bonding. In reality, chemical bonding is space propagating certain intensities of temperatures and pressures around nuclei configurations.

The nucleus's structure and its space-the paths, energies and the radiation through the space around the nucleus-contribute to different properties, explaining why we have so many different types of materials. Whether these nuclei form trees, tires, dust, ashes, animals, air, water, or the different types of food we eat, and the things we see around us, depends on how their nuclei configurations are put together, and their properties of space confining them.

Nuclei behave differently, changing their characteristics with altering temperatures and pressures/force. Further, when different nuclei are put together, there are reactions forming compounds and mixtures.

Why do nuclei/elements possess their different characteristics?

A proton is proton plasma together with its confining space. In the same way, a nucleus is a proton and neutron plasma conjoined with their confining space. A proton and nucleus with their confining space, is further confined by atmospheric, or star field space.

Atmospheric or star field space is continuously waving or propagating temperatures and pressure/force. However, increases in degrees or intensities of temperatures and pressure/force, alter the space confining protons and nuclei, explaining why protons, nuclei, elements, molecules and substances continuously alter their characteristics, or disintegrate, and terminate.

All nuclei are the same thing (a proton and a plasma neutron). All nuclei are confined by space. Atmospheric space propagates temperatures and pressure/force, but the different degrees of heat/temperatures and pressure/force intensities acting upon each nucleus, and their combinations/configurations, is what makes nuclei/elements, molecules, or matter, possess their different characteristics.

Nuclei of the same characteristics conjoin together to form their element. Different nuclei comprise elements, molecules and compounds, explaining the myriad and varied forms of matter.

Every element engenders/exists, and endures, under certain temperatures and pressures. If there are increases or decreases in these forms of energy, nuclei/elements, molecules and substances alter, or disintegrate. Helium is a solid under great pressure, but then exists as a gas when pressures are decreased. It is a liquid at absolute zero, and normal pressures.

How is our world filled with nuclei/elements, molecules and compounds?

This question is answered within revelation eight.

How is hydrogen produced?

The different formation of protons

1. Protons are formed as a nascent star expels its outer core of plasma, commencing a cycle between the nascent star, and the nebula of protons generated.

2. Star field space confines plasma (the dense region), and this plasma expelled from a star into star field space, engenders protons. Protons or hydrogen are not found within stars.

3. Protons are also formed, as a terminating star field sheds off its atmosphere engendering a nebula of protons.

4. Elements and molecules disintegrate into nuclei, and protons.

Why is an element light or heavy? Why do we say hydrogen is the lightest element?

At a functioning star, energy continuously condenses producing particles, and with further concentrations, quarks, and plasma, or proton plasma engender. As protons plasma is continuously produced it is expelled from the star, and confined by star field space, protons or hydrogen form.

Hydrogen is incorrectly termed an element. A nucleus of condensed energy (proton) conjoined with neutron plasma (disintegrating condensed energy) is in reality, the first element. Deuterium is the first element followed by tritium, helium, and the rest.

As elements in the periodic table in ascending order contain more protons and neutron plasma, they would possess more condensed energy, and are termed heavy. Hydrogen, or a single proton, or deuterium with a single proton and neutron plasma are the lightest elements. Simply, heavier elements have more protons and neutron plasma and therefore, more condensed energy, or mass, or matter, and weigh more.

Why are some elements plentiful and some rare?

Heavier elements engendered with increases in temperatures, pressures/force and space. As iron engendered temperatures and pressure/force were at extreme levels. As iron disintegrated explosively with sudden precipitation, heavier elements than iron engendered with nuclei blasts. As these heavier elements were instantaneously created with variations in temperatures and pressures/force intensities, some are plentiful, and some rare.

Over millions of years, some unstable or radioactive elements have decayed, or disintegrated into other nuclei. Platinum, palladium and rhenium are rare. Copper is common, whereas zinc, cadmium are less common than uranium. Silver is 40 times less common than uranium.

Molecules

Molecules are made up of two or more nuclei, either of the same element, or of two or more different elements, joined by one or more bonds. H2 is the hydrogen molecule.

As space contains a nucleus, and also a molecule, and as space propagates different intensities of temperatures and pressure/force, nuclei in molecules forms different kinds of bonds. At certain temperatures and pressures/force, or at room temperatures, nitrogen, oxygen, fluorine and chlorine exist as gases.

If a molecule contains nuclei of different kinds (as the water molecule which consist of an oxygen nucleus, and two hydrogen nuclei) it is a molecule of a chemical compound. A molecular compound is represented by its molecular formula; for example water is represented by the formula H_2O. A more structural formula is sometimes used to show the arrangements of nuclei in the molecule.

A molecule is the ultimate indivisible unit of a substance, as it cannot be broken down without losing its identity. A molecule of sugar, or water, can be divided into single nucleus or groups, but then it is no longer sugar, or water. Inorganic molecules generally consisted of a small number of different nuclei in definite proportions.

A molecule is the smallest particle of a compound having all the chemical properties of that compound. A single nucleus is not referred to as a molecule, and ionic compounds such as common salt are not made up of molecules.

The molecules of a substance are in constant motion. The solid, liquid, or gaseous state of matter, depend on the speed, and separation of the molecules in the matter. Substances differ according to the structure, and composition of their molecules.

Molecules differ in size and molecular weight, as well as in structure. In chemical reactions between molecular substances, the molecules are often broken into nuclei that recombine to form other molecules, or other substances. Molecules can combine to form a single large molecule, or a large one will be broken up into several smaller molecules.

Molecules assume many shapes and sizes. Molecules of Hydrogen gas H_2 are very small, each consists of two nuclei of hydrogen, but a water molecule H_2O is much larger containing a nucleus of oxygen, and two of hydrogen.

Complex molecules form the key units of living tissues, the proteins and nuclei acids. Molecules in living organisms are very complex, such as RNA and DNA resembling giant helices.

A large number of small molecules, are joined together to form a single large polymer molecule. Typical polymers include synthetic resins, rubbers and plastic.

One volume of hydrogen reacts with one volume of chlorine to form two volumes of hydrogen chloride gas. One nucleus of hydrogen cannot combine with one nucleus of chlorine to produce two nuclei of hydrogen chloride gas, as each nuclei of hydrogen chloride must have equal hydrogen and chlorine. Therefore, there must be two or more nuclei of one, to combine with two, or more, nuclei of the other. This combination of nuclei is called a molecule.

An assumption that there are two nuclei in a molecule of gaseous hydrogen or chlorine, and one nucleus each of hydrogen and chlorine in a molecule of hydrogen chloride, preserves both the hypothesis of indivisible nuclei, and the hypothesis of equal number of nuclei in equal volumes of gases. Similar reasoning would allow a larger even number of nuclei in the molecules of hydrogen or chlorine.

Are there such things as Ions?

As there is no such thing as an electron (read following) atoms are hypothetical structures. Ions are postulated to be atoms that 'loose/gain' 'electrons'. Space containing a nucleon, or nucleus/nuclei, continuously propagates temperatures and pressure/force, which alter the nucleus, or nuclei in an element altering its characteristics.

What are isotopes?

Any of the forms of an element that differ chiefly in the amount of neutron plasma is an isotope. Isotopes of an element are the same, but they display different physical properties, evidence that continuously propagating temperatures and pressure/force around their nuclei configurations determines their characteristics. Deuterium is a type of hydrogen containing proton plasma and equal amounts of neutron plasma. Tritium which is another type of hydrogen contains extra amounts of neutron plasma, and these are called isotopes of hydrogen.

As atmospheric space confining nuclei continuously propagates different degrees of temperatures and pressure/force intensities, nuclei alter their characteristics. Therefore, atmospheric nuclei are continuously disintegrating. Helium disintegrates/decays into a stream of particles called Alpha radiation. This radiation collides with other nuclei (or alter the characteristics of their confining space) resulting in a lot of 'ionization'. As protons are extracted from earth's surface, ocean and the atmosphere, they interact with the incoming radiated heat and pressure/force, and protons from the sun at our upper atmosphere. These interactions/disintegration processes generate a lot of alpha radiation resulting in a lot of ionization. This produces our visible light.

Beta radiation is intense, as it passes through solid matter without much ionization. This radiation is long range, passing easily through thin sheets of aluminum or steel, but is stopped by lead. There is also Gamma radiation which propagates quickly, weakening with distance.

OUR WORLD OF MATTER

What is matter?

Why does matter exist and endure?

What is matter, as/if it is a form of energy and energy is everything everywhere?

Humans hypothesize particles, quarks and plasma. It is only when condensing energy, as proton, or nuclei, is observable by humans, we call it matter.

If star fields did not terminate, there would be no neutron stars. Without neutron stars, existential energy with star fields/galaxies would consist only of protons, and there would have been no matter. Some type of star fields must terminate for neutron stars to form. Fragments of an exploding neutron star must dart through a nascent star field, engendering nuclei. Fragments turned planets must exist in close proximity to their star, and there must be an ensuing catastrophe that must deplete their atmospheres exposing their interiors to begin nuclei disintegration, as well as nuclei integration into helium and heavier elements. There must be precipitation to engender heavier elements, and matter. If there was no matter, humans would not exist to think/imagine, query, and try to explain matter.

Matter is a human concept. Matter is visible condensed energy. Existences, or physical phenomena created by humans have been given names to explain them. With incorrect explanations there are confusions, paradoxes and problems. Most of our words describe the same forms of matter differently. Water is also ice, sleet and snow, and water as vapor constitutes air. Air is also breeze, wind and the atmosphere. A tree is also wood and a table. A stick is also a piece of wood, a tool to dig, and be thrown, a weapon, and as a thing to burn to/for heat.

Matter is anything that has mass and volume, occupies space, and can be detected and measured. All matter is made up of nuclei. Matter is material existing as a solid, liquid, or gas.

Air, water, coffee, fire and human beings are matter. Light, X-rays, space, information and love aren't matter (within human understanding). A grain of sand, and an elephant all have one thing in common-they are composed of matter. A beam of light, the motion of a falling stone, and the explosion of a stick of dynamite all have one thing in common-they are expressions, or processes, or states of transmuting energy.

We are informed that matter is any physical substance. In contrast, energy is a non-physical quantity. But the word 'physics' is also our concept. Matter exists as humans created it.

As nuclei is confined by space, which continuously propagates different temperatures and pressure/force, each element, molecule and substance have different chemical and physical properties, explaining why there are myriad types, and forms of matter. Different nuclei integrate together to form molecules and substances, explaining myriad and different types of matter.

The same nuclei are used again and again. Structure determines not only the appearance of materials, but also their properties. When a pencil lead can become a diamond, we can understand the structure of materials. Disintegrate matter, and you will get nuclei, and particles.

Some nuclei take billions of years to decay, while others disappear in a matter of seconds. It all depends on the integration of nuclei, and the influence of altering temperatures and pressure/force.

Buildings, animals, sand, organs, air, butter, tofu, Mars, beer, paper, glue, wax and so many other things are all made up of different nuclei configurations bonded differently.

Different types of nuclei integrate together to form molecules, genes, chromosomes, nucleus and cytoplasm of cells, cells, tissues, organs, muscles, body structures, organisms-plants and animals. Within organic structures nuclei integrate into molecules, and yet further engender carbohydrates, fats, lipids, proteins, amino acids and other metabolic compounds. Within inorganic structures, nuclei cohere together to form wax, clay, salt, water, coal tar, asphalt, plastic, rubber and all the various substances and material we see around us. Most of these are human engendered. All these things eventually disintegrate, but exist for some duration, as environments alter (temperatures and force/pressure), or with human tinkering.

Radiation, radioactive decay, exposure to air, water, gases and other substances also instantaneously, or gradually, alter the structures, forms, processes or states of matter. We can however disintegrate matter by motion, burning/heating, explosion and other ways.

When is energy a particle, and when is it matter?

When enough nuclei of protons and neutron plasma conjoin, we begin to get matter that we can see. This full stop is made from about 10 million millions nuclei. However, protons and neutron plasma are made up of many quarks, and quarks are made up of still smaller elementary particles. A particle is a form/state of condensing energy and therefore, matter can be described as observable energy.

Humans create existences, and then try describing them, explaining why we have to theorize, and/or postulate, always running into problems/difficulties.

How do we describe existences as matter, and as energy?

Momentum or force or pressure, temperatures, forms of radiation, light, space, atmospheric particles or atmospheric air or nuclei, and sounds are some forms of energy, or existences we cannot observe, as our visual systems are constituted of matter. Objects, astronomical bodies, organisms, water are observable forms/states/structures of energy, or existences, which are observed, are described as matter.

We see within, and by visible light, but cannot see light per se. We can see the incandescence at the sun, and at the stars, and at tube lights. In reality, we see matter, as it stands out by light. Without light we would not be able to see matter, and would have evolved receptors to sense it.

Matter is condensed energy. Completely disintegrate any of the following matter: galaxy, star fields, planets, moons, satellites, objects, organisms, cells, elements, molecules, nuclei, and protons, and you will get virtual particles.

Why do we see matter for duration if energy is continuously transmuting?

As energy continuously transmutes there is flux, and every existence continuously alters, at every moment of time. Non living things gradually alter or decay, and eventually terminate. Living things continuously alter in/as a process we call living, or growth. Living things also eventually terminate.

Living things comprised of cells, process different forms and types of environments, and grow, or alter much faster than non-living things, explaining why we observe inert things as matter for longer durations. Both living and non-living things are comprised of the atmosphere/oceans, and the earth's materials, and alter, decay and eventually terminate, as environments continuously change.

Humans understand the concept/word 'termination', as the 'end' of something that once existed. In reality, this is not so, as existences, or states, processes, structures, or forms of energy, continuously transmute into yet other states, processes, structures or forms of energy. Therefore, nothing 'begins' or 'ends' and furthermore, nothing really 'engenders', 'exists' or 'endures', as existences are concepts within human illusory language minds.

Why is there no such thing as matter?

Only with nearly equal amounts of integrating condensed energy, and disintegrating condensing energy a nucleon engenders, which is the basic unit of all matter. Without tritium matter would not engender. It is the release of neutron plasma, concurrently with integrations of nuclei that elements, molecules and matter engendered. As a human concept and/or creation, matter is condensed energy that is observable. Everything everywhere is energy. As stated above, nothing really 'engenders', 'exists', or ‘endures’, as existences are concepts within human illusory language minds.

Is matter found in other solar systems or star fields?

We have to first describe, or explain our own concepts. Physicists talk of mass different from matter, explained following. Matter is condensed energy, and as more and more of energy condense, it becomes visible to human eyes, and detectors. We have to draw the line describing mass and matter. We have also to decide at what stage condensing energy, as particles, plasma, protons, and nuclei, can be described as matter.

At star processes, energy is continuously condensing into particles and plasma. Plasma expelled from a star immediately transforms into protons. Nuclei only form, as fragments of a neutron star are confined by space, as/with those processes that took place within our solar system. Therefore, if these events occur at other nascent star fields, nuclei and matter may form. Even if nuclei form, integrations may stop at helium, engendering bodies like our outer planets.

A new atmosphere (of the inner planets) generating precipitation, falling down on the fiery interiors solidified the gaseous elements, engendering the mantle and crust, and the myriad forms of matter.

There are many conditions to be met, not only for matter to engender, but also to exist, and endure. Our oceans now sustain the depleting atmosphere, so matter on earth, and earth per se also endures. At other places, nuclei and matter may engender, but without the right conditions, will soon disintegrate into free energy again.

MASS MATTER AND WEIGHT

What is mass? How do particles acquire their mass?

What is the definition of a mass unit?

Energy is everything everywhere. We cannot see virtual particles, and their confining space, and so we will never know how much energy transmutes into the tiniest virtual particle (structured energy).

Even if we theorize a particle's mass, we also have to take into account the space confining it, as space is also a form of energy, and an integral part of its particle. Our atmospheric space is continuously propagating temperatures and pressures/force. These forms of energy, or the waving of space, continuously alter the characteristics of every particle and therefore, we cannot measure, or calculate a particle's mass. Moreover, as energy continuously transmutes, everything is altering its characteristics, at every moment of time. The particle is always changing, as space confining it is in flux. Space confining a particle, and a nucleus is always transmuting within interplanetary space, or atmospheric space. We can only calculate the mass of a quark, or protons, if they are observable, taking into account the confining space.

As we shall learn, the atom, or protons, and the postulated neutrons, with their 'orbital' uncertain localized electrons are hypothetical. Space is invisible, and we cannot calculate waves of space, propagating forms of energy and therefore, cannot measure a particle's mass.

The incorrect explanations of mass and matter

It cannot be determined when a particle is said to have mass, as everything everywhere is energy. A particle can only be understood to have mass, as its mass can first be determined then calculated. No existence is an entity, and as a particle has space as its integral part, which cannot also be observed, invisible particles, and also their space confining them cannot be studied.

Is mass the amount of energy (matter) that can be weighed?

The tiniest existence exists, as an 'unknown' amount of energy is condensed into a particle (structure of energy). This unit of condensed energy would have mass, but it would possess no weight, as particles are extracted toward the sun as replenishment material. We can only weigh things within the pathways of the energy transmuting shuttling to-and-fro cycle, between the sun and a planet/body. We cannot observe a particle, and also space confining it. We can only observe particles with mass, when our visual systems can observe it through detectors-then it is called matter. Humans conceptualized/created energy, mass and matter, and can never come up with the correct explanation/descriptions.

We are informed that anything that takes up space, or has a mass is matter. Objects take up space. Objects have mass. Mass is how much there is of an object. Mass is related to how much something weighs, but mass, and weight, are two different things. The unit for mass is a gram. A nickel has the mass of about one gram. Objects that take up space, and have mass are called matter. Space is also a form of energy, and has weight if it could be weighed. In reality, these are incorrect explanations.

Einstein's famous formula, E=MC2 introduced a revolutionary relationship between matter and energy. Before his discovery, matter and energy were assumed to be completely separate quantities. But Einstein theorized that at the smallest levels, matter and energy are interchangeable. This is incorrect, as at all levels everything everywhere is energy. Nothing else other than energy exists, and matter is condensed energy.

Energy transmutes into forms, states, processes or structures which humans call existences. Some existences are invisible, and some that are observable are named matter. Matter and energy is the same thing, but humans conceptualize them distinctively.

As a particle engenders, space-as a form of energy-concurrently engenders. In other words, energy has only transmuted into a particle (with mass) and space-other forms of energy. In reality, an area/volume of energy has transmuted into a particle (an observed and/or conceptualized existence), and concurrently space (a form of energy) confines it, and vice versa. Both particle and space (are existences 'separately' or 'together') are integral parts of one another. Mass and matter are human words/constructs/concepts, to explain and understand the transmutations of energy. Even energy per se is a human concept/construct.

There is no distinction between energy and matter. Matter is visible transmuting energy, and energy is virtual matter. Matter is observable energy, and energy is invisible matter- it all depends how you want to comprehend it. Everything is energy, or energy transmutations.

E=MAC2 and E=MC2-both incorrect equations

This Einstein's equation did not speak of matter, but of mass. Presently, as the moon has little or no atmosphere, to continue an energy transmutation cycle, between it and the sun, there is no 'incoming' or 'outgoing' forms of energy. Therefore, things cannot be weighed on the moon. With little atmosphere (as density alters) things show incorrect weight. Again, things can be weighed on earth in relation to a given constant.

If things have weight, and as weight and mass are related, do things on the moon have mass, and/or weight? Science has hidden meanings behind most of its theories/concepts. Both mathematics, and also equations are abstract concepts.

Mass is just a theoretical construct, and it is designed to eliminate local forces from equations. We conceive some real-world method of measuring mass, and this is enough to render the construct mathematically valid. In the world of scientific units, the units are artifacts and true. A particle will weigh differently on the moon and the earth, but its mass will not.

Einstein states, that as the speed of a particle approaches that of light its mass increases. The matter is unchanging whilst the mass increases, and so it is the mass to matter ratio that is influenced by speed. Now we have to eliminate matter, and use only the word mass, as only a single variable can reflect Einstein's speed influence. It would be valid for small or large quantities of matter, and independent of local forces. Therefore, scientists using this equation who speak of dark matter actually mean dark mass. Mass can go missing, as it is invisible, but matter is more difficult to hide.

The mathematicians got the mass to matter ratio wrong by a factor of about ten to one. They state that the postulated 'universe' is ten times 'heavier' than everything they see. But that 'heaviness' is due to wrongly computed masses, not missing matter. If you see through a telescope you see matter. You also see light at all its wavelengths or energy. You do not see mass. What then is mass?

Redefining E=MC2

Equal is a wrong human word/concept. The word equal creates distinctions. Nothing equals anything. Energy and mass are not equivalent, but the same thing, as different energy transmutations-states, process, structures or forms. There are no entities, and everything is the same thing-energy.

In reality, mathematics has its own illusory independent existence. E=MC2 is only a hypothetical formula within our illusory language minds (revelation 33 and 35). It is shown in complex mathematical derivations, and describes matter mysteriously.

According to Einstein's E=MC2, nuclear fusion and fission convert ‘matter’ into energy. During fission heavy atoms such as uranium break up into lighter particles. With fusion two light atoms such as hydrogen are fused into a heavier element. In each case, some ‘matter’ gets converted into energy. This energy described is actually force/pressure-a form of energy.

Energy, matter, space and many other phenomena described distinctly, are human words and concepts. Everything is energy. Matter is visible energy, and invisible matter is energy. With fusion, fission or any experiment, we are only transmuting energy from one form, state, process or structure, into its other various forms, states, processes and structures. In other words, fusion is matter as condensed energy disintegrating into/as space, light or heat and pressure/force-all forms of transmuting energy. How can E=MC2 be correct when space-as a form of energy-propagating these forms of energy, is not taken into consideration.

Redefining Einstein’s theories

Einstein buried the ‘ether’ stating that absolute time was unnecessary, and introduced space-time. With space-time how can units be constant? How can we define, observe, or explain the motion of ‘variables’ to a unit, as all measurements are taken as dimensionless ratios with respect to a unit? Why do we need units to measure things in the physical world? What are 'dimensions' (length time mass etc), and where do they come from?

There are no entities, but only energy transmutations. There are existences/beings, or in physical terms, structures/processes/states or forms of transmuting energy (according to human concepts and minds), existing within/as the fabric of energy/space, and therefore, nothing can be 'equal' to anything.

Mass or matter is not equal to energy. Mass or matter is condensed energy. Matter is mass that can be weighed and observed. You can only weigh things within the energy transmuting shuttling to-and-fro cycle on the surface of the earth, with relation to a human created constant.

The transmuting of some forms of energy manifests force, or power or pressure, or momentum, and temperatures. All forms of radiation including light are transmuting states, processes, structures, and forms of energy. All forms of energy transmute or propagate through, or within/as space-also a form of energy.

Since space is not a void, but a form of energy, and gravity does not exist, the special theory of relativity does not hold. Gravity is neither attractive nor does it exist, and so the postulated universe is neither expanding nor contracting. If the theory of relativity is incorrect, then the postulated universe would not have a beginning, and we would not have to worry about its 'end'. General relativity is an incomplete theory, and moreover, it cannot tell us how the universe started off, as all theories break down at the beginning of the Big bang. Revelation three describes the non-existence of the human notion/construct called 'universe'.

MASS AND WEIGHT

Mass is not what you weigh. Mass and weight are different things. Mass is the measure of a quantity of matter-how much stuff there is. Matter is mass that can be weighed. Things can only be weighed within the energy transmuting shuttling to-and-fro pathway within earth's atmosphere, and at other areas with similar cycles. Obviously, weight will be dependant on the object's mass. The more mass an object possesses the more it will weigh, as the incoming forms of energy 'forces' it down.

In reality, 'incoming' energy is not more than 'outgoing' energy. As energy continuously transmute, different forms, states, processes and structures of energy continuously alter. Some forms of energy as pressure/force, and radiated heat, 'propagate' as space waves, and so there is momentum/force, explaining why things (structured with integrated nuclei heavier than air) are forced to the surface of the earth, and also why with a given constant they can be weighed.

Organisms and inert object are comprised of the atmospheric and water nuclei and molecules, and as they are heavier than atmospheric air they are weighed down, and remain on the surface of the earth. Nuclei configurations of the same weight as atmospheric nuclei will obviously float like air, and they will not weigh anything. If objects are completely disintegrated, the separated particles will be lighter than air, and are then susceptible within the energy transmuting cycle, and would therefore, have no 'weight'. When an astronaut floats in space we say that the astronaut is weightless, but the astronaut still has a body (matter), and so has mass.

Why is the astronaut weightless in 'space' above our troposphere?

The astronaut, his space vehicle, and everything within it, are comprised of atmospheric nuclei, and material of the troposphere. These nuclei configurations comprising them are designed, or patterned, conforming to/by/from that environment. At higher altitudes, as the atmosphere spreads out, it is tenuous, and as its rotation is faster the atmosphere is less dense. Without density things cannot be weighed within the energy transmuting shuttling cycle, and cannot be ‘forced down’, and are weightless floating aimlessly.

Water will float out of its container inside a shuttle in space. Its shape will vary, but its volume will remain constant if the air pressure and temperature acting upon it are constant. Water is constituted within our troposphere, subjected to our atmospheric temperatures and pressure it collects in a glass, and also on earth's surface, as it is weighed down.

Mass and weight in physics

Mass is a physical concept. In physics mass is the quantity of matter in a body regardless of its volume, or of any force acting on it. But is not mass matter?

They inform us that mass should not be confused with weight, which is a measure of the force (incoming energy) acting on a body. But how do you weigh something if you are not weighing its mass? They state, that under ordinary conditions the mass of a body can be considered to be constant; its weight however is not constant, since the postulated force of 'gravity' varies from place to place. As we have learnt, there is no such thing as the force of gravity. Things weigh as the incoming forms of energy 'weighs' down on things. In other words, as our atmosphere is continuously interacted, integrated and disintegrated within the cycle, there are alterations in temperature, density and pressure/force. At different altitudes, and at different places, or with altering temperatures, pressure/force and densities, 'weight' of objects will vary from place to place. However, their mass or matter always remains the same.

The postulated gravitational mass, and inertial mass is also incorrect. A body should weigh (the same at any location) according to its mass, as weighing is a human construct with human artifacts. A body cannot resist acceleration by some external force (in this case the energy cycle). If weighing is done out of the pathway of the energy transmuting cycle, there would be no mass and therefore, no weight.

Energy transmutes shuttling to the earth, and from the earth to the sun, and is not a one way phenomenon. In reality, the weight of any object is not real, but a human construct for human planning and living. We only consider the weight, or the incoming energy acting upon the object that is weighed. We do not calculate the effects of the changing densities, pressure and temperatures influencing that weight, and also the outgoing energy extracted from the oceans, and the atmosphere within that same energy transmuting pathway. Therefore mass or weight can be considered, or calculated, or measured only within the energy propagating or transmuting pathway.

As everything everywhere is energy transmuting, mass of a body cannot increase with speed. Actually, within the energy transmuting to-and-fro pathway, nuclei in a body in momentum will expand and therefore, susceptible to the extractions processes of the energy cycle, and in reality, get lighter. With this theory mass is being considered a distinct entity than energy. Energy is everything everywhere. Mass, matter, light and speed/momentum are forms/states/structures or process of transmuting energy, and not 'equal' to it. E=MC2 is totally incorrect.

Why cannot the physicists' standard model predict a particle's mass?

We create existences of the real world from within our illusory language world, and then have to theorize to explain our creations. In the same way, finding some elementary particles physicists created a 'standard model'.

Protons and neutron plasma are not fundamental, as they can be broken down into smaller parts. Each proton and neutron plasma is postulated to contain 3 quarks. It is postulated, that electrons are fundamental particles, and cannot be made simple, and is a member of the lepton family. Physicists think all of matter is made up of quarks and leptons. The standard model describes quarks, leptons, and the forces between them.

We will never know how much of energy condenses into the tiniest virtual particle (structure of energy which is observed). We will also never know how many virtual particles integrate until they form a particle that contains mass or matter, which is calculable and observable, as we cannot study invisibility. We will never know the influences and the amount of space confining a particle, or a nucleon. Quarks-the smallest particle theorized-are made up of hundreds of smaller more fundamental and virtual particles.

While the standard model provides a good description of phenomena observed by experiments, it is still an incomplete theory. Although the model divulges why particles exist as they do, as we cannot observe virtual or invisible particles, we will never be able to explain a possible pattern for particle masses, nor why a particle as discussed above has a certain mass. Moreover, within our atmosphere, temperatures and pressure or force are continuously propagating as space, as space waves. Space waves interact with every particle, and or nuclei, continuously altering its characteristics at every moment of time.

We learnt that a certain area/volume of space condenses or concentrates, or transmutes into a particle, and its space that confines it. This existence occupies, or supplants, the same area/volume of energy from which it engendered. How can we calculate how much energy is required to generate the tiniest indivisible particle? How can we calculate it's mass, if its confining space is an integral part? How can we calculate or measure space-a form of energy? How can we calculate the force (a form of energy) generated by space confining the particle, and the reacting force of the particle facilitating the space to confine it? The particle, it's containing space, the two forces, temperatures manifested with the process are all forms, states, processes or structures of transmuting energy. How can we ever predict a particle's mass with invisibility? How can we ever finite an atom with invisible space?

Studying invisible particles by postulating its mass is impossible, and/or enormously difficult, and these exercises are futile. We will never know how many virtual particles integrate together to form quarks, or smaller elementary particles.

Physicists postulate a particle is an entity, but its confining space determines it's characteristics-mass and the energy/space confining it. Space propagates continuously altering temperatures and pressure/force. Moreover, every particle's space is interlinked with interplanetary or atmospheric space-nothing is an entity.

How do physicists find the smallest parts, or fundamental particles?

We can describe anything of matter only if it is observable. Scientists impact nuclei into an accelerator motioning them at high speeds creating more massive, and more exotic particles. With detectors (as small as a computer chip, or as big as a house) they count particles, visualize their tracks, measure particle energies, and record times of flight, and identify different parts.

But even with the best of accelerators and sophisticated detectors, we will never be able to study invisible or virtual particles-the tiniest blocks of quarks-the building blocks of protons, or the building block of all observable matter. We cannot study invisibility, and above all, cannot measure invisible space-a form of energy.

Are there smaller building blocks then quarks?

Some time ago, atoms were thought to be the smallest indivisible units that could also not be cut further. Now scientists have found protons, the postulated neutrons, and other sub-atomic nuclei particles, which are postulated to be made up of quarks. We will never know how much volume/area of energy condenses to form the smallest particle. Also, we will never know whether the smallest particle is an observable particle, a particle with mass, or a particle that can be weighed, calculated or measured. As we will never be able to study invisible or virtual particles, we will never know how many of these smallest particles integrate to form the postulated quark.

At functioning stars energy continuously condenses into particles, and particles integrate to form quarks. Quarks further integrate to form plasma (even particles and quarks can be plasma). So the star's atmosphere contains plasma-naked protons. Plasma or 'particles' are not confined by their 'individual' space. Their space amalgamates collectively as star field space-a star field. Therefore, it would be difficult to study these particles. Particles can only be studied by disintegrating nuclei or protons, but as they are quickly extracted (amalgamating into/as space), they can be observed for very short duration

A particle's mass can only be correctly calculated if it can be measured together with its confining space, as space is also a form of energy.

TEMPERATURES AND FORCE/PRESSURE ON NUCLEI OR MATTER

Important note:

Every existence engenders, exists and endures, as a result of certain degrees of temperatures and pressure/force intensities. Increases or decreases of these forms of energy will alter or terminate existences.

A proton is proton plasma together with its confining space. In the same way, a nucleon is a proton and neutron plasma together with their confining space. A proton, and a nucleus with their space, exists within atmospheric or interplanetary space. Atmospheric or star field space is continuously waving/propagating temperatures and pressure/force. However, increases or decreases in temperatures and pressure/force intensities alter the space confining protons and nuclei, explaining why protons, nuclei, elements and molecules continuously alter their characteristics, or terminate.

Fusion and Fission

How is energy obtained from fusion and fission? Why is more energy 'obtained' during gaseous element fusion than with heavier element fission?

Neutron plasma is disintegrating condensed energy. A tritium has more of it than a deuterium. With fusion, deuterium and tritium are made to collide and combine producing helium. However, the extra neutron plasma 'released', or disintegrates, amalgamating into/as space/energy, concurrently producing heat and pressure/force, and space. This process 'releases' considerable more energy (force and heat) than a fission process.

With fission, neutron plasma is made to bombard uranium splitting it (lighter nuclei are formed), 'releasing' energy (heat and force). The neutron plasma 'released' may bombard and split other nuclei, and further fission can take place.

Deuterium, Tritium and Helium engendered with progressive increases in heat and pressure/force. Most of the heavier elements than iron engendered under varying degrees of heat and pressure/force, and as explained earlier, with more neutron plasma they are unstable. This explains why there is more heat and pressure/force or 'energy' obtained from fusion than fission.

In reality, energy 'released', or 'obtained' is force-a form of energy. With fission and fusion, force 'released' or 'obtained' is the same force, or heat intensities utilized/expelled to integrate, and confine those nuclei.

Energy is everything and everywhere. 'Obtain', 'release', 'fission' and 'fusion' are human concepts. In reality, energy continuously transmutes into yet other forms, states, processes or structures.

Will fusion always be the energy source of the future?

Earth's interiors of elements have, and continue to disintegrate, belching out magma to sustain our continuously depleting atmosphere and oceans. As a result of naturally occurring fission within earth's fiery core, heavier elements are broken down into lighter ones, and lighter ones into protons. Protons are continuously extracted toward the sun as replenishment material.

Although earth 'receives' energy in the form of radiated heat and pressure from the sun, these forms of energy have not yet been technologically transformed to meet our energy needs successfully. Most fusion research today is focused on fusing deuterium and tritium. As neutron plasma is 'released' from these interactions, heat and radioactive materials are produced just as in a fission reactor. This fusion produces must less radioactive waste than fission, but can be avoided using a better fuel. The fusion of hydrogen with boron also produces the same results, but researchers do not spend as much time on this fusion, as the deuterium-tritium fusion is easier to ignite. It requires temperatures of only 100 million Kelvin, while hydrogen-boron fusion requires 1 billion Kelvin.

Unfortunately, many fusion researchers have spent their entire careers developing a device called the tokomak, which cannot reach the temperatures required for hydrogen-boron fusion. Rather than looking for new ideas, the future research establishment has decided that the radioactive waste produced by deuterium-tritium fusion is acceptable.

By the 1955-1965 period several ZETAS or Zero Energy Thermonuclear Assemblies were built first in Russia, and then in UK. Zetas were built with the aim of achieving zero energy. But attention was soon shifted from the Zeta concept to the toroid field concept or Tokomak, a metal doughnut where very powerful electromagnets are used to hold/confine a plasma field, preventing or limiting the formation of eddies, and currents in the very hot environment.

The ITER or thermo-nuclear energy research center being built in France between 6 partners countries hopes that zero energy would be achieved within 3 to 5 years after the site has been completed. However, the ITER project as with all previous and failed attempts to achieve 'controlled' fusion will be faced with a huge number of constraints, extending from the postulated gravitational and thermal, to mechanical and quantum mechanical, and metallurgical and physical.

If the ITER project would lead to controlled and safe, economically feasible and permanent fusion, it would enable the construction, and operations of commercial fusion reactors producing electricity, and only electricity.

Commercial fusion reactors would be possible only after 2040. From about 2006 world oil production will start its inexorable decline, triggering massive price rises for oil and therefore, vast expansion of natural gas consumption. Whether or not ITER controls fusion in 30 or 40 years, there is definitely going to be a real energy crisis from 2007 onwards.

Electricity is a form of energy used for some urban transport and trains in some countries, but is useless for supplying the raw materials such as fertilizers, plastics and pharmaceuticals, which is supplied by the petrochemical, and gas industries. Moreover, electricity is impossible to cheaply and easily stock or store, unlike hydrocarbon fuels. With the rapid decline of oil and gas supplies, and with little time, cheap fusion energy is a far away dream. Fusion will always be the energy (heat and force/power-electricity) source of the future.

States of Matter - Solids Liquids Gases and Plasma

Matter exists in four different states or phases: solids, liquids, gas and plasma. Solids, liquids, gases and plasma are forms, states, processes or structures of transmuting energy, and are described as different existences having different properties/characteristics, only for human understanding. Volume and shape defines the states of matter.

Solids

In solid molecules vibrate slowly, as nuclei configurations are confined, or integrated more closely. That means nuclei and molecules become fixed or trapped in their places, or move in the same place. A solid has a regular arrangement of nuclei, ions or molecules close together, and cannot move, and often line up in a crystalline arrangement. At the very lowest temperature possible-absolute zero- all motion stops, and the nuclei do not vibrate/spin. The nuclei inside a solid are not allowed to move around much, no matter how small the pieces are.

One of the main characteristics of solids is that they hold their own shape. The shape and volume of a solid is fixed. It cannot be compressed. Matter is solid when it has a definite volume or size, and a distinct shape at a given temperature/pressure. Iron at room temperature has a shape and size, and does not change. Ice is a solid, but its temperature must be below 0 C. It will turn into a liquid, and a gas, as temperatures and pressures alter.

Liquids

The nuclei in liquids and gases move around in all directions. In a solid nuclei and molecules are trapped in their places. In a liquid nuclei are moving or vibrating fast enough, and break out of the constraints of a defined structure. The temperatures and pressures are strong enough to hold nuclei together in the form of a liquid.

Liquids have definite volume, and take the shape of the container. It starts filling at the bottom subjected to the incoming energy transmutation cycle, and incoming forms of energy also creates the flat top surface of liquids.

As the volume of liquids is fixed, they cannot be compressed. Gases are much easier to compress than liquids. A liquid has its nuclei close together, but they are free to move, and this is why a liquid will take the shape of its container.

Gases

With a gas the nuclei are more energetic vibrating rapidly. Temperatures and pressures easily propagate through the nuclei configuration's space explaining why it is a gas. A gas spreads out more easily explaining why it does not stay in an open container. The atmosphere is a big layer of gas that envelopes the solid earth (land/oceans).

Solids have compacted nuclei and molecules. In liquids they are slightly spread out, but in gases they spread out more, bouncing around constantly.

Gases fill containers of any size or shape. Liquids can fill only the bottom of the containers, while gases can fill it entirely. The volume of a quantity of gas is dependent on its temperature and surrounding pressure. A gas has no order and/as its nuclei are arranged at random. A gas will fill the whole volume of its container.

Plasma

Physicists describe plasma as a fourth state of matter. Matter is condensed energy that we can observe, but we cannot observe plasma, which is also condensing energy. Matter and plasma are human concepts explaining the states of transmuting energy, therefore, explanations, or descriptions, are inaccurate and/or incomplete.

Frozen water is a solid, melting it is a liquid, and with heat it turns into a gas or vapor. With further heat over 1500 degrees C, it is plasma. Sufficient heat and or pressure/force disintegrates a nucleon, and further heat knocks off the space confining a proton. The naked proton is plasma. This is how electricity (heat and momentum) directs neon and florescent lights.

Gases can be made into plasmas by breaking the nucleon from its confining space. In a florescent tube or a neon tube, a form of energy or electricity (pressure/momentum or heat) interacts with the space of certain type of nuclei generating plasma. Different nuclei produce different colors.

The solar wind is comprised of plasma. However, as plasma is propagated through star field or interplanetary space, it is confined by space, and transmutes to protons.

Temperatures and/or pressures on the nucleus/matter

Energy is everything everywhere, and matter is a structure (observable energy) of condensed energy. The matter observed around us engendered, exists, and endures, as a result of certain degrees of temperatures and pressure/force intensities. With increases or decreases of these forms of energy matter alters, and/or disintegrates.

Integrating nuclei engender elements, molecules, compounds, mixtures and so on. Some elements will be solids, liquids or gases. Some molecules and or compounds will form glue, rubber, metal, wood, stone coal, and so on. The density, hardness, softness, elasticity, tenacity, texture, fluidity or liquidity, gaseous states, lubricating properties of matter, all depend on how nuclei combined and structured. The influence of temperatures and/or pressure/force on nuclei determines how nuclei integrate or disintegrate.

The pressures/force or momentum, and temperatures/radiation generated at the sun, or a star, continuously propagates throughout the solar system, or star fields in a shuttling to-and-fro, energy transmutation cycle. As a result, every existence, or every nucleus, element or molecule comprising all states of matter, subjected to these forms of energy, alter their physical and chemical properties, at every moment of time.

Matter exists in different states, solids, liquids, gases, and has many types of physical and chemical properties, as a result of continuously altering temperatures or pressures/force acting on the nucleus. Matter can change physically and chemically. Matter constantly changes, as ice melts, glass breaks and paper are ripped. Chemical changes are spoiling milk or burning toast. Temperatures and/or pressure/force on nuclei also determine matter to have appearances, properties and behavior.

Mass, weight, color, shape, volume and density are some physical properties of matter. Matter reacts and combines. It has physical properties as melting, boiling and freezing points, and chemical properties such as the way elements combine with each other in reactions. Chemical properties of matter can be observed, as the chemical properties change, as a burning candle. Chemical properties describe the way that an element or a compound reacts chemically with other substances.

Water can be a solid as ice, a liquid or a vapor, depending on the prevailing temperatures and or pressures. Vaporization and evaporation takes place with altering temperatures and/or pressures. The atmosphere is moist, humid, or dry, or there is a wind/breeze, depending on altering temperatures and/or pressures and densities.

Color, odor, or smells, or taste of substances, depend on how nuclei configured to compose that substance, and manifest with those same temperatures and/or pressures.

Our atmospheric particles or air, the ocean molecules, the nuclei, elements and molecules in our crust, mantle and the inner and outer cores of earth are continuously altering to pressures or temperatures of the energy transmuting cycle between the sun and the earth. As energy is continuously transmuting, and everything is flux, everything is decaying, altering or terminating.

All elements are either gaseous, liquids or solids engendering, existing and enduring at those temperatures and pressures/force that generated them. In other words, every existence exists, or has its characteristics, or alters its characteristics with certain temperatures and pressures/force. Therefore, existences alter their characteristics, or disintegrate/terminate with increases or decreases of temperatures and pressure/force.

With reactions between nuclei, subjected to certain temperatures and forces/pressures, there are different types of molecules. With reactions between nuclei and molecules subjected to certain temperatures and force/pressures, there are different types of substances, and chemical compounds.

Why and how do materials change from solid to liquid to gaseous states and vice versa?

As a proton engenders, proton plasma spins in one direction, and its confining space spins in the opposite direction. This is the only way a proton engenders and exists. This is the same with a nucleon or nuclei. Nuclei in/as elements, molecules, substances and compounds are eternally spinning.

As nuclei configure in different patterns, and space propagates varying temperatures and pressures, there are different types of nuclei/elements and molecules, and different types of matter. Moreover, matter exists in different states, has different characteristics, or behavior, or properties.

For example two substances might have exactly the same composition, and yet show distinctly different properties. Ethyl alcohol and Methyl ether have same composition, yet one is a liquid, and the other is a gas. This is evidence that the characteristics of space containing nuclei, propagating as temperatures and or pressures, determines whether nuclei in elements or molecules exist/endure as a solid, liquid or a gas with their different characteristics.

Matter changes from solids to liquids, or liquids to gases, or vice versa, as nuclei is confined by space. Space continuously propagates temperatures and pressure/force, and increases or decreases of these forms of energy alter nuclei configurations, changing the characteristics of matter.

Solids

Freezing point is a measure when liquid turn to solids. When pressures surrounding a substance rise, the freezing point also rises, explaining why it is easier to freeze the substance at higher pressures. When it gets colder most solids shrink in size, and there are a few which expand, but most shrink. Solids of pure elements or compounds freeze into a very special structure called a crystal lattice-a very exact organization where every nuclei or molecule has a specific place. It is very neat and very compact-a diamond is a crystal lattice.

Liquids

Heat changes solid into a liquid. There is a melting point for every solid to turn it into a liquid. For a gas to turn into a liquid the reverse is true, as surrounding temperatures must be lowered. Condensation point is when gases become a liquid. Steam hitting a cool wall would quickly become into a liquid.

Gas

Boiling liquids change them into gases. Vapor and gas is the same thing, but the word vapor is used to describe gases that are usually found as liquids (like water). Liquids become a gas due to evaporation.

A material will change from one state to another at specific combinations of temperatures and surrounding pressure. Typically the pressure is atmospheric pressure, so temperature is the determining factor to the change in state in those cases. The changes in state are melting, freezing, boiling, condensation, sublimation and disposition. The temperature of a material will increase until it reaches the point where the change takes place. It will stay at that temperature until the change is completed.

Materials engender and maintain their state at certain atmospheric pressure. Changing the surrounding pressure will change the temperature, at which a material will go from one state to another. Some solids can change directly to gas, skipping the liquid phase (sublimation).

Materials expand, or contract with the change in pressure and temperature. If the pressure is altered, the temperature also alters in that material. The higher the pressure, the higher the temperature needed to change the state of the material. This explains that temperatures and or pressure/force are intimately linked. Both propagate together within/as space, as space waves.

Increased pressure will contract gases. A small volume of gas has trillions of nuclei. Compress the gas, and its temperature and pressure will go up in an orderly manner, adjusting to the new volume by establishing itself in a new equilibrium state. Gases can be compressed and expanded, as gases are composed of widely spaced nuclei.

The effect of pressure on liquids and solids is considerably less. Most substances expand when heated, and contract when cooled. This continues even after they have changed their states. Carbon dioxide becomes a liquid on compression, and if allowed to expand rapidly to atmospheric pressure, it becomes cooler, and freezes to dry ice-a white snow like solid. Dry ice sublimes-passes directly into vapor-without melting at-78.5 degrees C, and is used to produce stage fog, and is also used as a refrigerant.

Steam will contract with pressure and cooling. The volume of water is not really affected by pressure, but it does contract with cooling. Once water becomes ice it actually expands. It is one of the few materials that expand when it changes to the solid state (explained following).

Water in the cracks in the road form ice, and as ice forms it expands, and the crack gets bigger. As the temperature of the ice decreases, it starts to contract. But it still has a greater volume than it does in a liquid state. The water molecule is interacted and disintegrated within the energy transmuting cycle, and as the hydrogen (proton) is released, it is extracted toward the sun, as replenishment material. This process can only take place with normal temperatures and pressures, explaining why water is a liquid, and why it is found in our biosphere with those temperatures and pressures. As temperatures drop, extraction processes cannot disintegrate the water molecule. The water molecule becomes solidly bonded without interactions, and this process expands it, and ice forms. It has a greater volume as ice, as the proton without interactions from/by processes of the energy transmuting cycle 'absorbs' space. In reality, the water molecule consists of 1 part of oxygen and two parts of proton plasmas (protons without space).

Things move from one phase to another by physical means. Increasing temperatures or increasing pressure, or freezing something, or decreasing pressure change the states of matter.

Physicists talk of adding energy, energy being absorbed, or taking it away, but this is incorrect. Everything everywhere is energy. Energy only transmutes into any combinations of states, forms, structures or processes of energy changing the states of matter.

Superconductors

Scientists are trying to find out an element, or a metallic alloy, or compound that will conduct electricity without resistance below a certain temperature. Resistance is undesirable, as it produces losses in electricity flowing through the material.

Once set in motion electric current will flow forever in a closed loop of superconductive material making it the closest thing to perpetual motion in nature. Scientists refer to superconductivity as a macroscopic quantum phenomenon.

As explained within revelation 9, electricity (force and heat) is a form of energy, or energy in momentum. All nuclei are confined by space, which is continuously propagating temperatures and pressure/force enduring the nuclei, or altering it depending on increases, or decreases of these forms of energy.

Conductive materials have nuclei configured, freely allowing the propagation of electricity or pressure/force with heat, and will therefore, offer some resistance to these forms of energy. When mercury was cooled by means of liquid helium to a few degrees above absolute zero, the resistance vanished-this is superconductivity.

Scientists have found that ceramics (which are normally insulators) could be used as a superconductor, but again, temperatures and force must concurrently alter nuclei, and also propagate through it.

We observe and experience perpetual motion in nature, as energy continuously transmutes shuttling to-and-fro between the sun and the planets. Researchers think/agree serendipity will discover a superconductor, but it is only a dream. How can we conduct electricity without resistance at lower temperatures of conductive materials, and concurrently allow electricity (heat and force) to propagate through these materials?

Super fluids

As certain temperatures and force/pressure alter nuclei, helium can be solidified with pressure at very low temperatures. It can also become into a super fluid. Helium one of the light gases exists in two forms-two isotopes-one is helium with two protons, and two parts of neutron plasma, and the other two protons, and one part of neutron plasma, so this one is lighter.

As the heavier helium is more abundant than the lighter one (by a factor of about 10 million) it has been possible to produce the lighter helium at nuclear power reactors. As the heavier helium is cooled to even further temperatures than described above, liquid helium nuclei ceases internal movement, and becomes super fluid-helium nuclei flow freely out from the vessels they are kept in.

A super fluid has very complex properties exhibiting a mixture of three different phases. These phases have different properties, and the proportions in the mixture are dependant on temperatures, pressures, and external propagating temperatures and pressure or force.

Why are liquids and solids more soluble at higher temperatures?

Heat or temperatures and pressures continuously propagate through/as the nucleus's space. Higher temperatures increase the nuclei momentum, facilitating and abetting faster reactions.

Chemical reactions

During a chemical reaction, the nuclei constituents of the substances that react together are rearranged producing different substances. Chemical reactions are the means by which new substances are formed from old ones-fuels are burnt, and life processes are some chemical reactions.

Chemical Activity

Sugar glucose molecules dissolve in water. When more glucose is added it forms maltose, and further additions of glucose to maltose, cellulose forms. This long chain of glucose molecules is not soluble at all.

Plants cells start with glucose molecules in solution, and connect them into long chains. From the sugary solution emerge fibers of cellulose out of which a tree trunk is formed. This is chemical activity.

Chemical compounds

Nuclei of more than one element come together to form either molecules or infinite structures-salt-sodium chloride-this is a chemical compound. Chemical compounds are pure microscopic substances consisting of nuclei or ions, or two of more different elements.

The great variety of known chemical compounds results from the different ways in which space-with different characteristics-of nuclei are able to interact with each other, either with nuclei of the same kind, or with nuclei of a different kind. Compounds cannot be broken down by physical forces. There are millions of compounds everywhere. When elements join together, and become compounds, they loose their individual traits-sodium chloride. Water has a finite structure based on oxygen, chemically bonded to two hydrogen nuclei H2O.

Chemical mixtures

A mixture is different than a compound, where the substances react and combine chemically. The difference between a mixture and a compound concern how they combine, the result, and how they separate.

When elements or compounds are mixed together, but not chemically bonded, they are a chemical mixture. A chemical mixture can be of two solids, salt and sand, two liquids, or two gases or permutations of these. Mixtures combine physically in no definite proportions. They just mix.

Compounds combine chemically to form molecules. With a mixture no new substance forms, but with a compound new substances with new properties form. Combination of solids, liquid, or gas, can be involved in simple mixtures. Compounds can be separated by chemical means, and mixtures by physical and mechanical means, for example, magnetism, boiling, and sifting separate mixtures.

Chemical solution

A chemical solution is the result of dissolving a substance in a fluid solvent, where the materials do not react chemically. Here the solution's molecules or ions mix homogeneously, and do not separate by mechanical means, rather they are usually separated by changing the temperature.

Alcohol and water molecules will mix evenly together to form a solution, as the molecules are spread throughout the solution, but yet they do not chemically combine to form another substance. It is interesting to note, that you can combine a liter of water with a liter of alcohol, but you do not get two liters of the mixture. Instead you only get about 1 ½ liters. Here one plus one is not equal to two. In this case, the molecules mix in between each other, and fill in spaces which would normally be empty in the liquid by itself. Each molecule of water and alcohol exists with their different space characteristics. Normally with mixing we should obtain two liters, but the amount of space amalgamating when these two different molecules conjoin alters.

As volume/weight is dependant on atmospheric pressures and temperatures generating a substance, it decreases the amount of space. This is evidence, that the confining space (propagating temperatures and pressure/force) dictates and influences the characteristics of a nucleus, an element and a molecule.

Salt and sugar dissolve in water. When salt dissolves in water the sodium chloride molecules split into ions, as water molecules of hydrogen and water separate. This amalgamation generates a new amount of space confinement. When heat is applied during the mix, it dissolves quicker and evenly. This is also evidence that a nuclei or a molecules' space, continuously alters to propagating temperatures and pressures.

Carbon dioxide gas can be dissolved in a liquid as in carbonated drinks. Water as a liquid can dissolve into air forming humidity. The amount of a material, and the rate at which it dissolves in a solvent, depends on temperatures and pressure, and the altering characteristics of the nuclei space.

Bonds

Nuclei combine in countless ways to form materials with every imaginable property. The properties of all materials be they stocky, magnetic, brittle or green, arise from the infinite different ways nuclei can be arranged and linked together. Two nuclei form a chemical bond when their amalgamating space rearranges itself, so that each nucleon is subjected to external temperatures and pressures/force.

Although there are only 109 known elements, there are millions of chemical substances found in nature or made artificially. These substances are not simply mixtures of two or more elements, and they are specifically determined chemical compounds, formed by combining two or more elements together in a chemical reaction or chemical bonding. The nucleus architecture dictates/determines physical properties. There are several different kinds of bonds each producing dramatically different properties.

Temperatures and or pressures determine how substances are bonded. Each type of element, molecule or substance has been engendered with different degrees of temperatures, and varying intensities of pressures. Each also alters its characteristics or transmute into another form, state or structure with changing temperatures and pressures.

Although two carbon nuclei do not form a quadruple bond to one another, carbon nuclei combine to form a giant crystal lattice in which each nucleon is bonded to four others by single covalent bonds. This is the structure of diamond.

Many other elements and compounds exist as giant covalent crystal lattices, and containing millions of nuclei are held together by strong covalent bonds. As these hard solids have high melting points having formed with high temperatures and pressures, high temperatures and pressures are also required to disintegrate them.

Water H2O has the same amount of mass as the neon nuclei. Neon has a much higher melting point, as there are unusually strong intermolecular forces between the water molecules (the extraction processes of the energy cycle). Although oxygen and hydrogen nuclei share the same space in a covalent bond, the oxygen nuclei exerts a stronger 'pull' on these spaces, and so becomes space rich, leaving the hydrogen nuclei space-poor. As a result, there is a force of attraction between the hydrogen and oxygen nuclei on neighboring molecules. This is known as hydrogen bonding.

As well as accounting for the surprisingly high melting point of water, hydrogen bonding is responsible for the rigid open structured of ice crystals, and is very important in influencing the structures and properties of biological molecules. Water is also very unusual, as ice it is less dense than the liquid at the same temperature. The water molecule anywhere is continuously interacted and disintegrated, as the proton is extracted within the energy transmuting cycle, and therefore, oxygen nuclei are continuously 'made' to 'exert' a pull keeping the molecule in existence.

Temperatures and Pressures/force on Water

As temperatures and pressure/force propagate within/as space, and within/as space confining nuclei, these forms of energy alter the chemical and physical properties of gases, liquids and solids and therefore, water can exist as any one of these states.

Water is a gas (vapor), liquid or solid, depending on the prevailing degrees of temperatures and pressure/force intensities. Water is liquid at room temperature, becoming a solid or ice at 0C, and a vapor with heat or temperatures at 100C. Heating or boiling water disintegrates the nuclei of the water molecule, as the protons get excited and become susceptible, and are easily extracted within the energy transmuting cycle. When hot water and cold water are mixed, temperatures and pressures of both the hot and cold molecules interchange, and the water alters to a new thermal equilibrium.

Liquid water is denser than ice, as liquid water molecules are continuously interacted, and disintegrated, molecules continuously keep conjoining, forming new molecules. With ice, without interactions and disintegrations, molecules hold their configurations.

Only certain amounts of solids dissolve at a given temperature. Likewise, a certain amount of water can dissolve in air, at a given temperature. The percentage allowed at a given temperature, is called the relative humidity. The saturation point for adding water to air at a given temperature means the relative humidity is 100%. If the relative humidity is very high, and temperatures decrease, or cools down, less water vapor can be dissolved in the air, so the water will precipitate in the form of rain. If humidity and temperature are high, liquid water is not readily dissolved in the air, and we feel sweaty and uncomfortable. If the temperature is high, but the humidity is low, your perspiration evaporates into the air, and you will not feel as uncomfortable.

When water evaporates from a solution, the concentration of the other material becomes greater, until it reaches the saturation point at which it will precipitate out. Extraction of the water molecule from our oceans (evaporation) breaks down the salt water leaving the salt, increasing the salt contents in the oceans.

Boiling can result in a more volatile material to vaporize first leaving the other material. Boiling an alcohol-water solution will vaporize the alcohol before the water.

How can freezing water expand, even bursting metal pipes, with no energy to explain it? Is energy continuously drained from the water it cools toward freezing? Does hot water freeze faster than cold water?

Why does water expand at 4 degree C? Why does it form ice?

Why does ice crystal? Why is each snowflake different?

It is important to note, that atmospheric temperatures and pressure/force are continuously propagating around 'independent' nuclei, and also around nuclei that constitutes elements, molecules, objects/things.

The covalently bonded water molecule of two nuclei of hydrogen, and one of oxygen, is the strongest type of intermolecular bonding. This results from the continuous extraction of hydrogen (proton), from the continuously disintegrating water molecule, within the energy transmuting cycle.

When water is cooled, it contracts down to 4 degrees centigrade, and if cooled further, it begins to expand. This results, as increases or decreases of temperatures and pressure/force alter the characteristics of nuclei. At four degrees centigrade, propagating temperatures and pressure/force alter the nuclei configurations of water, and water reaches its maximum density, as the liquid contracts due to reduced molecular motion. As the nuclei configurations in the water molecule alter, there is a decrease in these forms of energy, and hydrogen bonds form in great quantity. These H bonds being long tend to force the two hydrogen protons in a water molecule away from each other, reducing the density. As there are fewer molecules in a given volume when water freezes, it subsequently expands by about 9 percent. This is why water turns into ice (solid) at zero degrees C. This also explains why freezing water bursts metal pipes.

As matter changes its states, there is an increase or decrease of temperatures and force/pressure. In other words, it is as a result of altering temperatures and pressure/force that matter exists, endures, or changes its states. Therefore, when water alters its state to ice, the resulting increase in force, or in other words, the expansion of the structure of nuclei in molecules from within, burst metal pipes. This is also why rain falls as snow, and sometimes as ice in areas where the temperatures are less than zero degrees C.

It is incorrect to state that freezing water has no energy input, and that energy is continually drained from the water, as it cools towards freezing. Energy is everything everywhere, and it is only forms, states, processes or structures that continuously alter to yet other forms, states, processes or structures of energy.

We do not experience fragments of clouds, or blocks of snow, or ice falling down, as there are variations of temperatures, pressure/force and densities within the troposphere. Disintegrating and contracting molecules subjected to different temperatures, and or forces, generate the crystalline structure. This is why each snow flake has a different structure.

As hot water engenders water vapor there are less molecules, and expansion increases, and hot water will cool faster than when cool water is set to freeze. But there is another point to note. Unless it is blast frozen, hot water that is set to freeze takes time (it would return to cool temperatures), and this also depends on the freezing temperatures. This answer is not complete if the freezing conditions are not known.

It is wrong to state, that water does not need energy to 'expand', and energy is not 'drained' from water as it freezes. Energy is everything everywhere. Energy transmutes into existences, as states/processes/structures or forms, and concurrently these processes generate momentum/pressure/force and/or temperatures, and all forms of radiation.

Why does ice melt?

Increases or decreases in temperatures and force/pressure on the nuclei constituting the water molecule will alter its nuclei's space configurations. With ice-forming temperatures these nuclei motion less, but with warmer temperatures these nuclei alter to liquid-motion intensities, and the ice melts to a liquid.

What is absolute zero?

A theoretical temperature marked by a complete absence of heat, and equivalent to exactly -273.15 C or -459.67F, or zero K.

Temperature is a measure of the momentum of nuclei comprising matter.

When the molecules are hot nuclei move faster, and this momentum increases temperatures. Colder molecules have lesser velocities, and lesser momentum is decreased temperatures. You cool an object by decreasing the velocity of its molecules.

By combining laser and evaporative cooling, scientists have been able to achieve temperatures in clouds of air (gases) below absolute zero. Cooling large crystals to absolute zero restricts the motions of nuclei. Lowering the momentum of nuclei we can control them, and study superconductivity.

The eternal unavoidable fluctuation (spin) in the position of a particle, no matter how tightly confined cannot be stopped. We can never cool a particle to such an extent that it is completely frozen in space. We can reduce it to the lower energy/momentum state called the ground state. Even though a virtual particle integrates with other virtual particles, it is eternally in momentum.

Why do balloons left in the sun expand and burst?

Balloons filled with lighter material such as hydrogen/helium will rise within the energy transmutation (extraction) cycle. Eventually, as pressures and temperatures change with altitude they will burst.

Balloons will stay in their form for a long time if they are filled with atmospheric air, and left in the same place. However, continuously altering atmospheric temperatures, and or pressures will alter the characteristics of the balloon's internal air or nuclei, and the balloon will eventually shrink in size, even if the neck of the balloon is tightly shut. Furthermore, altering temperatures and pressures also alter the nuclei comprising the balloon's membrane, explaining why it may also burst.

There are more changes in temperatures and pressure/force as a balloon is exposed to the sun, altering both the nuclei of the membrane, and the internal air which burst the balloon.

Why does food take a longer time to spoil under saran wrap or a plastic covered bag?

Materials continuously transmute with continuously altering temperatures and pressures. Atmospheric air, or nuclei, are continuously altering, and disintegrating, and food and organic matter nuclei also continuously alter. Saran wrap or plastic bag prevents the normal duration altering processes. Spoil is our concept.

Why does tea stay hotter than soup in a flask?

Soup is made up of many different types of molecules. There is quicker thermal equilibrium, as temperatures propagate through these different types of nuclei. Tea has fewer types of nuclei in its molecules, and heat remains for longer durations.

What is fire made up of? Why does smoke come from a fire?

Fire involves a chemical reaction between fuel and atmospheric oxygen. A fire started is self-sustaining, and releases a combination of heat, light, noxious gases, and particulate matter. The burning flame (or flaming combustion to occur) is the chemical process where solids or liquids are being transmuted into a gaseous state.

Heat disintegrates solids, and liquids transmute into a gaseous state, and into invisibility, or invisible energy particles. Smoke is a gaseous state, and disappears as protons are swept away.

THE HYPOTHETICAL ATOM

THE ATOMIC AND MOLECULAR THEORY

QUANTUM MECHANICS

As everything is energy there are no entities, or 'existences'. A particle, and its confining space, described herein are only for explanation purposes.

As particles integrate, their space also amalgamates forming a larger particle and space. As particles integration continues particles get large enough to be observed. This observable particle is called matter. The observable particle as matter-as a form of energy-can be observed, but its integral part-space (another form of energy) confining it-cannot be observed. We cannot observe space, or an elementary particle, as we have not evolved visual systems to observe those virtual particles making up quarks, protons, and neutron plasma, or a nucleon.

A proton and a postulated neutron, together with a postulated electron orbiting the nucleus is what physicists call an atom. The hydrogen atom consists only of a single proton. Deuterium with a proton, and the postulated neutron, and tritium with a proton, and two postulated neutrons are two isotopes of hydrogen.

Physicists assume there are 'electrons' that 'orbit' the nucleus of the hypothetical atom with a negative charge. The electrons are often described as revolving about the nucleus, as the planets revolve about the sun. This picture is misleading. The quantum theory shows that all particles in motion also have certain wave properties. For a particle the size of an electron, these properties are of considerable important. As a result, the electrons in an atom cannot be pictured as localized in space, but rather should be viewed, as smeared out over the entire orbit so that they form a cloud of charge. The electron clouds around the nucleus represent regions in which the electrons are most likely to be found. The shape of these clouds can be every complex in marked contrast to the simple elliptical orbits of planets.

The nucleus of the hypothetical atom is postulated to contain the proton, and the postulated neutron, together with the electron 'orbiting' this nucleus is what physicists call an atom. The proton is said to have a positive charge, and the neutron is neutral. This theory is preposterous, as protons (with a positive charge) cannot attract to each other forming elements. This contradicts the electromagnetic theory stating like charges repel each other, and unlike charges attract each other.

There is no such thing as an 'electron' explained following. Space is an integral part of the particle, and also a nucleus, and as a form of energy continuously propagating temperatures and pressure or force, influences the characteristics of the particle(s)/nucleus.

We learnt, that as energy condenses, a particle engenders concurrently with space containing it. There are no such things as electrons, as space is an integral part of the particle, or nucleus of the hypothetical atom. The hypothetical atom's nucleus is divisible. The question of the basic structure of matter has been carried beyond the atom, and now centers on the nature of, and relations between the hundreds of elementary particles that have been discovered in addition to the proton, the postulated neutron, and the hypothesized electron. New types of exotic atoms, such as positronium and muonium have been made from these new elementary particles.

No such thing as an electron

We are informed that the postulated electrons orbit about outside the atomic nucleus having a negative charge. They are held in the atom by the positive charge of the nucleus. This electrostatic attraction is part of one of the basic forces of nature-electromagnetism. It seems that electrons can be knocked off the atomic nucleus.

There is no such thing as an electron, as it is assumed to be fundamental, and cannot be split further. The postulated electron is supposed to have less mass (1800 times less than protons and neutrons). The protons and neutrons have the same mass, and the neutron helps keep the nucleus together. The diameter of the nucleus is 100000 times smaller than the diameter of the complete postulated atom. It is like a whole earth being squashed into a football stadium. All the mass of the postulated atom is in the nucleus. As space (not an electron) confines a nucleon, it has no mass.

We have been informed that when light or radiation shine on certain metals they emit sparks, which is postulated as 'electrons'. The sparks seen is light (heat and pressure/force), not the postulated ‘electrons’, propagating from the alteration of nuclei, as space waves (as light or heat and force is apparent). By itself this is not particularly remarkable, as metals have the property that some of their space binds their nucleus loosely-which is why they are good conductors of electricity. In other words, their unique nuclei configurations together with their space, can easily motion, or propagate temperatures/pressures, or electricity.

When light strikes the metallic surface, it relinquishes energy (transmutes energy as light), much as it does when it strikes the surface of your skin, causing you to feel warmer. This transferred/transmuted energy can agitate/alter the characteristics of the containing space in the metal (nuclei configurations), and some of loosely bound space (postulated 'electrons') can be knocked clear off the surface, as it happens with friction.

As temperatures/force and pressures propagate through/as space, there are interactions, integrations and disintegrations, changing or altering the characteristics of particles, or nuclei and their confining space. There are no entities, and an electron cannot be a particle without its confining space. Every particle within atmospheric space is confined within/with its own space.

Heavier elements with more nuclei have more confining space, explaining why heavier elements are believed to have more postulated electrons. The hypothetical electron is space and therefore, cannot be localized. The electron is postulated to be here, and there, taking different paths. The electron does not exist. The degrees of temperatures and pressure/force intensities on nuclei, and not the arrangements of the postulated electrons about the nucleus, explains the structure of the elements of the Periodic table.

The Atomic Theory and the Molecular theory

The atomic theory states that particles known as the electron rotates around the nucleus (protons and neutrons), and this existence is postulated as an atom. It is theorized, that when electrons break down from their nuclei, their motion results in electricity.

The molecular theory states that matter consists of small molecules in motion, and this explains heat. Molecules are constituted of smaller particles called atoms. These atoms are called elements in chemistry, as they are elementary to molecules. This explains chemistry.

The new sub-atomic theories have even smaller particles, having discovered that the protons and neutrons themselves are made of even smaller particles called quarks. These particles are then held together by particles called gluons.

Now new theories have emerged that these sub-atomic particles are not particles at all, but really vibrating strings, or membranes. New theories to explain the old unsuccessfully explained ones.

Is our atomic and molecular theory wrong?

Absolutely! We are taught to accept models of the inner atom, but the endless internal strong nuclear force, the weak nuclear force, the internal, and external electromagnetic energy bonding between atoms, the endless gravitational energy, and the endless external magnetic energy all at work, have no known power source driving these varied forces, and cannot be explained.

The weird world of Quantum Mechanics

QM is weird. QM states that at the sub atomic scale the world is governed by probabilities, not certainties. An electron may be in this location, but it would just be in that one. The postulated electrons cannot be 'found', and are anywhere around the nucleus of the atom at any time. This explains why there are vigorous philosophical debates concerning the uncertainty principle, and the statistical nature of all the predictions of the theory. Everything is subjective, including reality. The mere act of observing a particle alters the objective reality. Thoughts do not affect one's reality, thoughts shape, or create it. Science is full of theories or stories. As one theory cannot be fully explained another one is introduced to try and explain it. Within our illusory language (human thinking), this is the only method by which scientists are trying to find reality.

Older classical theories of physics state, that energy is a continuous phenomenon, while matter is assumed to occupy a very specific region of space, and to move in a continuous manner. According to the quantum theory, energy is held to be emitted and absorbed in tiny, discrete amounts called a quantum (plural quanta). A packet, or bundle of energy, or quanta, thus behaves in some situations, much like particles of matter, as particles are found to exhibit certain wavelike properties when in motion, and are no longer viewed as localized in a given region, but rather as spread out to some degree.

It is stated for example, that light, or other radiation is given off, or absorbed by an atom, has only certain frequencies (or wavelengths), as can be seen from the line spectrum associated with the chemical element represented by that atom. The quantum theory shows, that those frequencies correspond to definite energies of the light quanta or photons, and results from the fact that 'electrons' of the atom can only have a certain allowed energy values or levels. When an 'electron' changes from one allowed level to the other, a quantum of energy is emitted, or absorbed, whose frequency is directly proportional to the energy difference between the two levels.

Modern physical theory is concerned with the emission and absorption of energy by matter, and with the motion of material particles. Therefore, the quantum theory, and the theory of relativity together form the theoretical basis of modern physics. The theory of relativity assumes importance in situations where very large speeds are involved. The quantum theory is necessary for the special situation where very small quantities are involved for example, on the scale of molecules, atoms and elementary particles.

The quantum theory is a 'story' of the examined interior of the atom. The scientific examination of that interior is, according to quantum theory, an intrusive disturbance of what is going on there. It follows that further development of quantum theory, and additional scientific investigation cannot tell us about the character of the unexamined atomic interior. Quantum mechanics deals exclusively with what can be observed, and cannot describe what is happening during measurements. Therefore, there are only possibilities, probabilities and no certainties. Events can occur in different ways, and what is measured depends on whether it is known which of the possible paths were taken. Thus any additional knowledge which can be gained by making an additional measurement changes the outcome of the first experiment.

As we ideate and create phenomena, we have to invent such theories, and keep on refining it into experimental agreement. Quantum Mechanics is a very bizarre and mysterious hypothesis, and we continue to invent some sort of explanation, often mistaking hypotheses for reality. As we now know, ours is a very simple common sense world.

Quantum theory is an incomplete account of subatomic reality, and that incompleteness cannot be remedied by any means at the disposal of science. Why? Because sub atomic reality is invisibility, and our eyes, or any of our sophisticated instruments now, or in future cannot go there. More importantly, space is a form of energy, and an integral part of any particle, quark, nucleus or element. How can we measure space in our calculations or analysis?

With invisibility we cannot study sub atomic reality and therefore, cannot, and will never be able to know the building blocks of matter. We will never know how a particle forms, how nuclei integrate, exist/endure and function, and how nuclei form elements, and molecules, or the different types of matter we see all around us. We will never be able to create a proton, or a nucleon-the basic unit of all matter.

Micro-biologists will never know how the postulated genes form, how cells replicate, and all the complex functions taking place in an organism's cell. We will never be able to regenerate mutated cells, repair/replace damaged tissues, or completely cure illnesses, forget about creating even primitive unicellular life, or a cell. How can stem cells work the way we want them to? As particle physics is at the end of the road with invisibility, all related disciplines of science are at a standstill and forever will be.

Science says protons and positively charged neutrons are tightly clustered in the nucleus, but like-charges would strongly repel in such close proximity. Why does the nucleus not fly apart?

As they could not find the answer, scientists simply decided that a mysterious attracting force called the strong nuclear force binds protons. As scientists could not find the reason why particles in the nuclei decay, they suggested that there must be a weak nuclear force contradicting the strong nuclear force theory.

Protons or particles/nuclei should not be able to endlessly attract or repel each other, and without even a power source in sight. This is the problem with 'positively' charged protons in the nucleus, and also with the concept of postulated negatively charged 'electrons', held in orbit about the nucleus by an endless unknown power source.

We have two scientifically unexplained forces behind the stability of the postulated atoms-the electro magnetic force between charged particles, and the strong nuclear force, both acting endlessly, and with no known power source. Oops! We forgot another force-the weak nuclear force that 'decays' the bonding of the other two forces over time.

Protons and the postulated neutrons are not positively or negatively charged-there is no such thing as charges. We learnt how nuclei engender, endure and disintegrate.

Does the proton decay?

In a theory of everything, quarks which make up protons should somehow be convertible to leptons (such as electrons), so catching a proton decaying into something else might reveal new laws of particle physics.

A naked proton (without its confining space) is plasma, and plasma can be quarks, or smaller particles. Many postulated quarks make up a proton, and even if it decays, what is plasma? How would we present new laws in particle physics? A proton engenders, as plasma is expelled from a star through star field space. A proton also engenders, as elements are disintegrated with interactions from radiated heat and pressure/force. However, disintegrating protons result in plasma and/or particles, which immediately transmute, or amalgamate into/as space.

The physicist's world of particles

Physicists have identified 12 building blocks that are the fundamental constituents of matter. Three quarks with different directions make a proton, and also a postulated neutron. Together with the electron they make atoms, and atoms integrate together to form molecules.

Physicists also tell us that there are other leptons (an electron is a lepton) and force carrier particles called bosons, that carry discrete amount of energy called quanta, from one particle to another. Then there are gluons-a strong force carrier particle-that glues quarks together, and then the photon, which carries the electromagnetic force-it seems it also transmit light. There are the W and Z bosons that represent the weak force introducing different types of decay.

To top up this all, they are suggesting a gravitational force exists, associating it with a boson particle called the graviton-a hypothetical boson.

The smallest ‘virtual’ particles are the fundamental building blocks of existences, and not quarks, or leptons, or force carrier particles. There is no such thing as a lepton. W, Z, electrons, positrons, muons, taus, gluons, bosons are all hypothetical particles invented to either convey an electrical charge (which does not also exist) to an atom, or to quantify, and measure mass of particles, and the nucleon.

Most of these particles are invisible, as space is a form of energy, and seems to appear (ghostly) in particle accelerators, as particles are made to collide producing the postulated 'rays', light, and other radiation, that propagate through/as space. Neutrinos, and photons, and the above particles as leptons do not exist, and this is why they cannot be detected. Millions of dollars have been spent to build observatories underground, and millions more are spent conducting experiments to detect a theoretical neutrino.

Light (heat and force) propagates within/as a nuclei’s space, and also within/as atmospheric, and interplanetary space. As a nucleon is interacted with forms of energy, disintegrating into particles to light exhibiting intensity, this light is propagated through space, and this phenomenon suggests the possible virtual existence of particles as a neutrino, or a photon. Light propagates within/as space as it waves and therefore, there is no such thing as a photon. This explains why physicists have theorized photons to have no mass, and ‘no charge’.

As different forms of radiation are observed or calculated, particles are postulated to possess different amount of energy calculated in electron volts-muons being heavier than electrons, and taus being heavier than muons. This explains why the elusive hypothetical neutrino has no mass, has no electrical charge (as physicists put it), cannot interact with matter, and is difficult to detect.

No such thing as a photon

A photon or a hypothetical particle of light cannot exist, as light-a form of energy-propagates only within/as space, as space transmutes to light and waves, and light is apparent. The different degrees of temperatures and pressure/force generate different space wavelengths. Light cannot be considered as a particle, or a photon, as/and if space did not exist, light (heat and pressure or force) would not engender in the first place.

No such thing as a 'charge'

Elements should not form, as protons of an atom integrate with protons of another atom, with the same postulated charge.

Space is an integral part of a particle, and/or a nucleus, and the propagation of any form, state, or process of energy within/as space, alters the characteristics of the particle, and its confining space. The assumed positive, and negative charges, only describe opposite reactions. Every existence exists with another existence of a different characteristic, as parts of the same cycle, sustaining each other.

No such thing as a force carrier particle-boson

Quantum mechanics theorizes strong and weak forces that bind, and integrate particles together. However, gluon, weak gauge bosons and graviton particles do not exist.

In the summer of 2007 if all goes according to plans, many protons will accelerate in either direction around an oval 27 kilometer circumference tunnel, inside a particle accelerator called the Large Hadrons Collider, being built under the French and Swiss countryside. After months of fine tuning, many scientists from all over the world expect to find the God particle, or the Higgs boson, a particle so-called, as it may be the key to unlocking the ultimate nature of matter.

An estimated 10000 Higgs bosons (postulated) will be produced in the first year of operation. The smashing of protons is also expected to discover a group of new particles, and perhaps miniature black holes. Before the decade is over, scientists expect to explain how our universe (postulated) was born, and what holds it together.

As with other experiments this one will not produce all the answers, and over the years, there will be yet more questions, and whole new vistas will be opened up. With new vistas, scientists will want to build more powerful accelerators, and even more sensitive detectors.

No such thing as neutrinos

We are informed that the transformation of hydrogen into helium in the sun gives rise to two neutrinos, for each helium nucleus formed by a series of reactions. The flux of neutrinos from the sun was estimated to be very large; thousands of solar neutrinos were reckoned to pass through our bodies every second without our noticing them. This is incorrect.

Neutrinos like many other hypothetical particles are enigmatic candidates, introduced to modify the gaps in the Physicists' Standard Model. We were informed that neutrinos were particles without mass, but now they have a mass different from zero. We have been recently informed that they can change to any type of particle.

As explained elsewhere, at star processes, free energy condenses to plasma, and there are no protons at the sun. Particles/plasma, and different forms of energy, are continuously expelled, and concurrently extracted toward the sun, as replenishment material continuing the sun's processes.

OTHER MATTERS

What is dark Energy?

After developing the quantum theory of matter, physicists realized (within their own concepts) that 'empty space' was full of temporary ('virtual') particles, continuously forming and destroying themselves. They began to suspect, that indeed the 'vacuum' ought to have a dark form of energy, but they could not predict its magnitude.

In the 1990's, they discovered (within their concepts) that their postulated 'universe' is accelerating. The source of this mysterious force opposing the postulated gravity is what they call dark energy.

From recent measurements of the expansion of the postulated 'universe', astronomers discovered that some form of dark energy does indeed appear to dominate the total mass-energy content of the 'universe', and its weird repulsive 'gravity' is pulling the 'universe' apart. Now it seems that the WMAP satellite found early evidence of dark energy.

We are informed, that dark energy is the anti gravity force causing the ‘universe’ to rapidly expand. It makes up the bulk of the energy in the universe, and it has been changing over time growing stronger, as the postulated universe matures. We are also informed, that the 'atoms' that make up people, planets, stars, galaxies, quasars and black holes, and all the other visible objects, make up only 4 percent of the total content of the universe. At the beginning of the Big Bang, or the universe, there was very little dark energy, but it was growing. Dark energy has grown into a monster, and now accounts for 73 percent of the content in the universe.

So dark energy has reversed the slowing expansion, and is now driving the ‘universe’ to expand at faster rates. It will cool to the point at which it no longer has energy. The space between things is getting bigger. The energy is becoming less dense. As things get farther and farther apart, the temperature gets lower and lower, and we start seeing fewer and fewer galaxies in the sky. They are moving out away, and getting out of our vision, and so the universe dies in a kind of slow death? Fantastic theories being introduced to support the old unexplained ones!

Is there such thing as dark energy?

There is no such thing as 'universe', and there is no such thing as 'gravity’. Matter is visible transmuting energy, or 'visible' existences. There are only star fields conjoined as galaxies throughout existential energy's expanse. Each galaxy is kept apart from each other being confined by dark energy, and vice versa. This energy is dark, as it is invisible without star fields (producing light), and static, as it cannot transmute into existences, or star fields/galaxies.

Dark Matter

Where and what is the missing matter?

Why the missing mass?

No matter which direction astronomers point their telescopes the light from the centre of galaxies is red shifted, implying expansion in all directions-an expanding universe. The expansion of the universe is connected with the missing mass. The more mass there is the more 'gravity' exists. Whether the universe is theorized to be closed, open, or flat depends on how much mass there is. This is where dark matter comes into the picture.

How are cosmologists looking for dark matter, and how does it fit in their theories? Only large amounts of hidden mass could explain many of their observations. The existence of some unseen mass would also support theories regarding the structure of the universe. The mysterious dark matter would explain the gravitational motions of galaxies, and also validate current theories about the origin, and fate of the postulated 'universe'.

Without dark matter we live in an open universe. If there is a whole lot of dark matter we could live in a closed universe, and just the right amount, then we live in a flat universe. The amount of dark matter that exists, determines the fate of the universe. Scientists estimate that 90 to 99 percent of the total mass of the 'universe' is missing matter.

Dark matter is postulated to account for at least 90% of the mass of the postulated 'universe', which has not been detected so far. The evidence for dark matter comes mainly from observations of the dynamics of stars in galaxies, and of galaxies in cluster of galaxies from gravitational lenses, and weakly-interacting elementary particles such as low mass, fast moving neutrinos, or massive slow-moving cold dark matter particles.

Dark matter cannot be seen or touched; its existence is implied. Dark matter may consist from tiny subatomic particles-WIMPS (too small to be seen)-weakly interacting mass particles, and MACHOS (too far away)-small stars to black holes with masses millions times than that of the sun. Astronomers search MACHOS and particle physicists look for WIMPS, and as expected, both disagree what dark matter really is.

MACHOS are postulated to be non-luminous objects that make up the halos around galaxies, primarily brown dwarfs and black holes. Brown dwarfs are theorized to be made out of hydrogen, and lack the intense pressure to collapse under 'gravity' to initiate a nuclear reaction, emitting light and energy as a normal star-not a 'real' star.

The Hubble telescope did not reveal the large numbers of brown dwarfs astronomers hoped to find. Astronomers use a technique called gravitational lenses to search for dark matter halo objects. It seems intense gravity 'bends' the light rays, as MACHOS comes in between us and a star indicating their presence. It is 'proved' that gravity bends light rays as stated by Einstein (see the explanation 'does light bend'?)

Black holes are theorized to have an over abundance of matter that collapses under its own enormous 'gravity' into a relatively small area, emitting no light, and are truly black. When astronomers see stars circling around something, but cannot see what that something is, they suspect a black hole. And by observing the circling objects the astronomers can conclude that indeed a black hole exists.

Particle physicists theorize the existence of tiny particles that are 'different' from what we call ordinary matter. Smaller than atoms, WIMPS are 'thought' to have mass, but interact with other matter. Since each WIMP make up a small amount of mass, they need to be a large number to make up the bulk missing matter.

Like the Big Bang story, proponents of this story now suggest the 'missing matter' may be misleading, as it is really the light that is missing. They state they can tell that the dark matter is there, but they cannot see it. It is a fairly embarrassing situation to admit they cannot find 90 percent of the 'universe'. Scientists are scrambling to try and find not only where it is, but also what is it. If it exists, where it is, is any body's guess. What is it is also any body's guess. We create existences, and try to find them, or try to explain our own generated concepts.

Is there Missing matter?

The existences of a 'universe', 'dark matter', 'dark energy' and missing matter are all theories conceptualized within human illusory language. As we cannot theorize phenomenal reality within our language world, theories generate more theories, and more questions.

Postulating the concept 'universe' originated from a point, as stated in the Big Bang theory, has sent thousands of physicists banging their heads. Unable to explain many phenomena within the Big bang theory, there are many other theories that have cropped up, as discussed in revelation three. Presently, there are theories like the membranes, multi dimensional universe, and parallel universes.

How do you determine the mass of galaxies? How do we measure the mass of the 'universe'? As stated in revelation 3, there is no such thing as the 'universe'. Since there is no universe there are no boundaries, and the actual totality of mass can never be known. Scientists talk of the missing mass of the postulated 'universe' in percentages, not real numbers.

We are informed, that since the majority of the matter they can see is clumped together into galaxies, the total mass of all the galaxies should be a good indication of the mass of the 'postulated' universe. But they have not seen the boundaries of that postulated universe. Again, they state that although it is not possible to add up an infinite number of galaxies, they can infer the percentage of the postulated universe's missing mass from estimates of the missing mass in galaxies, and cluster of galaxies using different techniques. One of the tools is the Doppler shift. As the Doppler shift works with sound, they postulate that it also works with light.

Using the Doppler shift scientists observe, that light from one side of the galaxy is blue shifted, and light from the other side is red shifted. One side is moving toward the earth, and the other is moving away. They can also determine the speed at which the galaxy is rotating from how far the light is shifted, and with this speed they can figure out the mass of the galaxy mathematically. But there is something wrong here. As they look closer at the speeds of galactic rotation they find something strange. The individual stars in galaxies should act like planets in our solar system. The further away from the centre the slower they should move. But the Doppler shifts reveal that the stars in galaxies do not slow down at further distances. And on the top of that, stars should move at speeds that should rip the galaxies apart; there is not enough measured mass to supply the 'gravity' to hold the galaxy together. High rotating speeds suggest that the galaxies contain more mass than was calculated. Scientists therefore theorized that if the galaxies were surrounded by a halo of unseen matter, the galaxies could remain stable at such high rotating speeds. This explains why the Doppler shifting and/or the blue and red shifting calculations are wrong. Revelation 3 explains why blue and red shifting is incorrect.

Another method astronomers use to determine the mass of a galaxy, or cluster of galaxies, is simply to look how much light there is. By measuring the amount of light reaching the earth, scientists can estimate the number of stars in the galaxy. Knowing the number of stars in the galaxy, scientists can then mathematically determine the mass of the galaxy.

10% of what we call the universe is in a form that we can see, but the elusive dark matter-the 90% we cannot.

Scientists are bouncing theories back and forth. The dark matter problem discredits the Big Bang theory. Theories change minds every year, and experimentalists will still go ahead with their experiments

The search for dark matter is about more than explaining discrepancies in galactic mass calculations. The missing matter or the dark energy problem has people questioning the validity of current theories on how the postulated 'universe' formed, and how it will ultimately end.

What is cold dark matter?

The star/sun continuously propagates shuttling to-and-fro temperatures/pressures or force, and plasma/protons into a star field, or inter-planetary space. Some of the plasma/protons not interacted at atmospheres of bodies, or bodies within star fields, settle on the outskirts, or our solar system, as interstellar mediums. As radiated heat and pressure/force from star dwindle with distance, these areas are cold and icy-cold dark matter. Some regions of dark matter have temperatures of 3 degrees above absolute zero.

This matter is dark, as light from a star, as well as other adjacent star fields dwindle with distance, altering to other forms of energy. Matter around these areas is slow moving, and therefore cold. Cold dark matter surrounds each star field interweaving between stars in a galaxy. In these neutral zones, nuclei come together to form complicated molecules and these molecules manage to endure, as the flood of plasma/protons from stars do not have intense temperatures and pressures/force to smash them apart.

Anti-Matter

The antimatter that physicists seek

Anti-matter is one of the most bizarre concepts in science. It is the opposite of matter. It is assumed, that matter and anti-matter are supposed to destroy one another on contact.

A theory suggests that particles and anti-particles are always created in equal amounts, or destroyed in equal numbers, in explosions of energy and gamma rays. This balance between matter, and anti-matter, is known as symmetry.

Stars, dust, and everything observed, is made up of matter. As physicists theorized, that the 'entity', or the concept 'universe' oozed out from a point in the Big Bang theory (which we have proved is incorrect) there are many questions, as a result of incomplete, and incorrect explanations to speculations/theories.

Accepting the Big Bang theory, many scientists continue to research and source explanations, for example: Most theories of the evolution of the universe suggest, or even require that antimatter and matter were equally common in the earliest stages- so why is antimatter so uncommon today? If the cosmos began with matter with equal amounts of matter and anti-matter, where is the antimatter?

Scientists cannot observe or find their hypothesized antimatter, and propose unexplainable puzzles, and more theories for the supposed imbalance between antimatter and matter. They suppose, that the concept, or entity 'universe' would be a much less interesting place, as there would have been no matter left, as annihilations between matter and anti-matter would have converted everything into radiation by now.

Antimatter candidates are neutrinos, or/and two kinds of particles neutralinos and axions-predicted, but never detected. The 'wisp', axion, neutrino, neutralino, or muon, are all theorized unobtrusive light particles that physicists are trying to find. By speculating antimatter they have yet another problem stating, that this imbalance is a key property of the world they know.

So as has always been the scenario, there are a number of non-constructive experiments around the world, focusing on differences in the decay of a particle called the B-mesons, and their anti-particle partners, which physicists think could solve this imbalance problem. Matter and anti-matter do not exist in equal amounts. Anti matter is in reality, space-a form of invisible matter/energy discussed in revelation two.

Why is it that there is so much more matter than anti-matter as physicists call it?

Why do we observe matter and almost no antimatter?

To a particle physicist matter and antimatter are almost the same. Some subtle difference must explain why matter is common and antimatter rare. First of all there is no antimatter. Antimatter-as physicists are trying to search-as explained above, is the space around a particle, an integral part of the particle, and also a form of energy. Matter is condensed energy and therefore, energy can be explained as invisible matter.

The space surrounding a particle is invisible, and unless some particles integrate together to form mass or matter, which our visual system can observe, they are also invisible.

We always have this problem of describing physical reality, as we have conceptualized and named existences, before observing and experiencing physical phenomena.

It is incorrect to speculate, that only ten percent of matter can be observed, the rest being antimatter or dark energy, or dark matter. We can observe distant galaxies only as far as our telescopes can take us. Cosmologists believe that those distant galaxies are the first that ever flew out from the big bang. But as technology advances, they will see more distant galaxies, proving the Big bang theory, or the existence of the concept 'universe' incorrect.

Why do physicists assume that each type of particle has a corresponding antiparticle?

This theory is to quantify, and measure the entirety of matter in their concept 'universe'.

By adding the concept of anti-matter as distinct from matter, physicists narrowed the definition of matter to apply to only certain types of particles including however, all those found in every day experiences.

Particle physicists colliding protons, and other particles, in mass accelerators, and sophisticated detectors, observe hypothetical 'antimatter' that appears, and instantaneously disappears. They assume, that anytime a particle meets its matched antiparticle the two can annihilate one another-they both disappear-leaving the energy transformed into some other form. In reality, nuclei are disintegrated sending sparks of energy in the form of light, and other forms of radiation, and amalgamate with/as space.

There is no corresponding antiparticle to a particle. Every existence has another existence to sustaining it, and vice versa and therefore, would obviously possess different characteristics. As space is an integral part of a particle, it possesses the opposite characteristic of a particle. This is the antimatter. As space is invisible, it is speculated that anti-matter cannot be stored. Obviously a particle condensing from energy engenders its surrounding space, and on disintegration will amalgamate together with its confining space to 'disappear' in a flash back to energy. There is no antimatter. The antimatter that physicists seek is space surrounding any particle, nuclei or astronomical object.

Annihilation is a wrong word. Nothing annihilates. Energy transmutes, and cannot be destroyed or created.

Are neutrinos their own antiparticles?

Neutrinos are postulated to exist to try and find out the origins of matter. This is one more theory explaining other unsuccessful stories. There are some theories involving millions and millions of dollars, years of research, and untold humans hours. Millions of dollars have been spent on underground experiments to search for the so far elusive neutrinos.

Can the laws of physics be unified?

Physicists created the laws of physics from within their illusory world. How can they unify them?

CONCLUSION

These eleven revelations elucidated physical reality, or existential energy, or nothingness.

Existential Energy

There is no such thing as a universe, and therefore, questions on its existence, expansion or contracting is unconstructive thinking. Only existential energy exists, as everything, everywhere, and it continuously transmutes into forms, states, processes, or structures, or existences (human understanding). As everything is flux, virtual, or observable existences endure for certain duration, ranging from a millionth of a second, or for million/billions of years.

Energy transmutes engendering, and enduring our world. As energy transmutes, star fields structure lighting up worlds, but eventually terminate, structuring up again and again, existing/enduring for a moment, or for millions of years..

All existences/beings, whether they are galaxies, star fields generating momentum, or pressure or force, and temperatures, or radiation, or light, astronomical bodies, atmospheres, objects, humans/organisms, or cells, are like clouds for duration. They randomly appear, or exist, or endure, but eventually terminate, or disintegrate into particles again, into/as/within the fabric of space, or energy. Particles may float aimlessly, or once again be integrated with other particles to exist/endure, as another observable or invisible existence.

There is no such thing we conceptualize as 'space'-a void and therefore, there are no entities, as everything exists within/as the fabric of existential energy. Newton's dynamics are wrong.

Without a 'universe', and 'space'-as a void, and time as an abstract (generated from within human illusory minds), Einstein's theories of relativities, and space-time are also incorrect.

Without protons, and temperatures and pressures/force propagating to, and from the sun/star, nothing would exist.

Every existence engenders, or exists, or endures, as a result of prevailing temperatures and pressure/force, and every existence alters or terminates, as a result of changes in those temperatures and pressure/forces.

The formation of our Solar System

As many fragments from an exploding neutron star darted through our nascent star field, planets could engender. However, with this abnormal configuration of bodies, they could have eventually terminated one by one, terminating our solar system. They could have also terminated, after one of those planets exploded. But this event transformed the inner planets, engendering new atmospheres, with heavier elements in the interiors. These processes steadfastly set the sun, and all the prevailing energy transmuting cycles with bodies within the solar system, as the sun could easily replenish its expended energy in different forms from the inner planets.

The present state of our solar system

Even though everything is flux, our world endures, as a result of the several firmly set energy transmuting cycles within our solar system, explaining why it has endured for millions of years. As the sun is easily able to replenish its expended material to continue functioning, everything is smooth sailing.

With so many energy transmuting cycles with as many planets, and as each planet (with different characteristics) rotating around its imaginary axis, and orbiting the sun at different orbits, and at different speeds, there is continuous energy increases and decreases (within the energy transmuting cycle) with the expulsion and extraction of energy. As a result, there are continuous disruptions/alterations, both in the functioning, and the characteristics of bodies within each cycle. These situations alter existing cycles, with changing orbits, orbital motions, rotation speeds, and other disruptions. Correcting these abnormal processes, meteorites, asteroids, comets, and other bodies are perturbed from the localization wandering around the solar system, often crashing, or colliding with other bodies.

Over millions of years, as replenishment material from the atmospheres of inner bodies and planets is depleting, the sun sources its replenishment energy from bodies, and planets, next in proximity. There is plenty available at the large Jovian planets. As our solar system has endured for 4 billion years (as informed), we may conclude that our solar system can last for another 4 or 5 billions years.

This is the only unique solar system within any galaxy. There may be other solar systems that have formed, and are forming elsewhere, but our system has as many cycles, as there are planets, and with plenty of replenishment material, facilitating the processes of/at the sun, all the bodies continue to exist/endure. Our solar system will continue to exist, as long as there are no extreme energy transmuting variations internally, or externally, and energy expended from the sun is replenished from the bodies within the solar system.

Why and how will our solar system terminate?

Our solar system's multiple energy transmuting cycle system is a rare event. Even if this happens in one of the billions of star fields within our galaxy, or within one of other star fields, among billions and billions of other galaxies, this unique configuration would not last long. As one of the planets exploded, the inner planets somehow quickly generated new atmospheres, which confined the disintegrating interiors, and endured the solar system, otherwise, the planets would have exploded, or terminated one by one.

As energy continuously transmutes, and everything is flux, there are continuous alterations in everything, everywhere. If there is a disruption in a cycle, other cycles must immediately alter to accommodate the energy replenishment at the sun, if not planets can explode, or there will be other types of catastrophes.

Energy transmuting variations could terminate an energy cycle, between the sun, and any one of the other planets. Planets could disintegrate, and the sun processes could go haywire. Any major disruptions in one, or many energy cycles would terminate our entire solar system. There could also be disruptions in adjacent star fields, altering any one of our energy cycles within our solar system. There are no neutron stars, which are close enough to worry about. However, one or a group of exploding neutron stars can trigger the end of our solar system. Any galaxy, any star field, our solar system, any planet, our earth, and life on earth can terminate any time.

It is indisputable that over the years, Mercury, Mars, our moon, Mar's moons, and the asteroids in the asteroid belt, have degassed their interiors, sustaining their depleting atmospheres. Without an active energy transmuting cycle with the sun, they are now impotent bodies within the solar system. As the sun has completely depleted Mercury's atmosphere (and the interior has completely solidified) it is now wobbling. We can envisage Mercury's disintegration in the near future. If earth and its inhabitants survive this disintegration, the next one to follow will be Mars, or our moon.

Mercury and Venus initially may have had moons, and or gas rings, which may have disintegrated. Earth and Mars also may have had many more moons, as well as gas rings like the Jovian planets. If this was the case, these rings, and some of the moons may have already disintegrated.

The asteroids will also disintegrate, and then Venus and Earth will follow, but we have yet to ascertain, how much longer will the unique atmospheres of Venus and Earth continue enduring, having contained their molten core to some extent, not to have completely belched out their interior gases, as Mercury and Mars. It is a possibility that with Mercury disintegrating, all the inner planets may immediately soon follow in a chain reaction, or each over periods of time.

Stars exist, and endure, processing particles into plasma. All matter is comprised of protons. Protons are continuously being extracted to the sun from every nuclei, molecule, substance, compound, objects and organism from within the inner planets. This explains why earth's elements are continuously disintegrating into nuclei and protons. This also explains why every organism and object alters, decay, and everything eventually disintegrates, or terminates. We grow old, as protons gradually leave our cells during our lifespan, and although we replenish some extraction of protons with food and water intake, our metabolism mechanisms have not been fine tuned to avoid, or repair chromosome degeneration. We cannot redesign, or restructure our genes to replace departing protons. Incidentally, growth, cell replication, or life, is the consequent of proton extraction, and other atmospheric processes (revelation 12 and 13). If protons were, and are not continuously extracted, there would be no cell replication, or growth, or life. Eventually, as all the protons or hydrogen is extracted, earth and the inner planets will disintegrate. The sun will then source its particles, or plasma, or protons from the outer planets. Our solar system will endure, as long as protons can be extracted by/to the sun, within energy transmuting cycles. If this cannot happen, our solar system, or any star field terminates.

Over millions of years, as plasma/protons are expelled to the far reaches of our solar system, the interstellar medium grows larger. After all the planets solidify, and eventually disintegrate, all energy transmuting cycles will end. The sun will exist for some more time, processing the protons available at the interstellar medium, but this cycle does not last long. The star/sun will shut down its processes, as energy cycles cannot form between the star and the interstellar medium, as all forms of energy dwindle with distance. The sun will then expand into a red giant, and subsequently into a white dwarf, and/or collapse/drown, amalgamating with space, into/as a black hole. The swirling black hole transmutes all terminating material into free energy, from which another star field may engender.

The present state of Earth

Protons with radiated heat and pressure/force from the sun interact, disintegrate and integrate our atmospheric nuclei. Protons from these interactions are extracted to the sun, transforming into plasma enroute. The interactions within our atmosphere generate altering temperatures, pressures and densities, resulting in the disintegration of the water molecules on land, and our oceans. Through many volcanoes and other openings, earth's molten core belches out to sustain the continuously regenerating depleting atmosphere, and oceans.

As the inner planets and bodies are solidifying, as their nuclei deplete, they get closer to the sun. Altering solar cycles imply more solar activity, which in turn means more solar radiation is reaching earth's atmosphere than ever before. Floods, draughts, forest fires, light reduction, increase or decrease of rainfall, tidal waves, hurricanes, volcanoes, tsunamis, earthquakes etc, are all the result of alterations in the energy transmuting cycle.

Why and how will earth terminate?

Energy continuously transmutes, and everything is flux, and so is every existence or being. Everything has duration-a moment, a day, a month, a year, or millions of years.

As fragments of a neutron star, all our planets are extremely compressed fragments. There are plenty of protons to be extracted from earth over the next billions of years. However, fragments from an exploding neutron star can terminate this thought.

There may have been a planet, or planets closer to the sun than Mercury, which may have disintegrated, as it solidified, terminating its cycle with the sun. There might have been a moon orbiting Venus, or another moon orbiting earth, which may have disintegrated after degassing its interiors. There may have been more moons around the inner planets, and it can be postulated that an exploding body, or alterations in energy transmuting cycles within the solar system inner planets, or elsewhere, may have altered earth's environment, which may have resulted in mass extinctions of large creatures, perhaps terminating the dinosaur era. Some catastrophe like this could alter our atmosphere, which may result in the termination of all life forms.

If Mercury terminates, all the other planets will change their orbits, and move closer to the sun. As Earth moves closer to the sun, our days will either be longer or shorter, depending on the availability, and facilitation of replenishment material from other bodies. Accordingly, earth will rotate faster, or slower than normal. If earth survives Mercury's explosive disintegration everything may alter. Alterations in the earth's rotation, and/or orbit around the sun will change its climate, weather, seasons, currents, tides, and the hydrological, nitrogen, carbon dioxide and other cycles. As the many layered atmosphere will alter their densities, pressures/force and temperatures, there could be an ecological disaster, and living forms may terminate.

If there is a major disruption within the sun-earth energy cycle, or within any other cycles within our solar system, our atmospheres would disintegrate, and immediately depleted, exposing the troposphere to radiated heat. There would be instantaneous changes in weather, tides and ocean currents. Our oceans could rise, and magma would erupt exposed, and will disintegrate. The exposed surface would tear the ecological systems apart, and this would be the end of living things. Disruptions will also alter the composition of the earth's atmosphere, with changing pressures and temperatures, altering plate tectonics, resulting in earthquakes, and the resurrecting of dormant volcanoes, and myriad forms of catastrophes will ensue.

We live in these times, assuming everything is smooth sailing. As our atmosphere is depleting, so are the oceans, and the molten core. As the molten core erodes, the outer core, mantle and crust sink. Presently, the increasing in flooding, and the sinking of low level surfaces all over the globe are evidence. Plate tectonics happen, as the mantle and crust sink, generating under water earthquakes, land quakes, and increase in volcanic activity Tsunamis experienced around Indonesia during December 2004, and July 2006 is evidence.

Our atmosphere, oceans and organisms including ourselves will endure, as long as earth's molten core continues to disintegrate, and belch out materials. As the molten core depletes, the atmosphere also will completely deplete, and the oceans will rise up, and disintegrate. All life forms will terminate, and earth will turn into an inert ball waiting to be disintegrated. Earth would disintegrate retaining no evidence whatsoever, that it was once a garden teeming with myriad forms of plants and creatures, or organisms/living things, and self proclaimed 'intelligent' human beings. One by one all our planets will follow the same fate, until our solar system collapses into/as a black hole from where another star field may engender.

Within this new solar system, and without our type of atmosphere, and oceans to bring forth cells to evolve into multi cellular organisms, humans will not evolve to think/ideate, imagine and query. Then there will be nothingness, or existential energy, as it was, and ever will be.

Matter

The key to protons and neutron plasma integrating into/as matter, and enduring, incredibly, is the concurrent disintegration of matter into nuclei, and nuclei into plasma. Plasma is expelled from the sun, and also extracted to the sun. In other words, if plasma is not expelled from the sun, nuclei or matter would not disintegrate, and also if nuclei or matter does not disintegrate, plasma would not be available to be extracted to sustain the cycle, and continue the sun's processes.

It is incorrect to postulate we are functioning star material. Matter and we do not engender at, or from functioning star material. A functioning star must explode, and the core should transform into a neutron star. Furthermore, a neutron star must also explode, and also exist in close proximity to a nascent star field, in order for its blasted away fiery fragments to engender nuclei, concurrently inciting radiation from the star/sun, to engender atmospheres confining the disintegrating interiors. From the disintegration of the interiors of the planets, and also the interactions of radiated heat and pressure from the sun, matter, and we can then engender, exist, and endure.

All the inner planets took the same pathways, but earth at a certain distance from the sun, produced a unique mix of elements. Eventually, water vapor constituted the new atmosphere producing clouds, and precipitation. Precipitation on the fiery interiors engendered the outer core, the mantle and crust, which in turn, concurrently contained the precipitation into/as the rivers, lakes and oceans. From these oceans life has evolved, subsequently bringing forth humans to think, imagine and query about existences, or physical reality.

Every particle is confined by space, and as we cannot study space, or invisibility, or virtual energy particles, we will never understand how energy transmutes into a particle to exist/endure, and engender matter-the stuff that makes us, or the things we see around us.

We will also never understand other forms of invisible virtual states, processes, forms or structures of transmuting energy. We have researched into how most things work, but with invisibility cannot ascertain why things work, and/or engender, or exist, or endure, and eventually terminate.

In reality, quarks, or protons, or neutrons, the postulated atoms or molecules, or compounds, objects, planets or astronomical bodies, or our sun, or the stars, or a cell, or organisms including us, are constituted of particles/protons, or the same thing-matter or condensed energy.

Matter and Life

It is incorrect to postulate that it was the distance of earth from the sun, which dictated temperatures, facilitating life like ours to evolve, and proliferate for millions of years. In reality, our continuous disintegrating, and integrating atmospheres, sustained by the disintegration, and regeneration of our oceans, and the belching out from the molten core, are reasons why our biosphere experiences those temperatures. Our multi layered continuously altering atmosphere, facilitates life like ours to emerge, evolve and endure. With so many considerations, and conditions presented within these revelations, it is impossible to find life like ours anywhere else. Earth still has a very active energy transmuting cycle with the sun, and many cycles endure a unique ecological biosphere favorable for life. It is possible that ours is a unique solar system, and only earth, as a unique planet, could have brought forth life-as we know it.

Who or what created Energy?

Some billion years ago, within one spiral arm called the Orion, in our Milky Way galaxy, a neutron star exploded. Being in close proximity, this event transformed our previous nascent solar system into our present one.

A fragment of that exploded star with radiation from our nascent sun, produced an atmosphere enveloping itself, and transformed into planet Earth. With another event, a new atmosphere formed, engendering water vapor and clouds. Precipitation on the exposed fiery interiors transformed parts of it into the mantle and crust, containing the rains, as lakes and oceans.

Within/with this atmosphere, and the oceans, life evolved, and proliferated to eventually bring forth primates. Their utterances or sounds eventually unfolded into a language, engendering humans, and/or our illusory world, within which we think/ideate, imagine, and query about existences. Within our thoughts we created energy, and ask questions about existences. These profound questions can only be asked during our lifetime, or within the span of human existence. As our thoughts disintegrate when we die, these questions become meaningless, or nothingness. Humans think/ideate, and query the purpose, aim and design of living, and non-living things, within their own engendered vocabulary of sounds, or language. Why and how should these questions, within an illusory language-distinct from reality-be constructively answered?

Particles/plasma/protons/nuclei/molecules constituting our thought neurons, and our bodies, will eventually disintegrate. These may lie down as earth's soil for generations. They may integrate together to engender leaves, flowers, fruits, food for worms, or float aimlessly as air, or atmospheric nuclei. These nuclei may disintegrate further, extracted as lighter particles toward our sun, where they may integrate together to form plasma/protons, and may be propagated to another planet. However, energy particles will exist eternally, somewhere, within something, until they may become part of neurons of another being like us humans, who think/ideate, imagine, and query who, or what, created energy.

If a planet at our original solar system did not explode, and if this event did not alter our original atmosphere, nuclei heavier than helium, or matter would not have engendered, and exist. Well, life and we would not have existed. Like all existences, the engendering of matter, life, and us was an accident.

Revelation 8 detailed that if those fragments turned planets did not experience a sudden complete extraction of their atmospheres, some time before the advent of multi cellular organisms, the inner planets would have endured today with the same constituents as the outer planets, and matter, and us would never have existed.

Humanity dawned only a few thousand years ago, and within illusory language minds, or their world, they think/ideate, and query about beginnings, ends, after life, and durations. We created the gods to explain to ourselves, who and why all existences that we have also created exist. We just happen to live within this smooth period, without any major catastrophe for these thousands of years. In reality, energy transmutations continuously alter everything, explaining why everything is flux, and duration. Humankind, our earth, and the solar system, as an ephemeral cloud, or a water bubble can terminate any time, or will eventually disintegrate.

Some insects (cicadas) sleep for 17-20 years, waking up for only a few weeks till their death. Some creatures live for a few days, and some for a few hours. Some flies live for only five minutes. We must understand that each night when we are asleep, we die, and every morning when we awake, it is the 'beginning' of a new life. Enjoy every minute, as if it is a year, and every day, as if it is a lifetime. We could probably have just spent our last minute. The end can come anytime.

Only existential energy exists eternally, continuously transmuting, cycling eternally, continuously engendering, enduring, and disintegrating existences/beings. There are no creators. There is no causation. There are no entities, no beginnings, and no ends. There are no purposes, no aims, and no goals, but only random spontaneously engendering existences from particles, which eventually terminate, or disintegrate, into energy particles again.

Matter is our word/concept. We are matter, or visible energy. We are here, and something engendered us, and that something is energy, everything, and everywhere.

As we can think, or conceptualize, we create existences within our thoughts. In reality, every existence as transmuting energy is nothingness. We live (think, feel, and action) within/with our thoughts, or language. In reality, this is nothingness. Within/with nothingness, there are no such things, as something, or anything, or nothing.

BOOK 2

(Revelations 12-27)

Part Three elucidates physio-ecological existences-cellular and multi cellular structuring, and evolution.

Finally, we will learn why, and how a mysterious process we call 'life' 'originated' or engendered, and proliferates/continues.

Revelation 13 and 14 details how the hypothetical genes engender, and how multi cellular organisms structure. A revelation explains why and how creatures endure, as structural processes of the environment.

Multi cellular organisms could engender and evolve, and pass on their experiences and information to their progenies, only by possessing a significant unique structure. Furthermore, multi cellular life is possible only with dual bodies, and sex.

Multi-cellular organisms can endure, and continue to exist, and proliferate, as they are their parents all over again, and more, and better.

Darwin was incorrect with his theory of Natural Selection-there is no such thing. Evolution 'began', and continues, conforming to changing environments, and other processes that concurrently continuously restructure (evolve).

Creatures are the environments, and action or behave, endure or survive, as, or with changing environments, and more.

Elucidating why, and how creatures move (locomotion), and action.

Would you be surprised if you are told that you see with sounds?

We will now understand why and how some mammals including humans are 'intelligent' and 'creative'.

The last revelation of this part details, why our altered anatomy of walking upright brings about a world full of suffering/privations.

BOOK 3

(Revelations 28-40)

Part Four elucidates human psychological, and physiological structuring, existence, and deterioration.

Whether there are such things as minds, or brains, and consciousness will astonish readers.

Finally, we will learn why we dream, what they are, and what they mean.

Section B 'Humans new illusory world/existence' describes how humans created themselves, concurrently as they created their gods. Revelation 34 discusses how and whether gods exist.

Learning why Pasteur's germ disease theory is wrong and what cancer, diabetes, and other diseases really signify will shock readers.

Is obesity an evolutionary alteration, or are we directly accountable? We can easily stay fit and healthy without excessive physical activity, or exercises, or diets. Learn why diets work only for some time. If exercises or physical activity is to keep you fit, why are we unfit during these exercises, or activity?

Other revelations detail how humans live within their illusory 'minds', indoctrinating, and influencing themselves.

We will understand what happiness is, and why it cannot be prolonged.

Is there good or bad, right or wrong?

Why we grow old? Why do we have to face the inevitable? What and why there is death, and whether there is an afterlife? These and many other questions are gently explained.

Envisions and Contemplations at the end recapitulates the 40 revelations, and may perhaps assist you in your future thinking, and lifestyle.

Made in the USA
Monee, IL
08 November 2026

68113597R00273